Power System Dynamics and Stability

PETER W. SAUER

M. A. PAI

Department of Electrical and Computer Engineering
University of Illinois at Urbana-Champaign

Prentice Hall
Upper Saddle River, New Jersey 07458

Library of Congress Cataloging-in-Publication Data

Sauer, Peter W.
    Power system dynamics and stability / Peter W. Sauer and M. A. Pai.
        p.   cm.
    Includes bibliographical references and index.
    ISBN 0–13–678830–0
    1. Electric power system stability. 2. Electric machinery,
Synchronous—Mathematical models. 3. Electric power stystems—
Control. I. Pai, M. A.   II. Title.
TK1010.S38   1998
621.31'01'1—dc21   97–17360
                                                                         CIP

Acquisitions editor: *ERIC SVENDSEN*
Editor-in-chief: *MARCIA HORTON*
Managing editor: *BAYANI MENDOZA DE LEON*
Director of production and manufacturing: *DAVID W. RICCARDI*
Production editor: *KATHARITA LAMOZA*
Cover director: *JAYNE CONTE*
Manufacturing buyer: *JULIA MEEHAN*
Editorial assistant: *ANDREA AU*

©1998 by Prentice-Hall, Inc.
Simon & Schuster / A Viacom Company
Upper Saddle River, New Jersey 07458

The author and publisher of this book have used their best efforts in preparing this book. These efforts include the development, research, and testing of the theories and programs to determine their effectiveness. The author and publisher make no warranty of any kind, expressed or implied, with regard to these programs or the documentation contained in this book. The author and publisher shall not be liable in any event for incidental or consequential damages in connection with, or arising out of, the furnishing, performance, or use of these programs.

This book was prepared using L^AT_EX and T_EX. It was published from T_EX files prepared by the authors.

T_EX is a trademark of the American Mathematical Society.

Printed in the United States of America

10 9 8 7 6 5 4 3 2 1

ISBN   0-13-678830-0

Prentice-Hall International (UK) Limited, London
Prentice-Hall of Australia Pty. Limited, Sydney
Prentice-Hall Canada Inc., Toronto
Prentice-Hall Hispanoamericana, S.A., Mexico
Prentice-Hall of India Private Limited, New Delhi
Prentice-Hall of Japan, Inc., Tokyo
Simon & Schuster Asia Pte. Ltd., Singapore
Editora Prentice-Hall do Brasil, Ltda., Rio de Janeiro

To

Sylvia and Nandini

Contents

PREFACE

The need for power system dynamic analysis has grown significantly in recent years. This is due largely to the desire to utilize transmission networks for more flexible interchange transactions. While dynamics and stability have been studied for years in a long-term planning and design environment, there is a recognized need to perform this analysis in a weekly or even daily operation environment. This book is devoted to dynamic modeling and simulation as it relates to such a need, combining theoretical as well as practical information for use as a text for formal instruction or for reference by working engineers.

As a text for formal instruction, this book assumes a background in electromechanics, machines, and power system analysis. As such, the text would normally be used in a graduate course in electrical engineering. It has been designed for use in a one-semester (fifteen-week), three-hour course. The notation follows that of most traditional machine and power system analysis books and attempts to follow the industry standards so that a transition to more detail and practical application is easy.

The text is divided into two basic parts. Chapters 1 to 6 give an introduction to electromagnetic transient analysis and a systematic derivation of synchronous machine dynamic models together with speed and voltage control subsystems. They include a rigorous explanation of model origins, development, and simplification. Particular emphasis is given to the concept of reduced-order modeling using integral manifolds as a firm basis for understanding the derivations and limitations of lower-order dynamic models. An appendix on integral manifolds gives a mathematical introduction to this technique of model reduction. Chapters 6 to 9 utilize these dynamic models for simulation and stability analysis. Particular care is given to the calculation of initial conditions and the alternative computational methods for simulation. Small-signal stability analysis is presented in a sequential manner, concluding with the design of power system stabilizers. Transient stability analysis is formulated using energy function methods with an emphasis on the essentials of the potential energy boundary surface and the controlling unstable equilibrium point approaches.

The book does not claim to be a complete collection of all models and simulation techniques, but seeks to provide a basic understanding of power system dynamics. While many more detailed and accurate models exist in the literature, a major goal of this book is to explain how individual component models are interfaced for a system study. Our objective is to provide a firm theoretical foundation for power system dynamic analysis to serve as a starting point for deeper exploration of complex phenomena and applications in electric power engineering.

We have so many people to acknowledge for their assistance in our careers and lives that we will limit our list to six people who have had a direct impact on the University of Illinois power program and the preparation of this book: Stan Helm, for his devotion to the power area of electrical engineering for over sixty years; George Swenson, for his leadership in strengthening the power area in the department; Mac VanValkenburg, for his fatherly wisdom and guidance; David Grainger, for his financial support of the power program; Petar Kokotovic, for his inspiration and energetic discussions; and Karen Chitwood, for preparing the manuscript.

Throughout our many years of collaboration at the University of Illinois, we have strived to maintain a healthy balance between education and research. We thank the University administration and the funding support of the National Science Foundation and the Grainger Foundation for making this possible.

Peter W. Sauer and M. A. Pai
Urbana, Illinois

Chapter 1

INTRODUCTION

1.1 Background

Power systems have evolved from the original central generating station concept to a modern highly interconnected system with improved technologies affecting each part of the system separately. The techniques for analysis of power systems have been affected most drastically by the maturity of digital computing. Compared to other disciplines within electrical engineering, the foundations of the analysis are often hidden in assumptions and methods that have resulted from years of experience and cleverness. On the one hand, we have a host of techniques and models mixed with the art of power engineering and, at the other extreme, we have sophisticated control systems requiring rigorous system theory. It is necessary to strike a balance between these two extremes so that theoretically sound engineering solutions can be obtained. The purpose of this book is to seek such a middle ground in the area of dynamic analysis. The challenge of modeling and simulation lies in the need to capture (with minimal size and complexity) the "phenomena of interest." These phenomena must be understood before effective simulation can be performed.

The subject of power system dynamics and stability is clearly an extremely broad topic with a long history and volumes of published literature. There are many ways to divide and categorize this subject for both education and research. While a substantial amount of information about the dynamic behavior of power systems can be gained through experience working with and testing individual pieces of equipment, the complex problems and operating practices of large interconnected systems can be better understood if this experience is coupled with a mathematical model. Scaled-model systems such as transient network analyzers have a value in providing a physical feeling for the dynamic response of power systems, but they are limited to small sizes and are not flexible enough to accommodate complex issues. While analog simulation techniques have a place in the study of system dynamics, capability and flexibility have made digital simulation the primary method for analysis.

There are several main divisions in the study of power system dynamics and stability [1]. F. P. deMello classified dynamic processes into three categories:

1. Electrical machine and system dynamics

2. System governing and generation control

3. Prime-mover energy supply dynamics and control

In the same reference, C. Concordia and R. P. Schulz classify dynamic studies according to four concepts:

1. The time of the system condition: past, present, or future

2. The time range of the study: microsecond through hourly response

3. The nature of the system under study: new station, new line, etc.

4. The technical scope of the study: fault analysis, load shedding, sub-synchronous resonance, etc.

All of these classifications share a common thread: They emphasize that the system is not in steady state and that many models for various components must be used in varying degrees of detail to allow efficient and practical analysis. The first half of this book is thus devoted to the subject of modeling, and the second half is devoted to the use of interconnected models for common dynamic studies. Neither subject receives an exhaustive treatment; rather, fundamental concepts are presented as a foundation for probing deeper into the vast number of important and interesting dynamic phenomena in power systems.

1.2   Physical Structures

The major components of a power system can be represented in a block-diagram format, as shown in Figure 1.1.

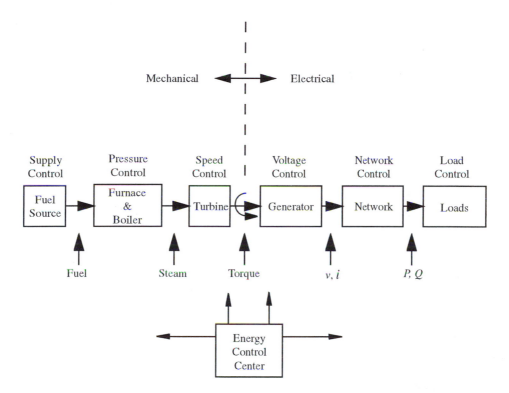

Figure 1.1: *System dynamic structure*

While this block diagram representation does not show all of the complex dynamic interaction between components and their controls, it serves to broadly describe the dynamic structures involved. Historically, there has been a major division into the mechanical and electrical subsystems as shown. This division is not absolute, however, since the electrical side clearly contains components with mechanical dynamics (tap-changing-under-load (TCUL) transformers, motor loads, etc.) and the mechanical side clearly contains components with electrical dynamics (auxiliary motor drives, process controls, etc.). Furthermore, both sides are coupled through the monitoring and control functions of the energy control center.

1.3   Time-Scale Structures

Perhaps the most important classification of dynamic phenomena is their natural time range of response. A typical classification is shown in Figure 1.2. A similar concept is presented in [6].

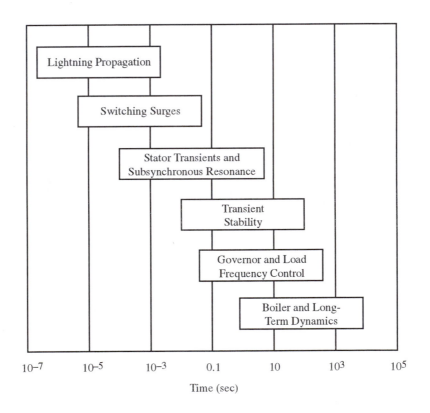

Figure 1.2: *Time ranges of dynamic phenomena*

This time-range classification is important because of its impact on component modeling. It should be intuitively obvious that it is not necessary to solve the complex transmission line wave equations to investigate the impact of a change in boiler control set points. This brings to mind a statement made earlier that "the system is not in steady state." Evidently, depending on the nature of the dynamic disturbance, portions of the power system can be considered in "quasi-steady state." This rather ambiguous term will be explained fully in the context of time-scale modeling [2].

1.4   Political Structures

The dynamic structure and time-range classifications of dynamic phenomena illustrate the potential complexity of even small or moderate-sized problems. The problems of power system dynamics and stability are compounded immensely by the current size of interconnected systems. A general system structure is shown in Figure 1.3. While this structure is not necessarily common to interconnected systems throughout the world, it represents a typical North American system and serves to illustrate the concept of a "large-scale system."

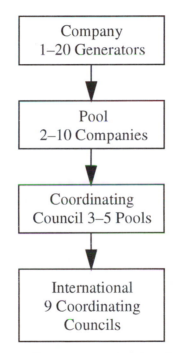

Figure 1.3: *System organizational structure*

If we speculate about the possible size of a single interconnected system containing nine coordinating councils, four pools per coordinating council, six companies per pool, and ten generators per company, the total possible number of generating stations can exceed 2000. The bulk power transmission network (138–765 kV) then typically consists of over 10,000 buses. Indeed, the current demand in the nine coordinating councils within the North American Electric Reliability Council (NERC) exceeds 500,000 MW [3]. At an average 250 MW per generator, this roughly confirms the estimate of over

2000 generators in the interconnected North American grid.

Dynamic studies are routinely performed on systems ranging in size from the smallest company to the largest coordinating council. These are made at both the planning/design and operating stages. These studies provide information about local capabilities as well as regional power interchange capabilities. In view of the potential size, dynamic studies must be capable of sufficiently accurate representation without prohibitive computational cost. The nature of system engineering problems inherent in such a complex task was emphasized in two benchmark reports by the U. S. Department of Energy (DOE) and the Electric Power Research Institute (EPRI) [4, 5]. These reports resulted in a meeting of international leaders to identify directions for the future of this technology. These reports set the stage for a whole new era of power system planning and operation. The volume of follow-on research and industry application has been tremendous. Perhaps the most significant impact of these reports was the stimulation of new ideas that grew into student interest and eventual manpower.

1.5   The Phenomena of Interest

The dynamic performance of power systems is important to both the system organizations, from an economic viewpoint, and society in general, from a reliability viewpoint. The analysis of power system dynamics and stability is increasing daily in terms of number and frequency of studies, as well as in complexity and size. Dynamic phenomena have been discussed according to basic function, time-scale properties, and problem size. These three fundamental concepts are very closely related and represent the essence of the challenges of effective simulation of power system dynamics. When properly performed, modeling and simulation capture the phenomena of interest at minimal cost. The first step in this process is understanding the phenomena of interest. Only with a solid physical *and* mathematical understanding can the modeling and simulation properly reflect the critical system behavior. This means that the origin of mathematical models must be understood, and their purpose must be well defined. Once this is accomplished, the minimal cost is achieved by model reduction and simplification without significant loss in accuracy.

Chapter 2

ELECTROMAGNETIC TRANSIENTS

2.1 The Fastest Transients

In the time-scale classification of power system dynamics, the fastest transients are generally considered to be those associated with lightning propagation and switching surges. Since this text is oriented toward system analysis rather than component design, these transients are discussed in the context of their propagation into other areas of an interconnected system. While quantities such as conductor temperature, motion, and chemical reaction are important aspects of such high-speed transients, we focus mainly on a circuit view, where voltage and current are of primary importance. While the theories of insulation breakdown, arcing, and lightning propagation rarely lend themselves to incorporation into standard circuit analysis [7], some simulation software does include a portion of these transients [8, 9]. From a system viewpoint, the transmission line is the main component that provides the interconnection to form large complex models. While the electromagnetic transients programs (EMTP) described in [8] and [9] are unique for their treatment of switching phenomena of value to designers, they include the capability to study the propagation of transients through transmission lines. This feature makes the EMTP program a system analyst's tool as well as a designer's tool. The transmission line models and basic network solution methods used in these programs are discussed in the following sections.

2.2   Transmission Line Models

Models for transmission lines for use in network analysis are usually categorized by line lengths (long, medium, short) [10]–[12]. This line length concept is interesting, and presents a major challenge in the systematic formulation of line models for dynamic analysis. For example, most students and engineers have been introduced to the argument that shunt capacitance need not be included in short-line models because it has a negligible effect on "the accuracy." Thus, a short line can be modeled using only series resistance and inductance, resulting in a single (for a single line) differential equation in the current state variable. With capacitance, there would also be a differential equation involving the voltage state variable.

Reducing a model from two or three differential equations to only one is a process that has to be justified mathematically as well as physically. As will be shown, the "long-line" model involves partial differential equations, which in some sense represent an infinite number of ordinary differential equations. The reduction from infinity to one is, indeed, a major reduction and deserves further attention.

Since this text deals with dynamics, it is important to be careful with familiar models and concepts. Lumped-parameter models are normally valid for transient analysis unless they are the result of a reduction technique such as Thevenin equivalencing. Investigation of the various traditional transmission line models illustrates this point very well. The traditional derivation of the "long-line" model begins with the construction of an infinitesimal segment of length Δx in Figure 2.1. This length is assumed to be small enough that magnetic and electric field effects can be considered separately, resulting in per-unit length line parameters R', L', G', C'. These distributed parameters have the units of ohms/mi, henries/mi, etc., and are calculated from the line configuration. This incremental lumped-parameter model is, in itself, an approximation of the exact interaction of the electric and magnetic fields. The hope, of course, is that as the incremental segment approaches zero length, the resulting model gives a good approximation of Maxwell's equations. For certain special cases, it can be shown that such an approach is indeed valid ([13], pp. 393–397).

The line has voltages and currents at its sending end (k) and receiving end (m). The voltage and current anywhere along the line are simply

$$v = v(x,t) \qquad\qquad\qquad (2.1)$$
$$i = i(x,t) \qquad\qquad\qquad (2.2)$$

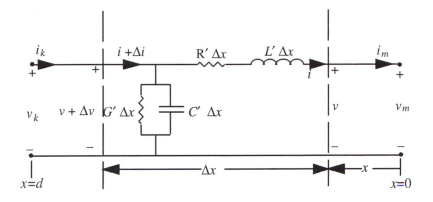

Figure 2.1: *Transmission line segment*

so that

$$v_m = v(o,t) \tag{2.3}$$
$$i_m = i(o,t) \tag{2.4}$$
$$v_k = v(d,t) \tag{2.5}$$
$$i_k = i(d,t) \tag{2.6}$$

where d is the line length. The dynamic equation for the voltage drop across the infinitesimal segment is

$$\Delta v = R'\Delta x i + L'\Delta x \frac{\partial i}{\partial t} \tag{2.7}$$

and the current through the shunt is

$$\Delta i = G'\Delta x(v + \Delta v) + C'\Delta x \frac{\partial}{\partial t}(v + \Delta v) \tag{2.8}$$

Substituting (2.7) into (2.8),

$$\Delta i = G'\Delta x v + G'\Delta x \left[R'\Delta x i + L'\Delta x \frac{\partial i}{\partial t}\right] + C'\Delta x \frac{\partial v}{\partial t}$$
$$+C'\Delta x \left[R'\Delta x \frac{\partial i}{\partial t} + L'\Delta x \frac{\partial^2 i}{\partial t^2}\right] \tag{2.9}$$

Dividing by Δx,

$$\frac{\Delta v}{\Delta x} = R'i + L'\frac{\partial i}{\partial t} \tag{2.10}$$

$$\frac{\Delta i}{\Delta x} = G'v + G'\left[R'\Delta x i + L'\Delta x \frac{\partial i}{\partial t}\right] + C'\frac{\partial v}{\partial t}$$

$$+C'\left[R'\Delta x \frac{\partial i}{\partial t} + L'\Delta x \frac{\partial^2 i}{\partial t^2}\right] \tag{2.11}$$

Now, the original assumption that the magnetic and electric fields may be analyzed separately to obtain the distributed parameters has more credibility when the length under consideration is zero. Thus, the final step is to evaluate (2.10) and (2.11) in the limit as Δx approaches zero, which gives

$$\lim_{\Delta x \to 0} \frac{\Delta v}{\Delta x} = \frac{\partial v}{\partial x} = R'i + L'\frac{\partial i}{\partial t} \tag{2.12}$$

$$\lim_{\Delta x \to 0} \frac{\Delta i}{\Delta x} = \frac{\partial i}{\partial x} = G'v + C'\frac{\partial v}{\partial t} \tag{2.13}$$

Equations (2.12) and (2.13) are the final distributed-parameter models of a lossy transmission line.

There are two special cases when these partial differential equations have very nice known solutions. The first is the special case of no shunt elements ($C' = G' = 0$). From (2.13), the line current is independent of x so that (2.12) simplifies to

$$v(x,t) = v(o,t) + R'xi + L'x\frac{di}{dt} \tag{2.14}$$

which has a simple series lumped R-L circuit representation. This special case essentially neglects all electric field effects.

The second special case is the lossless line ($R' = G' = 0$), which has the general solution [13, 14]

$$i(x,t) = -f_1(x - \nu_p t) - f_2(x + \nu_p t) \tag{2.15}$$

$$v(x,t) = z_c f_1(x - \nu_p t) - z_c f_2(x + \nu_p t) \tag{2.16}$$

where f_1 and f_2 are unknown functions that depend on the boundary conditions, and the phase velocity and the characteristic impedance

$$\nu_p = \frac{1}{\sqrt{L'C'}} \qquad z_c = \sqrt{L'/C'} \tag{2.17}$$

If only the terminal response (v_k, i_k, v_m, i_m) is of interest, the following method, often referred to as Bergeron's method, has a significant value in practical implementations. The receiving end current is

$$i_m(t) = i(o,t) = -f_1(-\nu_p t) - f_2(\nu_p t) \tag{2.18}$$

Now, $f_1(-\nu_p t)$ can be expressed as a function of $v(o,t)$ and $f_2(\nu_p t)$ from (2.16) to obtain

$$i_m(t) = -\frac{1}{z_c}v(o,t) - 2f_2(\nu_p t) \tag{2.19}$$

$$= -\frac{1}{z_c}v_m(t) - 2f_2(\nu_p t) \tag{2.20}$$

To determine $f_2(\nu_p t)$, it is necessary to evaluate the sending end current at time d/ν_p seconds before t as

$$i_k\left(t - \frac{d}{\nu_p}\right) = -f_1(d - \nu_p t + d) - f_2(d + \nu_p t - d) \tag{2.21}$$

Using (2.16) at $x = d$ and at time d/ν_p before t,

$$v_k\left(t - \frac{d}{\nu_p}\right) = z_c f_1(d - \nu_p t + d) - z_c f_2(d + \nu_p t - d) \tag{2.22}$$

so that (2.21) can be evaluated as

$$i_k\left(t - \frac{d}{\nu_p}\right) = -\frac{1}{z_c}v_k\left(t - \frac{d}{\nu_p}\right) - 2f_2(\nu_p t) \tag{2.23}$$

This solves for $f_2(\nu_p t)$ to obtain the expression for the current $i_m(t)$:

$$i_m(t) = -\frac{1}{z_c}v_m(t) + i_k\left(t - \frac{d}{\nu_p}\right)$$

$$+ \frac{1}{z_c}v_k\left(t - \frac{d}{\nu_p}\right) \tag{2.24}$$

This expression has a circuit model as shown in Figure 2.2, where

$$I_m = i_k\left(t - \frac{d}{\nu_p}\right) + \frac{1}{z_c}v_k\left(t - \frac{d}{\nu_p}\right) \tag{2.25}$$

A similar derivation (see Problem 2.1) can be made to determine the sending-end model shown in Figure 2.3, where

$$I_k = i_m\left(t - \frac{d}{\nu_p}\right) - \frac{1}{z_c}v_m\left(t - \frac{d}{\nu_p}\right) \tag{2.26}$$

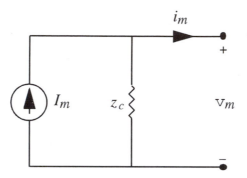

Figure 2.2: *Receiving-end terminal model*

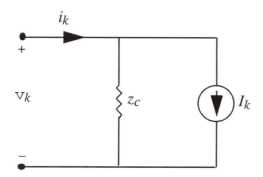

Figure 2.3: *Sending-end terminal model*

These circuit models are illustrated with other components in the next section.

Before leaving this topic, we consider the special case in which the voltages and currents are sinusoidal functions of the form

$$v(x,t) = V(x)\cos(\omega_s t + \theta(x)) \tag{2.27}$$

$$i(x,t) = I(x)\cos(\omega_s t + \phi(x)) \tag{2.28}$$

where ω_s is a constant. Substitution of these functions into the partial
differential equations yields the model of Figure 2.4, with phasors [10]

$$\overline{V}(x) = \frac{1}{\sqrt{2}}V(x)\angle\theta_v(x) \tag{2.29}$$

$$\overline{I}(x) = \frac{1}{\sqrt{2}}I(x)\angle\theta_i(x) \tag{2.30}$$

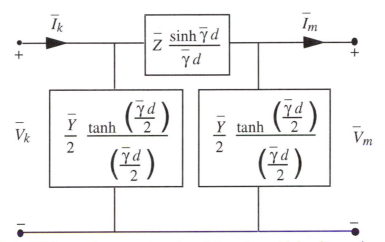

Figure 2.4: *PI lumped-parameter circuit for sinusoidal voltages/currents*

where

$$\overline{Z} = R'd + j\omega_s L'd,$$
$$\overline{Y} = G'd + j\omega_s C'd,$$
$$\overline{\gamma} = \sqrt{\frac{ZY}{d^2}} \tag{2.31}$$

There is always a great temptation to convert Figure 2.4 into a lumped-parameter time-domain R-L-C circuit for transient analysis. While such a circuit would clearly be without mathematical justification, it would be some approximation of the more exact partial differential equation representation. The accuracy of such an approximation would depend on the phenomena of interest and on the relative sizes of the line parameters. In later chapters, we will discuss the concept of "network transients" in the context of fast and slow dynamics.

2.3   Solution Methods

Since most power system models contain nonlinearities, transient analysis usually involves some form of numerical integration. Such numerical methods are well documented for general networks and for power systems [15]–[19]. The trapezoidal rule is a common method used in EMTP and other transient analysis programs. For a dynamic system of the form

$$\frac{dy}{dt} = f(y, t) \tag{2.32}$$

the trapezoidal rule approximates the change of state y over a change of time Δt as

$$y(t_i + \Delta t) \approx y(t_i) + \frac{\Delta t}{2}[f(y(t_i + \Delta t), t_i + \Delta t) + f(y(t_i), t_i)] \tag{2.33}$$

This is an implicit integration scheme, since $y(t_i + \Delta t)$ appears on the right-hand side of (2.33). To illustrate the method, consider the pure inductive branch of Figure 2.5. The state representation is

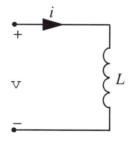

Figure 2.5: *Pure linear inductive branch*

$$\frac{di}{dt} = \frac{1}{L}v \qquad i(t_o) = i_o \tag{2.34}$$

The trapezoidal rule approximate solution is

$$i(t_i + \Delta t) \approx i(t_i) + \frac{\Delta t}{2}\left[\frac{v(t_i + \Delta t)}{L} + \frac{v(t_i)}{L}\right] \tag{2.35}$$

This approximation can be written as the circuit constraint of Figure 2.6. The circuit is linear for given values of $i(t_i)$, $v(t_i)$, and Δt. A similar circuit can be constructed for a pure linear capacitor (see Problem 2.2). Since these are linear elements, there is really no need to employ such an approximation, but recall the lossless-line circuit representation of Figures 2.2 and 2.3. The combination of the Bergeron lossless-line model and the trapezoidal rule for lumped parameters is appealing. There are two difficulties. First, it may be necessary to consider transmission line losses. Second, the lossless-line terminal constraints require knowledge of voltages and currents at a previous time $(t - \frac{d}{\nu_p})$, which may not coincide with a multiple of the integration step size Δt. The first problem is usually overcome by simply lumping a series or shunt resistance at either end of the transmission line terminal model. It is common to break the line into several segments, each using the circuits of Figures 2.2 and 2.3 together with a corresponding fraction of the line losses. The second problem is usually overcome either by rounding off the time d/ν_p to the nearest multiple of Δt, or by linear interpolation over one time step Δt (see discussion published with [14]). This is illustrated in the following example.

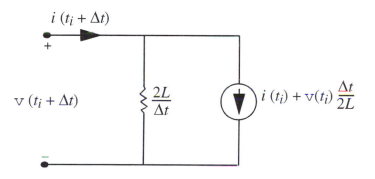

Figure 2.6: *Circuit representation of trapezoidal rule (linear L)*

Example 2.1

Consider a single-phase lossless transmission line connected to an R-L load, as shown in Figure 2.7:

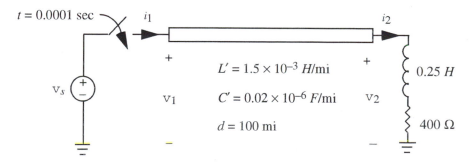

Figure 2.7: *Single lossless line and R-L load diagram*

$$v_s = \frac{230,000\sqrt{2}}{\sqrt{3}} \cos(2\pi 60 t)$$

Find i_1, i_2, and v_2 if the switch is closed at $t = 0.0001$ sec using the trapezoidal rule with a time step $\Delta t = 0.0001$ sec. Initial conditions are $i_1(0) = i_2(0) = v_1(0) = v_2(0) = 0$.

Solution:

For the parameters given,

$$z_c = 274 \text{ ohm}, \quad \frac{2L}{\Delta t} = 5000 \text{ ohm}, \quad \frac{d}{v_p} = 0.00055 \text{ sec}, \quad v_p = 182,574 \text{ mi/sec}$$

The circuit connection when the switch is closed is found using $t = t_i + 0.0001$ as shown in Figure 2.8.

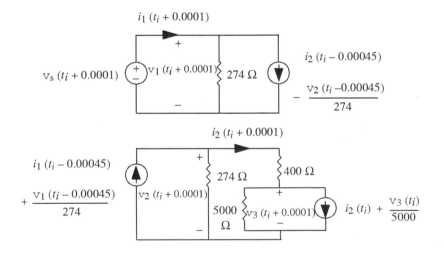

Figure 2.8: *Single line and R-L load circuit at* $t = t_i + 0.0001$

For $t_i = 0$ $(t = 0.0001$ sec$)$

From the initial conditions,

$$i_1(-0.00045) = 0$$

$$v_1(-0.00045) = 0$$
$$i_2(-0.00045) = 0$$
$$v_2(-0.00045) = 0$$
$$i_2(0) = 0$$
$$v_3(0) = 0$$
$$v_s(0.0001) = 187,661 \ V$$

The circuit to be solved at $t = 0.0001$ sec is shown in Figure 2.9.

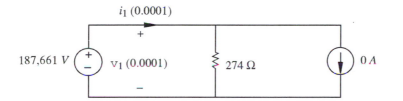

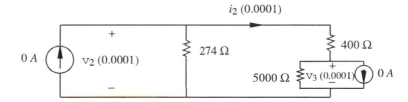

Figure 2.9: *Single line and R-L load circuit at t = 0.0001 sec*

Solving the circuits gives

$$i_1(0.0001) = 685 \ A$$
$$v_1(0.0001) = 187,661 \ V$$
$$i_2(0.0001) = 0$$
$$v_2(0.0001) = 0$$
$$v_3(0.0001) = 0$$

The sending-end current has changed instantaneously from zero to 685 A as the switch is closed.

For $t_i = 0.0001$ $(t = 0.0002$ sec$)$

From the initial conditions and the solution at time $t = 0.0001$ sec,

$$i_1(-0.00035) = 0$$
$$v_1(-0.00035) = 0$$
$$i_2(-0.00035) = 0$$
$$v_2(-0.00035) = 0$$
$$i_2(0.0001) = 0$$
$$v_3(0.0001) = 0$$
$$v_s(0.0002) = 187,261 \ V$$

The circuit for this time is the same as before, except that the source has changed. Solving the circuit gives

$$i_1(0.0002) = 683 \ A$$
$$v_1(0.0002) = 187,261 \ V$$
$$i_2(0.0002) = 0$$
$$v_2(0.0002) = 0$$
$$v_3(0.0002) = 0$$

This will continue until the traveling wave reaches the receiving end at $t_i = 0.00055$ sec. Since we are using a time step of 0.0001 sec, it will first appear at $t_i = 0.0006$ sec.

For $t_i = 0.0006$ $(t = 0.0007$ sec$)$

Need:

$$i_1(0.00015), \ v_1(0.00015), \ i_2(0.00015),$$
$$v_2(0.00015), \ i_2(0.0006), \ v_3(0.0006), \ v_s(0.0007)$$

There is now a problem, since i_1, i_2, and v_2 are not known at $t = 0.00015$. The voltage $v_1(0.00015)$ can be found exactly, since it is equal to the source v_s. The other "sources" in the circuits must be approximated. There are at least two approximations that can be used. The first approximation is to use

$$i_1(0.00015) \approx i_1(0.0002) = 683 \ A$$

The second approximation uses linear interpolation as

$$i_1(0.00015) \approx i_1(0.0001) + \frac{0.00015 - 0.0001}{0.0002 - 0.0001}(i_1(0.0002) - i_1(0.0001))$$
$$= 685 + 0.5 \times (683 - 685)$$
$$= 684 \ A$$

Using this approximation for $i_1(0.00015)$, the circuits to be solved at $t_i = 0.0006$ ($t = 0.0007$ sec) are shown in Figure 2.10.

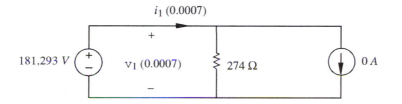

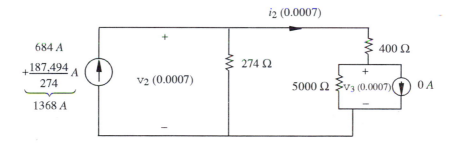

Figure 2.10: *Single line and R-L load at t = 0.0007 sec*

Solving these circuits,

$$i_1(0.0007) = 662 \ A$$
$$v_1(0.0007) = 181,293 \ V$$
$$i_2(0.0007) = 66 \ A$$
$$v_2(0.0007) = 356,731 \ V$$

The traveling wave has reached the receiving end and has resulted in nearly a doubling of voltage, because the receiving end is initially nearly an open circuit. The analysis continues with linear interpolation as needed. When a study contains only one line (like this example), the interface problem between Δt and d/ν_p can be avoided by choosing Δt to be an integer fraction of d/ν_p:

$$\Delta t = \frac{1}{N}\frac{d}{\nu_p}$$

Typical values of N range between 5 and 10,000.

$\square$

The unique feature of the combination of Bergeron circuits with trapezoidal rule circuits is the heart of most EMTP programs. This enables transmission line plus load transients to be solved using simple "dc" circuits. Most EMTP programs contain many other features, including three-phase representations and other devices. Its use is normally limited to small-sized systems in which the very fast transients of switching are the phenomena of interest.

The purpose of this chapter was to present the basic concepts for dealing with the fastest transients from a systems viewpoint. In most studies, these dynamics are approximated further as attention shifts to electromechanical dynamics and subsystems with slower response times.

2.4   Problems

2.1 Starting with (2.15) and (2.16), derive the circuit representation of Figure 2.3 for the sending-end terminals of a lossless transmission line.

2.2 Given the continuous time-domain circuit shown:

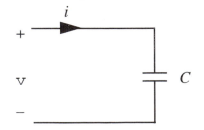

use the trapezoidal rule approximation to find an algebraic "dc" circuit representation of the relationship between $v(t_i + \Delta t)$ and $i(t_i + \Delta t)$.

2.3 Given the sinusoidal source and de-energized lossless transmission line shown:

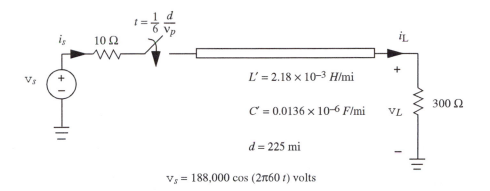

$$v_s = 188{,}000 \cos{(2\pi 60\,t)} \text{ volts}$$

draw the "Bergeron" algebraic "dc" circuit and find v_L, i_L, i_s for $0 \le t \le 0.04$ sec using a time step of $\Delta t = \frac{1}{6}\frac{d}{v_p}$. Plot v_L.

2.4 Given the sinusoidal source and de-energized lossless transmission line shown:

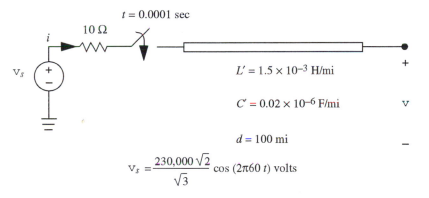

$$v_s = \frac{230{,}000\sqrt{2}}{\sqrt{3}} \cos{(2\pi 60\,t)} \text{ volts}$$

use Bergeron's method with linear interpolation to find v and i using a time step of $\Delta t = 0.0001$ sec. Solve for a total time of 0.02 sec. Plot the results.

2.5 Repeat Problem 2.4 using the lumped-parameter model shown:

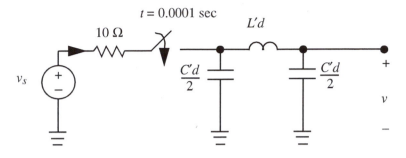

using the trapezoidal rule approximation with a time step of 10^{-6} sec.

2.6 Euler's forward integration scheme solves ordinary differential equations $(dx/dt = f(x))$ using a time step h as

$$x(t_i + h) = x(t_i) + (dx/dt)h$$

where dx/dt is evaluated at time t_i. Use this solution scheme to derive an algebraic "dc" circuit to solve for the current through a lumped-parameter R-L series circuit at each time step for any given applied voltage.

Chapter 3

SYNCHRONOUS MACHINE MODELING

3.1 Conventions and Notation

There is probably more literature on synchronous machines than on any other device in electrical engineering. Unfortunately, this vast amount of material often makes the subject complex and confusing. In addition, most of the work on reduced-order modeling is based primarily on physical intuition, practical experience, and years of experimentation. The evolution of dynamic analysis has caused some problems in notation as it relates to common symbols that eventually require data from manufacturers. This text uses the conventions and notations of [20], which essentially follows those of many publications on synchronous machines [21]–[27]. When the notation differs significantly from these and other conventions, notes are given to clarify any possible misunderstanding. The topics of time constants and machine inductances are examples of such notations. While some documents define time constants and inductances in terms of physical experiments, this text uses fixed expressions in terms of model parameters. Since there can be a considerable difference in numerical values, it is important to always verify the meaning of symbols when obtaining data. This is most effectively done by comparing the model in which a parameter appears with the test or calculation that was performed to produce the data. In many cases, the parameter values are provided from design data based on the same expressions given in this text. In some cases, the parameter values are provided from standard tests that may not precisely relate to the expressions given in this text. In this case, there is normally a procedure to convert the values into consistent data [20].

The original Park's transformation is used together with the "x_{ad}" per-unit system [28, 29]. This results in a reciprocal transformed per-unit model where 1.0 per-unit excitation results in rated open-circuit voltage for a linear magnetic system. Even with this standard choice, there is enough freedom in scaling to produce various model structures that appear different [30]. These issues are discussed further in later sections.

In this chapter, the machine transformation and scaling were separated from the topic of the magnetic circuit representation. This is done so that it is clear which equations and parameters are independent of the magnetic circuit representation.

3.2   Three-Damper-Winding Model

This section presents the basic dynamic equations for a balanced, symmetrical, three-phase synchronous machine with a field winding and three damper windings on the rotor. The simplified schematic of Figure 3.1 shows the coil orientation, assumed polarities, and rotor position reference. The stator windings have axes 120 electrical degrees apart and are assumed to have an equivalent sinusoidal distribution [20]. While a two-pole machine is shown, all equations will be written for a P-pole machine with $\omega = \frac{P}{2}\omega_{\text{shaft}}$ expressed in electrical radians per second. The circles with dots and x's indicate the windings. Current flow is assumed to be into the "x" and out of the "dot." The voltage polarity of the coils is assumed to be plus to minus from the "x" to the "dots."

This notation uses "motor" current notation for all the windings at this point. The transformed stator currents will be changed to "generator" current notation at the point of per-unit scaling. The fundamental Kirchhoff's, Faraday's and Newton's laws give

$$v_a = i_a r_s + \frac{d\lambda_a}{dt} \tag{3.1}$$

$$v_b = i_b r_s + \frac{d\lambda_b}{dt} \tag{3.2}$$

$$v_c = i_c r_s + \frac{d\lambda_c}{dt} \tag{3.3}$$

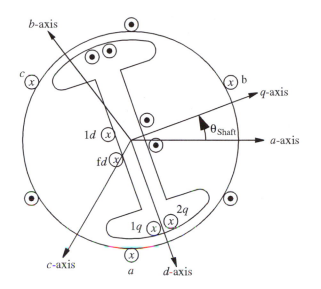

Figure 3.1: *Synchronous machine schematic*

$$v_{fd} = i_{fd}r_{fd} + \frac{d\lambda_{fd}}{dt} \tag{3.4}$$

$$v_{1d} = i_{1d}r_{1d} + \frac{d\lambda_{1d}}{dt} \tag{3.5}$$

$$v_{1q} = i_{1q}r_{1q} + \frac{d\lambda_{1q}}{dt} \tag{3.6}$$

$$v_{2q} = i_{2q}r_{2q} + \frac{d\lambda_{2q}}{dt} \tag{3.7}$$

$$\frac{d\theta_{\text{shaft}}}{dt} = \frac{2}{P}\omega \tag{3.8}$$

$$J\frac{2}{P}\frac{d\omega}{dt} = T_m - T_e - T_{fw} \tag{3.9}$$

where λ is flux linkage, r is winding resistance, J is the inertia constant, P is the number of magnetic poles per phase, T_m is the mechanical torque applied to the shaft, $-T_e$ is the torque of electrical origin, and T_{fw} is a friction windage torque. A major modeling challenge is to obtain the relationship between flux linkage and current. These relationships will be presented in later sections.

3.3   Transformations and Scaling

The sinusoidal steady state of balanced symmetrical machines can be transformed to produce constant states. The general form of the transformation that accomplishes this is Park's transformation [20],

$$v_{dqo} \triangleq T_{dqo} v_{abc}, \quad i_{dqo} \triangleq T_{dqo} i_{abc}, \quad \lambda_{dqo} \triangleq T_{dqo} \lambda_{abc} \qquad (3.10)$$

where

$$v_{abc} \triangleq [v_a v_b v_c]^t, \quad i_{abc} \triangleq [i_a i_b i_c]^t, \quad \lambda_{abc} \triangleq [\lambda_a \lambda_b \lambda_c]^t \qquad (3.11)$$

$$v_{dqo} \triangleq [v_d v_q v_o]^t, \quad i_{dqo} \triangleq [i_d i_q i_o]^t, \quad \lambda_{dqo} \triangleq [\lambda_d \lambda_q \lambda_o]^t \qquad (3.12)$$

and

$$T_{dqo} \triangleq \frac{2}{3} \begin{bmatrix} \sin \frac{P}{2}\theta_{\text{shaft}} & \sin(\frac{P}{2}\theta_{\text{shaft}} - \frac{2\pi}{3}) & \sin(\frac{P}{2}\theta_{\text{shaft}} + \frac{2\pi}{3}) \\ \cos \frac{P}{2}\theta_{\text{shaft}} & \cos(\frac{P}{2}\theta_{\text{shaft}} - \frac{2\pi}{3}) & \cos(\frac{P}{2}\theta_{\text{shaft}} + \frac{2\pi}{3}) \\ \frac{1}{2} & \frac{1}{2} & \frac{1}{2} \end{bmatrix} \qquad (3.13)$$

with the inverse

$$T_{dqo}^{-1} = \begin{bmatrix} \sin \frac{P}{2}\theta_{\text{shaft}} & \cos \frac{P}{2}\theta_{\text{shaft}} & 1 \\ \sin(\frac{P}{2}\theta_{\text{shaft}} - \frac{2\pi}{3}) & \cos(\frac{P}{2}\theta_{\text{shaft}} - \frac{2\pi}{3}) & 1 \\ \sin(\frac{P}{2}\theta_{\text{shaft}} + \frac{2\pi}{3}) & \cos(\frac{P}{2}\theta_{\text{shaft}} + \frac{2\pi}{3}) & 1 \end{bmatrix} \qquad (3.14)$$

From (3.1)–(3.9), Kirchhoff's and Faraday's laws are

$$v_{abc} = r_s i_{abc} + \frac{d}{dt}(\lambda_{abc}) \qquad (3.15)$$

which, when transformed using (3.13) and (3.14), are

$$v_{dqo} = r_s i_{dqo} + T_{dqo}\frac{d}{dt}(T_{dqo}^{-1}\lambda_{dqo}) \qquad (3.16)$$

After evaluation, the system in dqo coordinates has the forms

$$v_d = r_s i_d - \omega \lambda_q + \frac{d\lambda_d}{dt} \qquad (3.17)$$

$$v_q = r_s i_q + \omega \lambda_d + \frac{d\lambda_q}{dt} \qquad (3.18)$$

$$v_o = r_s i_o + \frac{d\lambda_o}{dt} \tag{3.19}$$

$$v_{fd} = r_{fd} i_{fd} + \frac{d\lambda_{fd}}{dt} \tag{3.20}$$

$$v_{1d} = r_{1d} i_{1d} + \frac{d\lambda_{1d}}{dt} \tag{3.21}$$

$$v_{1q} = r_{1q} i_{1q} + \frac{d\lambda_{1q}}{dt} \tag{3.22}$$

$$v_{2q} = r_{2q} i_{2q} + \frac{d\lambda_{2q}}{dt} \tag{3.23}$$

$$\frac{d\theta_{\text{shaft}}}{dt} = \frac{2}{P}\omega \tag{3.24}$$

$$J\frac{2}{P}\frac{d\omega}{dt} = T_m - T_e - T_{fw} \tag{3.25}$$

To derive an expression for T_e, it is necessary to look at the overall energy or power balance for the machine. This is an electromechanical system that can be divided into an electrical system, a mechanical system, and a coupling field [31]. In such a system, resistance causes real power losses in the electrical system, friction causes heat losses in the mechanical system, and hysteresis causes losses in the coupling field. Energy is stored in inductances in the electrical system, the rotating mass of the mechanical system, and the magnetic field that couples the two. Any energy that is not lost or stored must be transferred. In this text we make two assumptions about this energy balance. First, all energy stored in the electrical system inside the machine terminals is included in the energy stored in the coupling field. Second, the coupling field is lossless. The first assumption is arbitrary, and the second assumption neglects phenomena such as hysteresis (but not saturation). A diagram that shows such a power balance for a single machine using the above notation is given in Figure 3.2, with input powers for both the electrical and mechanical systems.

The electrical powers are

$$P_{\substack{\text{in}\\\text{elec}}} = v_a i_a + v_b i_b + v_c i_c + v_{fd} i_{fd} + v_{1d} i_{1d} + v_{1q} i_{1q} + v_{2q} i_{2q} \tag{3.26}$$

$$P_{\substack{\text{lost}\\\text{elec}}} = r_s(i_a^2 + i_b^2 + i_c^2) + r_{fd} i_{fd}^2 + r_{1d} i_{1d}^2 + r_{1q} i_{1q}^2 + r_{2q} i_{2q}^2 \tag{3.27}$$

$$P_{\substack{trans \\ elec}} = i_a \frac{d\lambda_a}{dt} + i_b \frac{d\lambda_b}{dt} + i_c \frac{d\lambda_c}{dt} + i_{fd} \frac{d\lambda_{fd}}{dt} + i_{1d} \frac{d\lambda_{1d}}{dt} + i_{1q} \frac{d\lambda_{1q}}{dt}$$

$$+ i_{2q} \frac{d\lambda_{2q}}{dt} \tag{3.28}$$

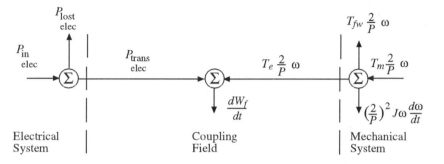

Figure 3.2: *Synchronous machine power balance*

The summation of the electrical system is simply Kirchhoff's plus Faraday's laws, and the summation of the mechanical system is Newton's second law. In terms of the transformed variables, since

$$v_a i_a + v_b i_b + v_c i_c = \frac{3}{2} v_d i_d + \frac{3}{2} v_q i_q + 3 v_o i_o \tag{3.29}$$

$$P_{\substack{in \\ elec}} = \frac{3}{2} v_d i_d + \frac{3}{2} v_q i_q$$

$$+ 3 v_o i_o + v_{fd} i_{fd} + v_{1d} i_{1d} + v_{1q} i_{1q} + v_{2q} i_{2q} \tag{3.30}$$

$$P_{\substack{lost \\ elec}} = \frac{3}{2} r_s i_d^2 + \frac{3}{2} r_s i_q^2 + 3 r_s i_o^2$$

$$+ r_{fd} i_{fd}^2 + r_{1d} i_{1d}^2 + r_{1q} i_{1q}^2 + r_{2q} i_{2q}^2 \tag{3.31}$$

$$P_{\substack{trans \\ elec}} = -\frac{3}{2} \frac{P}{2} \frac{d\theta_{shaft}}{dt} \lambda_q i_d$$

$$+ \frac{3}{2} i_d \frac{d\lambda_d}{dt} + \frac{3}{2} \frac{P}{2} \frac{d\theta_{shaft}}{dt} \lambda_d i_q + \frac{3}{2} i_q \frac{d\lambda_q}{dt}$$

$$+ 3 i_o \frac{d\lambda_o}{dt} + i_{fd} \frac{d\lambda_{fd}}{dt} + i_{1d} \frac{d\lambda_{1d}}{dt}$$

$$+ i_{1q} \frac{d\lambda_{1q}}{dt} + i_{2q} \frac{d\lambda_{2q}}{dt} \tag{3.32}$$

The power balance equation in the coupling field in these *dqo* coordinates gives the time derivative of the energy stored in the coupling field as

$$\frac{dW_f}{dt} = \left[\frac{3}{2}\frac{P}{2}(\lambda_d i_q - \lambda_q i_d) + T_e\right]\frac{d\theta_{\text{shaft}}}{dt} + \frac{3}{2}i_d\frac{d\lambda_d}{dt} + \frac{3}{2}i_q\frac{d\lambda_q}{dt} + 3i_o\frac{d\lambda_o}{dt}$$

$$+ i_{fd}\frac{d\lambda_{fd}}{dt} + i_{1d}\frac{d\lambda_{1d}}{dt} + i_{1q}\frac{d\lambda_{1q}}{dt} + i_{2q}\frac{d\lambda_{2q}}{dt} \tag{3.33}$$

For independent states $\theta_{\text{shaft}}, \lambda_d, \lambda_q, \lambda_o, \lambda_{fd}, \lambda_{1d}, \lambda_{1q}, \lambda_{2q}$ the total derivative of W_f is

$$\frac{dW_f}{dt} = \frac{\partial W_f}{\partial \theta_{\text{shaft}}}\frac{d\theta_{\text{shaft}}}{dt} + \frac{\partial W_f}{\partial \lambda_d}\frac{d\lambda_d}{dt} + \frac{\partial W_f}{\partial \lambda_q}\frac{d\lambda_q}{dt} + \frac{\partial W_f}{\partial \lambda_o}\frac{d\lambda_o}{dt} + \frac{\partial W_f}{\partial \lambda_{fd}}\frac{d\lambda_{fd}}{dt}$$

$$+ \frac{\partial W_f}{\partial \lambda_{1d}}\frac{d\lambda_{1d}}{dt} + \frac{\partial W_f}{\partial \lambda_{1q}}\frac{d\lambda_{1q}}{dt} + \frac{\partial W_f}{\partial \lambda_{2q}}\frac{d\lambda_{2q}}{dt} \tag{3.34}$$

For this total derivative to be exact [32], the following identities must hold:

$$\frac{\partial W_f}{\partial \theta_{\text{shaft}}} = \frac{3}{2}\frac{P}{2}(\lambda_d i_q - \lambda_q i_d) + T_e, \quad \frac{\partial W_f}{\partial \lambda_d} = \frac{3}{2}i_d \quad \text{etc.} \tag{3.35}$$

With appropriate continuity assumptions [33], the coupling field energy can be obtained from (3.33) as a path integral

$$W_f = W_f^o + \int \left[\frac{3}{2}\frac{P}{2}(\lambda_d i_q - \lambda_q i_d) + T_e\right]d\theta_{\text{shaft}} + \int \frac{3}{2}i_d d\lambda_d$$

$$+ \int \frac{3}{2}i_q d\lambda_q + \int 3i_o d\lambda_o + \int i_{fd} d\lambda_{fd} + \int i_{1d} d\lambda_{1d}$$

$$+ \int i_{1q} d\lambda_{1q} + \int i_{2q} d\lambda_{2q} \tag{3.36}$$

For this integral to be path independent, the partial derivatives of all integrands with respect to other states must be equal [34], i.e.,

$$\frac{3}{2}\frac{\partial i_d}{\partial \lambda_{fd}} = \frac{\partial i_{fd}}{\partial \lambda_d} \quad \text{etc.} \tag{3.37}$$

These constraints can also be obtained from those of (3.35) by taking the second partials of W_f with respect to states. The assumption that the coupling field is conservative is sufficient to guarantee that these constraints are satisfied. Nevertheless, these constraints should always be kept in mind when deriving the magnetic circuit relationships between flux linkage and current.

Assuming that these constraints are satisfied, the following arbitrary path is chosen from the de-energized ($W_f^o = 0$) condition to the energized condition:

1. Integrate rotor position to some arbitrary θ_{shaft} while all sources are de-energized. This adds zero to W_f since λ_d, λ_q, and T_e must be zero.

2. Integrate each source in sequence while maintaining θ_{shaft} at its arbitrary position.

With this chosen path, W_f will be the sum of seven integrals for the seven independent sources $\lambda_d, \lambda_q, \lambda_o, \lambda_{fd}, \lambda_{1d}, \lambda_{1q}, \lambda_{2q}$. Each integrand is the respective source current that must be given as a function of the states. Since this is not done until later sections, we make the following assumption. Assume that the relationships between $\lambda_d, \lambda_q, \lambda_o, \lambda_{fd}, \lambda_{1d}, \lambda_{1q}, \lambda_{2q}$ and $i_d, i_q, i_o, i_{fd}, i_{1d}, i_{1q}, i_{2q}$ are independent of θ_{shaft}. For this assumption, W_f will be independent of θ_{shaft} so that, from (3.35),

$$T_e = -\left(\frac{3}{2}\right)\left(\frac{P}{2}\right)(\lambda_d i_q - \lambda_q i_d) \tag{3.38}$$

To complete the dynamic model in the transformed variables, it is desirable to define an angle that is constant for constant shaft speed. We define this angle as follows:

$$\delta \triangleq \frac{P}{2}\theta_{\text{shaft}} - \omega_s t \tag{3.39}$$

where ω_s is a constant normally called rated synchronous speed in electrical radians per second, giving

$$\frac{d\delta}{dt} = \omega - \omega_s \tag{3.40}$$

The final unscaled model in the new variables is

$$\frac{d\lambda_d}{dt} = -r_s i_d + \omega\lambda_q + v_d \tag{3.41}$$

$$\frac{d\lambda_q}{dt} = -r_s i_q - \omega\lambda_d + v_q \tag{3.42}$$

$$\frac{d\lambda_o}{dt} = -r_s i_o + v_o \tag{3.43}$$

$$\frac{d\lambda_{fd}}{dt} = -r_{fd}i_{fd} + v_{fd} \tag{3.44}$$

$$\frac{d\lambda_{1d}}{dt} = -r_{1d}i_{1d} + v_{1d} \tag{3.45}$$

$$\frac{d\lambda_{1q}}{dt} = -r_{1q}i_{1q} + v_{1q} \tag{3.46}$$

$$\frac{d\lambda_{2q}}{dt} = -r_{2q}i_{2q} + v_{2q} \tag{3.47}$$

$$\frac{d\delta}{dt} = \omega - \omega_s \tag{3.48}$$

$$J\frac{2}{p}\frac{d\omega}{dt} = T_m + \left(\frac{3}{2}\right)\left(\frac{P}{2}\right)(\lambda_d i_q - \lambda_q i_d) - T_{fw} \tag{3.49}$$

It is customary to scale the synchronous machine equations using the traditional concept of per-unit [28, 29]. This scaling process is presented here as a change of variables and a change of parameters. We begin by defining new *abc* variables as

$$V_a \triangleq \frac{v_a}{V_{BABC}}, \quad V_b \triangleq \frac{v_b}{V_{BABC}}, \quad V_c \triangleq \frac{v_c}{V_{BABC}},$$

$$I_a \triangleq \frac{-i_a}{I_{BABC}}, \quad I_b \triangleq \frac{-i_b}{I_{BABC}}, \quad I_c \triangleq \frac{-i_c}{I_{BABC}},$$

$$\psi_a \triangleq \frac{\lambda_a}{\wedge_{BABC}}, \quad \psi_b \triangleq \frac{\lambda_b}{\wedge_{BABC}}, \quad \psi_c \triangleq \frac{\lambda_c}{\wedge_{BABC}} \tag{3.50}$$

where V_{BABC} is a rated *RMS* line to neutral stator voltage and

$$I_{BABC} \triangleq \frac{S_B}{3V_{BABC}}, \quad \wedge_{BABC} \triangleq \frac{V_{BABC}}{\omega_B} \tag{3.51}$$

with S_B equal to the rated three-phase voltamperes and ω_B equal to rated speed in electrical radians per second (ω_s). This scaling also converts the model to "generator" notation. The new *dqo* variables are defined as

$$V_d \triangleq \frac{v_d}{V_{BDQ}}, \quad V_q \triangleq \frac{v_q}{V_{BDQ}}, \quad V_o \triangleq \frac{v_o}{V_{BDQ}},$$

$$I_d \triangleq \frac{-i_d}{I_{BDQ}}, \quad I_q \triangleq \frac{-i_q}{I_{BDQ}}, \quad I_o \triangleq \frac{-i_o}{I_{BDQ}},$$

$$\psi_d \triangleq \frac{\lambda_d}{\wedge_{BDQ}}, \quad \psi_q \triangleq \frac{\lambda_q}{\wedge_{BDQ}}, \quad \psi_o \triangleq \frac{\lambda_o}{\wedge_{BDQ}} \tag{3.52}$$

where V_{BDQ} is rated peak line to neutral voltage, and

$$I_{BDQ} \triangleq \frac{2S_B}{3V_{BDQ}}, \quad \wedge_{BDQ} \triangleq \frac{V_{BDQ}}{\omega_B} \tag{3.53}$$

with S_B and ω_B defined as above, and again the scaling has converted to generator notation. The new rotor variables are defined as

$$V_{fd} \triangleq \frac{v_{fd}}{V_{BFD}}, \quad V_{1d} \triangleq \frac{v_{1d}}{V_{B1D}}, \quad V_{1q} \triangleq \frac{v_{1q}}{V_{B1Q}}, \quad V_{2q} \triangleq \frac{v_{2q}}{V_{B2Q}},$$

$$I_{fd} \triangleq \frac{i_{fd}}{I_{BFD}}, \quad I_{1d} \triangleq \frac{i_{1d}}{I_{B1D}}, \quad I_{1q} \triangleq \frac{i_{1q}}{I_{B1Q}}, \quad I_{2q} \triangleq \frac{i_{2q}}{I_{B2Q}},$$

$$\psi_{fd} \triangleq \frac{\lambda_{fd}}{\wedge_{BFD}}, \quad \psi_{1d} \triangleq \frac{\lambda_{1d}}{\wedge_{B1D}}, \quad \psi_{1q} \triangleq \frac{\lambda_{1q}}{\wedge_{B1Q}}, \quad \psi_{2q} \triangleq \frac{\lambda_{2q}}{\wedge_{B2Q}} \tag{3.54}$$

where the rotor circuit base voltages are

$$V_{BFD} \triangleq \frac{S_B}{I_{BFD}}, \quad V_{B1D} \triangleq \frac{S_B}{I_{B1D}}, \quad V_{B1Q} \triangleq \frac{S_B}{I_{B1Q}}, \quad V_{B2Q} \triangleq \frac{S_B}{I_{B2Q}} \tag{3.55}$$

and the rotor circuit base flux linkages are

$$\wedge_{BFD} \triangleq \frac{V_{BFD}}{\omega_B}, \quad \wedge_{B1D} \triangleq \frac{V_{B1D}}{\omega_B}, \quad \wedge_{B1Q} \triangleq \frac{V_{B1Q}}{\omega_B}, \quad \wedge_{B2Q} \triangleq \frac{V_{B2Q}}{\omega_B} \tag{3.56}$$

with S_B and ω_B defined as above. The definitions of the rotor circuit base currents will be given later, when the flux linkage/current relationships are presented. In some models, it is convenient to define a scaled per-unit speed as

$$\nu \triangleq \frac{\omega}{\omega_B} \tag{3.57}$$

where ω_B is as defined above.

This completes the scaling of the model variables. The model parameters are scaled as follows. Define new resistances

$$R_s \triangleq \frac{r_s}{Z_{BDQ}}, \quad R_{fd} \triangleq \frac{r_{fd}}{Z_{BFD}}, \quad R_{1d} \triangleq \frac{r_{1d}}{Z_{B1D}},$$

$$R_{1q} \triangleq \frac{r_{1q}}{Z_{B1Q}}, \quad R_{2q} \triangleq \frac{r_{2q}}{Z_{B2Q}} \tag{3.58}$$

where

$$Z_{BDQ} \triangleq \frac{V_{BDQ}}{I_{BDQ}}, \quad Z_{BFD} \triangleq \frac{V_{BFD}}{I_{BFD}}, \quad Z_{B1D} \triangleq \frac{V_{B1D}}{I_{B1D}},$$

$$Z_{B1Q} \triangleq \frac{V_{B1Q}}{I_{B1Q}}, \quad Z_{B2Q} \triangleq \frac{V_{B2Q}}{I_{B2Q}} \tag{3.59}$$

The shaft inertia constant is scaled by defining

$$H \triangleq \frac{\frac{1}{2}J(\omega_B \frac{2}{P})^2}{S_B} \qquad (3.60)$$

It is also common to define other inertia constants as

$$M = \frac{2H}{\omega_s}, \quad M' = 2H \qquad (3.61)$$

where the constants H and M' have the units of seconds, while M has the units of seconds squared. The shaft torques are scaled by defining

$$T_M \triangleq \frac{T_m}{T_B}, \quad T_{ELEC} \triangleq \frac{T_e}{T_B}, \quad T_{FW} \triangleq \frac{T_{fw}}{T_B} \qquad (3.62)$$

where

$$T_B \triangleq \frac{S_B}{\omega_B \frac{2}{P}} \qquad (3.63)$$

Using these scaled variables and parameters, the synchronous machine dynamic equations at this stage of development with $\omega_B = \omega_s$ are

$$\frac{1}{\omega_s}\frac{d\psi_d}{dt} = R_s I_d + \frac{\omega}{\omega_s}\psi_q + V_d \qquad (3.64)$$

$$\frac{1}{\omega_s}\frac{d\psi_q}{dt} = R_s I_q - \frac{\omega}{\omega_s}\psi_d + V_q \qquad (3.65)$$

$$\frac{1}{\omega_s}\frac{d\psi_o}{dt} = R_s I_o + V_o \qquad (3.66)$$

$$\frac{1}{\omega_s}\frac{d\psi_{fd}}{dt} = -R_{fd} I_{fd} + V_{fd} \qquad (3.67)$$

$$\frac{1}{\omega_s}\frac{d\psi_{1d}}{dt} = -R_{1d} I_{1d} + V_{1d} \qquad (3.68)$$

$$\frac{1}{\omega_s}\frac{d\psi_{1q}}{dt} = -R_{1q} I_{1q} + V_{1q} \qquad (3.69)$$

$$\frac{1}{\omega_s}\frac{d\psi_{2q}}{dt} = -R_{2q} I_{2q} + V_{2q} \qquad (3.70)$$

$$\frac{d\delta}{dt} = \omega - \omega_s \qquad (3.71)$$

$$\frac{2H}{\omega_s}\frac{d\omega}{dt} = T_M - (\psi_d I_q - \psi_q I_d) - T_{FW} \qquad (3.72)$$

It is important to pause at this point to consider the scaling of variables. Consider a balanced set of scaled sinusoidal voltages and currents of the form:

$$V_a = \sqrt{2}V_s \cos\left(\omega_s t + \theta_s\right) \tag{3.73}$$

$$V_b = \sqrt{2}V_s \cos\left(\omega_s t + \theta_s - \frac{2\pi}{3}\right) \tag{3.74}$$

$$V_c = \sqrt{2}V_s \cos\left(\omega_s t + \theta_s + \frac{2\pi}{3}\right) \tag{3.75}$$

$$I_a = \sqrt{2}I_s \cos\left(\omega_s t + \phi_s\right) \tag{3.76}$$

$$I_b = \sqrt{2}I_s \cos\left(\omega_s t + \phi_s - \frac{2\pi}{3}\right) \tag{3.77}$$

$$I_c = \sqrt{2}I_s \cos\left(\omega_s t + \phi_s + \frac{2\pi}{3}\right) \tag{3.78}$$

Using the transformation (3.13),

$$V_d = \left(\frac{\sqrt{2}V_s V_{BABC}}{V_{BDQ}}\right) \sin\left(\frac{P}{2}\theta_{\text{shaft}} - \omega_s t - \theta_s\right) \tag{3.79}$$

$$V_q = \left(\frac{\sqrt{2}V_s V_{BABC}}{V_{BDQ}}\right) \cos\left(\frac{P}{2}\theta_{\text{shaft}} - \omega_s t - \theta_s\right) \tag{3.80}$$

$$V_o = 0 \tag{3.81}$$

$$I_d = \left(\frac{\sqrt{2}I_s I_{BABC}}{I_{BDQ}}\right) \sin\left(\frac{P}{2}\theta_{\text{shaft}} - \omega_s t - \phi_s\right) \tag{3.82}$$

$$I_q = \left(\frac{\sqrt{2}I_s I_{BABC}}{I_{BDQ}}\right) \cos\left(\frac{P}{2}\theta_{\text{shaft}} - \omega_s t - \phi_s\right) \tag{3.83}$$

$$I_o = 0 \tag{3.84}$$

By the definitions of V_{BABC}, V_{BDQ}, I_{BABC}, and I_{BDQ},

$$\frac{\sqrt{2}V_s V_{BABC}}{V_{BDQ}} = V_s, \quad \frac{\sqrt{2}I_s I_{BABC}}{I_{BDQ}} = I_s \tag{3.85}$$

Using the definition of δ from (3.39),

$$V_d = V_s \sin\left(\delta - \theta_s\right) \tag{3.86}$$

$$V_q = V_s \cos (\delta - \theta_s) \tag{3.87}$$

$$I_d = I_s \sin (\delta - \phi_s) \tag{3.88}$$

$$I_q = I_s \cos (\delta - \phi_s) \tag{3.89}$$

These algebraic equations can be written as complex equations

$$(V_d + jV_q)e^{j(\delta - \pi/2)} = V_s e^{j\theta_s} \tag{3.90}$$

$$(I_d + jI_q)e^{j(\delta - \pi/2)} = I_s e^{j\phi_s} \tag{3.91}$$

These are recognized as the per-unit RMS phasors of (3.73) and (3.76).

It is also important, at this point, to note that the model of (3.64)–(3.72) was derived using essentially four general assumptions. These assumptions are summarized as follows.

1. Stator has three coils in a balanced symmetrical configuration centered 120 electrical degrees apart.

2. Rotor has four coils in a balanced symmetrical configuration located in pairs 90 electrical degrees apart.

3. The relationship between the flux linkages and currents must reflect a conservative coupling field.

4. The relationships between the flux linkages and currents must be independent of θ_{shaft} when expressed in the dqo coordinate system.

The following sections give the flux linkage/current relationships that satisfy these four assumptions and thus complete the dynamic model.

3.4 The Linear Magnetic Circuit

This section presents the special case in which the machine flux linkages are assumed to be linear functions of currents:

$$\lambda_{abc} = L_{ss}(\theta_{\text{shaft}})i_{abc} + L_{sr}(\theta_{\text{shaft}})i_{\text{rotor}} \tag{3.92}$$

$$\lambda_{\text{rotor}} = L_{rs}(\theta_{\text{shaft}})i_{abc} + L_{rr}(\theta_{\text{shaft}})i_{\text{rotor}} \tag{3.93}$$

where

$$i_{\text{rotor}} \triangleq [i_{fd} i_{1d} i_{1q} i_{2q}]^t, \quad \lambda_{\text{rotor}} \triangleq [\lambda_{fd} \lambda_{1d} \lambda_{1q} \lambda_{2q}]^t \tag{3.94}$$

If space harmonics are neglected, the entries of these inductance matrices can be written in a form that satisfies assumptions (3) and (4) of the last section. Reference [20] discusses this formulation and gives the following standard first approximation of the inductances for a P-pole machine.

$$L_{ss}(\theta_{\text{shaft}}) \triangleq$$

$$\begin{bmatrix} L_{\ell s} + L_A - L_B \cos P\theta_{\text{shaft}} & -\frac{1}{2}L_A - L_B \cos(P\theta_{\text{shaft}} - \frac{2\pi}{3}) \\ -\frac{1}{2}L_A - L_B \cos(P\theta_{\text{shaft}} - \frac{2\pi}{3}) & L_{\ell s} + L_A - L_B \cos(P\theta_{\text{shaft}} + \frac{2\pi}{3}) \\ -\frac{1}{2}L_A - L_B \cos(P\theta_{\text{shaft}} + \frac{2\pi}{3}) & -\frac{1}{2}L_A - L_B \cos P\theta_{\text{shaft}} \end{bmatrix}$$

$$\begin{matrix} -\frac{1}{2}L_A - L_B \cos(P\theta_{\text{shaft}} + \frac{2\pi}{3}) \\ -\frac{1}{2}L_A - L_B \cos P\theta_{\text{shaft}} \\ L_{\ell s} + L_A - L_B \cos(P\theta_{\text{shaft}} - \frac{2\pi}{3}) \end{matrix} \Bigg] \tag{3.95}$$

$$L_{sr}(\theta_{\text{shaft}}) = L_{rs}^t(\theta_{\text{shaft}}) \triangleq$$

$$\begin{bmatrix} L_{sfd} \sin \frac{P}{2}\theta_{\text{shaft}} & L_{s1d} \sin \frac{P}{2}\theta_{\text{shaft}} \\ L_{sfd} \sin(\frac{P}{2}\theta_{\text{shaft}} - \frac{2\pi}{3}) & L_{s1d} \sin(\frac{P}{2}\theta_{\text{shaft}} - \frac{2\pi}{3}) \\ L_{sfd} \sin(\frac{P}{2}\theta_{\text{shaft}} + \frac{2\pi}{3}) & L_{s1d} \sin(\frac{P}{2}\theta_{\text{shaft}} + \frac{2\pi}{3}) \end{bmatrix}$$

$$\begin{matrix} L_{s1q} \cos \frac{P}{2}\theta_{\text{shaft}} & L_{s2q} \cos \frac{P}{2}\theta_{\text{shaft}} \\ L_{s1q} \cos(\frac{P}{2}\theta_{\text{shaft}} - \frac{2\pi}{3}) & L_{s2q} \cos(\frac{P}{2}\theta_{\text{shaft}} - \frac{2\pi}{3}) \\ L_{s1q} \cos(\frac{P}{2}\theta_{\text{shaft}} + \frac{2\pi}{3}) & L_{s2q} \cos(\frac{P}{2}\theta_{\text{shaft}} + \frac{2\pi}{3}) \end{matrix} \Bigg] \tag{3.96}$$

$$L_{rr}(\theta_{\text{shaft}}) \triangleq \begin{bmatrix} L_{fdfd} & L_{fd1d} & 0 & 0 \\ L_{fd1d} & L_{1d1d} & 0 & 0 \\ 0 & 0 & L_{1q1q} & L_{1q2q} \\ 0 & 0 & L_{1q2q} & L_{2q2q} \end{bmatrix} \tag{3.97}$$

The rotor self-inductance matrix $L_{rr}(\theta_{\text{shaft}})$ is independent of θ_{shaft}. Using the transformation of (3.10),

$$\lambda_d = (L_{\ell s} + L_{md})i_d + L_{sfd}i_{fd} + L_{s1d}i_{1d} \tag{3.98}$$

$$\lambda_{fd} = \frac{3}{2}L_{sfd}i_d + L_{fdfd}i_{fd} + L_{fd1d}i_{1d} \tag{3.99}$$

$$\lambda_{1d} = \frac{3}{2}L_{s1d}i_d + L_{fd1d}i_{fd} + L_{1d1d}i_{1d} \tag{3.100}$$

and

$$\lambda_q = (L_{\ell s} + L_{mq})i_q + L_{s1q}i_{1q} + L_{s2q}i_{2q} \tag{3.101}$$

$$\lambda_{1q} = \frac{3}{2}L_{s1q}i_q + L_{1q1q}i_{1q} + L_{1q2q}i_{2q} \tag{3.102}$$

$$\lambda_{2q} = \frac{3}{2}L_{s2q}i_q + L_{1q2q}i_{1q} + L_{2q2q}i_{2q} \tag{3.103}$$

and

$$\lambda_o = L_{\ell s}i_o \tag{3.104}$$

where

$$L_{md} \triangleq \frac{3}{2}(L_A + L_B), \quad L_{mq} \triangleq \frac{3}{2}(L_A - L_B) \tag{3.105}$$

This set of flux linkage/current relationships does reflect a conservative coupling field, since the original matrices of (3.95)–(3.97) are symmetric, and the partial derivatives of (3.37) are satisfied by (3.98)–(3.104). This can be easily verified using Cramer's rule to find entries of the inverses of the inductance matrices.

In terms of the scaled quantities of the last section,

$$\psi_d = \frac{\omega_s(L_{\ell s} + L_{md})(-I_d I_{BDQ})}{V_{BDQ}} + \frac{\omega_s L_{sfd} I_{fd} I_{BFD}}{V_{BDQ}}$$
$$+ \frac{\omega_s L_{s1d} I_{1d} I_{B1D}}{V_{BDQ}} \tag{3.106}$$

$$\psi_{fd} = \frac{\omega_s \frac{3}{2} L_{sfd}(-I_d I_{BDQ})}{V_{BFD}} + \frac{\omega_s L_{fdfd} I_{fd} I_{BFD}}{V_{BFD}}$$
$$+ \frac{\omega_s L_{fd1d} I_{1d} I_{B1D}}{V_{BFD}} \tag{3.107}$$

$$\psi_{1d} = \frac{\omega_s \frac{3}{2} L_{s1d}(-I_d I_{BDQ})}{V_{B1D}} + \frac{\omega_s L_{fd1d} I_{fd} I_{BFD}}{V_{B1D}}$$
$$+ \frac{\omega_s L_{1d1d} I_{1d} I_{B1D}}{V_{B1D}} \tag{3.108}$$

$$\psi_q = \frac{\omega_s(L_{\ell s} + L_{mq})(-I_q I_{BDQ})}{V_{BDQ}} + \frac{\omega_s L_{s1q} I_{1q} I_{B1Q}}{V_{BDQ}}$$
$$+ \frac{\omega_s L_{s2q} I_{2q} I_{B2Q}}{V_{BDQ}} \tag{3.109}$$

$$\psi_{1q} = \frac{\omega_s \frac{3}{2} L_{s1q}(-I_q I_{BDQ})}{V_{B1Q}} + \frac{\omega_s L_{1q1q} I_{1q} I_{B1Q}}{V_{B1Q}}$$

$$+ \frac{\omega_s L_{1q2q} I_{2q} I_{B2Q}}{V_{B1Q}} \qquad (3.110)$$

$$\psi_{2q} = \frac{\omega_s \frac{3}{2} L_{s2q}(-I_q I_{BDQ})}{V_{B2Q}} + \frac{\omega_s L_{1q2q} I_{1q} I_{B1Q}}{V_{B2Q}}$$

$$+ \frac{\omega_s L_{2q2q} I_{2q} I_{B2Q}}{V_{B2Q}} \qquad (3.111)$$

$$\psi_o = \frac{\omega_s L_{\ell s}(-I_o I_{BDQ})}{V_{BDQ}} \qquad (3.112)$$

Although the values of I_{BFD}, I_{B1D}, I_{B1Q}, I_{B2Q} have not yet been specified, their relationship to their respective voltage bases assures that the scaled transformed system (3.106)–(3.112) is reciprocal (symmetric inductance matrices). The rotor current bases are chosen at this point to make as many off-diagonal terms equal as possible. To do this, define

$$I_{BFD} \triangleq \frac{L_{md}}{L_{sfd}} I_{BDQ}, \quad I_{B1D} \triangleq \frac{L_{md}}{L_{s1d}} I_{BDQ} \qquad (3.113)$$

$$I_{B1Q} \triangleq \frac{L_{mq}}{L_{s1q}} I_{BDQ}, \quad I_{B2Q} \triangleq \frac{L_{mq}}{L_{s2q}} I_{BDQ} \qquad (3.114)$$

and the following scaled parameters:

$$X_{\ell s} \triangleq \frac{\omega_s L_{\ell s}}{Z_{BDQ}}, \quad X_{md} \triangleq \frac{\omega_s L_{md}}{Z_{BDQ}}, \quad X_{mq} \triangleq \frac{\omega_s L_{mq}}{Z_{BDQ}} \qquad (3.115)$$

$$X_{fd} \triangleq \frac{\omega_s L_{fdfd}}{Z_{BFD}}, \quad X_{1d} \triangleq \frac{\omega_s L_{1d1d}}{Z_{B1D}}, \quad X_{fd1d} \triangleq \frac{\omega_s L_{fd1d} L_{sfd}}{Z_{BFD} L_{s1d}} \qquad (3.116)$$

$$X_{1q} \triangleq \frac{\omega_s L_{1q1q}}{Z_{B1Q}}, \quad X_{2q} \triangleq \frac{\omega_s L_{2q2q}}{Z_{B2Q}}, \quad X_{1q2q} \triangleq \frac{\omega_s L_{1q2q} L_{s1q}}{Z_{B1Q} L_{s2q}} \qquad (3.117)$$

It is convenient also to define the scaled leakage reactances of the rotor windings as

$$X_{\ell fd} \triangleq X_{fd} - X_{md}, \quad X_{\ell 1d} \triangleq X_{1d} - X_{md} \qquad (3.118)$$

$$X_{\ell 1q} \triangleq X_{1q} - X_{mq}, \quad X_{\ell 2q} \triangleq X_{2q} - X_{mq} \qquad (3.119)$$

Similarily, we also define

$$X_d \triangleq X_{\ell s} + X_{md}, \quad X_q \triangleq X_{\ell s} + X_{mq} \tag{3.120}$$

$$c_d \triangleq \frac{X_{fd1d}}{X_{md}}, \quad c_q \triangleq \frac{X_{1q2q}}{X_{mq}} \tag{3.121}$$

The resulting scaled $\psi - I$ relationship is

$$\psi_d = X_d(-I_d) + X_{md}I_{fd} + X_{md}I_{1d} \tag{3.122}$$

$$\psi_{fd} = X_{md}(-I_d) + X_{fd}I_{fd} + c_d X_{md}I_{1d} \tag{3.123}$$

$$\psi_{1d} = X_{md}(-I_d) + c_d X_{md}I_{fd} + X_{1d}I_{1d} \tag{3.124}$$

and

$$\psi_q = X_q(-I_q) + X_{mq}I_{1q} + X_{mq}I_{2q} \tag{3.125}$$

$$\psi_{1q} = X_{mq}(-I_q) + X_{1q}I_{1q} + c_q X_{mq}I_{2q} \tag{3.126}$$

$$\psi_{2q} = X_{mq}(-I_q) + c_q X_{mq}I_{1q} + X_{2q}I_{2q} \tag{3.127}$$

and

$$\psi_o = X_{\ell s}(-I_o) \tag{3.128}$$

While several examples [30] have shown that the terms c_d and c_q are important in some simulations, it is customary to make the following simplification [20]:

$$c_d \approx 1, \quad c_q \approx 1 \tag{3.129}$$

This assumption makes all of the off-diagonal entries of the decoupled inductance matrices equal. An alternative way to obtain the same structure without the above simplification would require a different choice of scaling and different definitions of leakage reactances [30]: Using the previously defined parameters and the simplification (3.129), it is common to define the following parameters [20]

$$X_d' \triangleq X_{\ell s} + \frac{1}{\frac{1}{X_{md}} + \frac{1}{X_{\ell fd}}} = X_{\ell s} + \frac{X_{md}X_{\ell fd}}{X_{fd}} = X_d - \frac{X_{md}^2}{X_{fd}} \tag{3.130}$$

$$X_q' \triangleq X_{\ell s} + \frac{1}{\frac{1}{X_{mq}} + \frac{1}{X_{\ell 1q}}} = X_{\ell s} + \frac{X_{mq}X_{\ell 1q}}{X_{1q}} = X_q - \frac{X_{mq}^2}{X_{1q}} \tag{3.131}$$

$$X_d'' \triangleq X_{\ell s} + \cfrac{1}{\frac{1}{X_{md}} + \frac{1}{X_{\ell fd}} + \frac{1}{X_{\ell 1d}}} \qquad (3.132)$$

$$X_q'' \triangleq X_{\ell s} + \cfrac{1}{\frac{1}{X_{mq}} + \frac{1}{X_{\ell 1q}} + \frac{1}{X_{\ell 2q}}} \qquad (3.133)$$

$$T_{do}' \triangleq \frac{X_{fd}}{\omega_s R_{fd}} \qquad (3.134)$$

$$T_{qo}' \triangleq \frac{X_{1q}}{\omega_s R_{1q}} \qquad (3.135)$$

$$T_{do}'' \triangleq \frac{1}{\omega_s R_{1d}} \left(X_{\ell 1d} + \cfrac{1}{\frac{1}{X_{md}} + \frac{1}{X_{\ell fd}}} \right) \qquad (3.136)$$

$$T_{qo}'' \triangleq \frac{1}{\omega_s R_{2q}} \left(X_{\ell 2q} + \cfrac{1}{\frac{1}{X_{mq}} + \frac{1}{X_{\ell 1q}}} \right) \qquad (3.137)$$

and the following variables

$$E_q' \triangleq \frac{X_{md}}{X_{fd}} \psi_{fd} \qquad (3.138)$$

$$E_{fd} \triangleq \frac{X_{md}}{R_{fd}} V_{fd} \qquad (3.139)$$

$$E_d' \triangleq -\frac{X_{mq}}{X_{1q}} \psi_{1q} \qquad (3.140)$$

Adkins [23] and several earlier references define X_q'' as in (3.131) and T_{qo}'' as in (3.135). This practice is based on the convention that single primes refer to the so-called "transient" period, while the double primes refer to the supposedly faster "subtransient" period. Thus, when a single damper winding is modeled on the rotor, this is interpreted as a "subtransient" effect and denoted as such with a double prime. Some published models use the notation of Young [35], which uses an E_d' definition that is the same as (3.140) but with a positive sign. In several publications, the terminology X_{fd} is used to define leakage reactance rather than self-reactance. The symbols X_{ad} and X_{aq} are common alternatives for the magnetizing reactances X_{md} and X_{mq}.

The dynamic model can contain at most only seven of the fourteen flux linkages and currents as independent state variables. The natural form of the state equations invites the elimination of currents by solving (3.122)–(3.128). Since the terminal constraints have not yet been specified, it is unwise to eliminate I_d, I_q, or I_o at this time. Since the terminal constraints do not affect I_{fd}, I_{1d}, I_{1q}, I_{2q}, these currents can be eliminated from the dynamic model now. This is done by rearranging (3.122)–(3.128) using the newly defined variables and parameters to obtain

$$\psi_d = -X_d'' I_d + \frac{(X_d'' - X_{\ell s})}{(X_d' - X_{\ell s})} E_q' + \frac{(X_d' - X_d'')}{(X_d' - X_{\ell s})}\psi_{1d} \tag{3.141}$$

$$I_{fd} = \frac{1}{X_{md}}[E_q' + (X_d - X_d')(I_d - I_{1d})] \tag{3.142}$$

$$I_{1d} = \frac{X_d' - X_d''}{(X_d' - X_{\ell s})^2}[\psi_{1d} + (X_d' - X_{\ell s})I_d - E_q'] \tag{3.143}$$

and

$$\psi_q = -X_q'' I_q - \frac{(X_q'' - X_{\ell s})}{(X_q' - X_{\ell s})} E_d' + \frac{(X_q' - X_q'')}{(X_q' - X_{\ell s})}\psi_{2q} \tag{3.144}$$

$$I_{1q} = \frac{1}{X_{mq}}[-E_d' + (X_q - X_q')(I_q - I_{2q})] \tag{3.145}$$

$$I_{2q} = \frac{X_q' - X_q''}{(X_q' - X_{\ell s})^2}[\psi_{2q} + (X_q' - X_{\ell s})I_q + E_d'] \tag{3.146}$$

and

$$\psi_o = -X_{\ell s}I_o \tag{3.147}$$

Substitution into (3.64)–(3.72) gives the dynamic model for a linear magnetic circuit with the terminal constraints (relationship between V_d, I_d, V_q, I_q, V_o, I_o) not yet specified. Since these terminal constraints are not specified, it is necessary to keep the three flux linkage/current algebraic equations involving I_d, I_q, and I_o. In addition, the variables E_{fd}, T_M, and T_{FW} are also as yet unspecified. It would be reasonable, if desired at this point, to make E_{fd} and T_M constant inputs, and T_{FW} equal to zero. We will continue to carry them along as variables. With these clarifications, the linear magnetic circuit model is shown in the following boxed set.

$$\frac{1}{\omega_s}\frac{d\psi_d}{dt} = R_s I_d + \frac{\omega}{\omega_s}\psi_q + V_d \tag{3.148}$$

$$\frac{1}{\omega_s}\frac{d\psi_q}{dt} = R_s I_q - \frac{\omega}{\omega_s}\psi_d + V_q \tag{3.149}$$

$$\frac{1}{\omega_s}\frac{d\psi_o}{dt} = R_s I_o + V_o \tag{3.150}$$

$$T'_{do}\frac{dE'_q}{dt} = -E'_q - (X_d - X'_d)[I_d - \frac{X'_d - X''_d}{(X'_d - X_{\ell s})^2}(\psi_{1d}$$
$$+(X'_d - X_{\ell s})I_d - E'_q)] + E_{fd} \tag{3.151}$$

$$T''_{do}\frac{d\psi_{1d}}{dt} = -\psi_{1d} + E'_q - (X'_d - X_{\ell s})I_d \tag{3.152}$$

$$T'_{qo}\frac{dE'_d}{dt} = -E'_d + (X_q - X'_q)[I_q - \frac{X'_q - X''_q}{(X'_q - X_{\ell s})^2}(\psi_{2q}$$
$$+(X'_q - X_{\ell s})I_q + E'_d)] \tag{3.153}$$

$$T''_{qo}\frac{d\psi_{2q}}{dt} = -\psi_{2q} - E'_d - (X'_q - X_{\ell s})I_q \tag{3.154}$$

$$\frac{d\delta}{dt} = \omega - \omega_s \tag{3.155}$$

$$\frac{2H}{\omega_s}\frac{d\omega}{dt} = T_M - (\psi_d I_q - \psi_q I_d) - T_{FW} \tag{3.156}$$

$$\psi_d = -X''_d I_d + \frac{(X''_d - X_{\ell s})}{(X'_d - X_{\ell s})}E'_q + \frac{(X'_d - X''_d)}{(X'_d - X_{\ell s})}\psi_{1d} \tag{3.157}$$

$$\psi_q = -X''_q I_q - \frac{(X''_q - X_{\ell s})}{(X'_q - X_{\ell s})}E'_d + \frac{(X'_q - X''_q)}{(X'_q - X_{\ell s})}\psi_{2q} \tag{3.158}$$

$$\psi_o = -X_{\ell s}I_o \tag{3.159}$$

Although there are time constants that appear on all of the flux linkage derivatives, the right-hand sides contain flux linkages multiplied by constants. Furthermore, the addition of the terminal constraints could add more terms when I_d, I_q, and I_o are eliminated. Thus, the time constants shown are not true time constants in the traditional sense, where the respective states appear on the right-hand side multiplied only by -1. It is also possible to define a mechanical time constant T_s as

$$T_s \triangleq \sqrt{\frac{2H}{\omega_s}} \qquad (3.160)$$

and a scaled transient speed as

$$\omega_t \triangleq T_s(\omega - \omega_s) \qquad (3.161)$$

to produce the following angle/speed state pair

$$T_s \frac{d\delta}{dt} = \omega_t \qquad (3.162)$$

$$T_s \frac{d\omega_t}{dt} = T_M - (\psi_d I_q - \psi_q I_d) - T_{FW} \qquad (3.163)$$

While this will prove useful later, in the analysis of the time-scale properties of synchronous machines, the model normally will be used in the form of (3.148)–(3.159). This concludes the basic dynamic modeling of synchronous machines if saturation of the magnetic circuit is not considered. The next section presents a fairly general method for including such nonlinearities in the flux linkage/current relationships.

3.5   The Nonlinear Magnetic Circuit

In this section, we propose a fairly generalized treatment of nonlinearities in the magnetic circuit. The generalization is motivated by the multitude of various representations of saturation that have appeared in the literature. Virtually all methods proposed to date involve the addition of one or more nonlinear terms to the model of (3.148)–(3.159). The following treatment returns to the original *abc* variables so that any assumptions or added terms can be traced through the transformation and scaling processes of the last section. It is clear that, as in the last section, the flux linkage/current relationships must satisfy assumptions (3) and (4) at the end of Section 3.3 if the results here are to be valid for the general model of (3.64)–(3.72). Toward this end, we propose a flux linkage/current relationship of the following form:

$$\lambda_{abc} = L_{ss}(\theta_{\text{shaft}})i_{abc} + L_{sr}(\theta_{\text{shaft}})i_{\text{rotor}}$$
$$- S_{abc}(i_{abc}, \lambda_{\text{rotor}}, \theta_{\text{shaft}}) \qquad (3.164)$$

$$\lambda_{\text{rotor}} = L^t_{sr}(\theta_{\text{shaft}})i_{abc} + L_{rr}(\theta_{\text{shaft}})i_{\text{rotor}}$$

$$-S_{\text{rotor}}(i_{abc}, \lambda_{\text{rotor}}, \theta_{\text{shaft}}) \qquad (3.165)$$

where all quantities are as previously defined, and S_{abc} and S_{rotor} satisfy assumptions (3) and (4) at the end of Section 3.3. The choice of stator currents and rotor flux linkages for the nonlinearity dependence was made to allow comparsion with traditional choices of functions. With these two assumptions, S_{abc} and S_{rotor} must be such that when (3.164) and (3.165) are transformed using (3.10), the following nonlinear flux linkage/current relationship is obtained

$$\lambda_d = (L_{\ell s} + L_{md})i_d + L_{sfd}i_{fd} + L_{s1d}i_{1d} - S_d(i_{dqo}, \lambda_{\text{rotor}}) \quad (3.166)$$

$$\lambda_{fd} = \frac{3}{2}L_{sfd}i_d + L_{fdfd}i_{fd} + L_{fd1d}i_{1d} - S_{fd}(i_{dqo}, \lambda_{\text{rotor}}) \qquad (3.167)$$

$$\lambda_{1d} = \frac{3}{2}L_{s1d}i_d + L_{fd1d}i_{fd} + L_{1d1d}i_{1d} - S_{1d}(i_{dqo}, \lambda_{\text{rotor}}) \qquad (3.168)$$

$$\lambda_q = (L_{\ell s} + L_{mq})i_q + L_{s1q}i_{1q} + L_{s2q}i_{2q} - S_q(i_{dqo}, \lambda_{\text{rotor}}) \quad (3.169)$$

$$\lambda_{1q} = \frac{3}{2}L_{s1q}i_q + L_{1q1q}i_{1q} + L_{1q2q}i_{2q} - S_{1q}(i_{dqo}, \lambda_{\text{rotor}}) \qquad (3.170)$$

$$\lambda_{2q} = \frac{3}{2}L_{s2q}i_q + L_{1q2q}i_{1q} + L_{2q2q}i_{2q} - S_{2q}(i_{dqo}, \lambda_{\text{rotor}}) \qquad (3.171)$$

and

$$\lambda_o = L_{\ell s}i_o - S_o(i_{dqo}, \lambda_{\text{rotor}}) \qquad (3.172)$$

This system includes the possibility of coupling between all of the d, q, and o subsystems. Saturation functions that satisfy these two assumptions normally have a balanced symmetrical three-phase dependence on shaft position.

In terms of the scaled variables of the last two sections, and using (3.129) ($c_d = c_q = 1$),

$$\psi_d = X_d(-I_d) + X_{md}I_{fd} + X_{md}I_{1d} - S_d^{(1)}(Y_1) \qquad (3.173)$$

$$\psi_{fd} = X_{md}(-I_d) + X_{fd}I_{fd} + X_{md}I_{1d} - S_{fd}^{(1)}(Y_1) \qquad (3.174)$$

$$\psi_{1d} = X_{md}(-I_d) + X_{md}I_{fd} + X_{1d}I_{1d} - S_{1d}^{(1)}(Y_1) \qquad (3.175)$$

and

$$\psi_q = X_q(-I_q) + X_{mq}I_{1q} + X_{mq}I_{2q} - S_q^{(1)}(Y_1) \qquad (3.176)$$

$$\psi_{1q} = X_{mq}(-I_q) + X_{1q}I_{1q} + X_{mq}I_{2q} - S_{1q}^{(1)}(Y_1) \qquad (3.177)$$

$$\psi_{2q} = X_{mq}(-I_q) + X_{mq}I_{1q} + X_{2q}I_{2q} - S_{2q}^{(1)}(Y_1) \qquad (3.178)$$

and

$$\psi_o = X_{\ell s}(-I_o) - S_o^{(1)}(Y_1) \qquad (3.179)$$

where

$$Y_1 \triangleq [I_d \ \psi_{fd} \ \psi_{1d} \ I_q \ \psi_{1q} \ \psi_{2q} \ I_o]^t \qquad (3.180)$$

$$S_d^{(1)} \triangleq S_d/\Lambda_{BDQ}, \quad S_{fd}^{(1)} \triangleq S_{fd}/\Lambda_{BFD}, \quad S_{1d}^{(1)} \triangleq S_{1d}/\Lambda_{B1D}$$

$$S_q^{(1)} \triangleq S_q/\Lambda_{BDQ}, \quad S_{1q}^{(1)} \triangleq S_{1q}/\Lambda_{B1Q}, \quad S_{2q}^{(1)} \triangleq S_{2q}/\Lambda_{B2Q}$$

$$S_o^{(1)} \triangleq S_o/\Lambda_{BDQ} \qquad (3.181)$$

with each S evaluated using λ_{dqo}, λ_{rotor} written as a function of Y_1. Using new variables E_d' and E_q', and rearranging so that rotor currents can be eliminated, gives

$$\psi_d = -X_d''I_d + \frac{(X_d'' - X_{\ell s})}{(X_d' - X_{\ell s})}E_q' + \frac{(X_d' - X_d'')}{(X_d' - X_{\ell s})}\psi_{1d} - S_d^{(2)}(Y_2) \qquad (3.182)$$

$$I_{fd} = \frac{1}{X_{md}}[E_q' + (X_d - X_d')(I_d - I_{1d}) + S_{fd}^{(2)}(Y_2)] \qquad (3.183)$$

$$I_{1d} = \frac{X_d' - X_d''}{(X_d' - X_{\ell s})^2}[\psi_{1d} + (X_d' - X_{\ell s})I_d - E_q' + S_{1d}^{(2)}(Y_2)] \qquad (3.184)$$

and

$$\psi_q = -X_q'' I_q - \frac{(X_q'' - X_{\ell s})}{(X_q' - X_{\ell s})} E_d' + \frac{(X_q' - X_q'')}{(X_q' - X_{\ell s})} \psi_{2q} - S_q^{(2)}(Y_2) \quad (3.185)$$

$$I_{1q} = \frac{1}{X_{mq}} [-E_d' + (X_q - X_q')(I_q - I_{2q}) + S_{1q}^{(2)}(Y_2)] \quad (3.186)$$

$$I_{2q} = \frac{X_q' - X_q''}{(X_q' - X_{\ell s})^2} [\psi_{2q} + (X_q' - X_{\ell s})I_q + E_d' + S_{2q}^{(2)}(Y_2)] \quad (3.187)$$

and

$$\psi_o = -X_{\ell s} I_o - S_o^{(2)}(Y_2) \quad (3.188)$$

where

$$Y_2 \triangleq [I_d \ \ E_q' \ \ \psi_{1d} \ \ I_q \ \ E_d' \ \ \psi_{2q} \ \ I_o]^t \quad (3.189)$$

and

$$S_d^{(2)} \triangleq S_d^{(1)} - S_{fd}^{(2)} - \frac{(X_d' - X_d'')}{(X_d' - X_{\ell s})} S_{1d}^{(2)},$$

$$S_{fd}^{(2)} \triangleq \frac{X_{md}}{X_{fd}} S_{fd}^{(1)}, \quad S_{1d}^{(2)} \triangleq S_{1d}^{(1)} - S_{fd}^{(2)},$$

$$S_q^{(2)} \triangleq S_q^{(1)} - S_{1q}^{(2)} - \frac{(X_q' - X_q'')}{(X_q' - X_{\ell s})} S_{2q}^{(2)},$$

$$S_{1q}^{(2)} \triangleq \frac{X_{mq}}{X_{1q}} S_{1q}^{(1)}, \quad S_{2q}^{(2)} \triangleq S_{2q}^{(1)} - S_{1q}^{(2)},$$

$$S_o^{(2)} \triangleq S_o^{(1)} \quad (3.190)$$

with each $S^{(1)}$ evaluated using Y_1, written as a function of Y_2.

Elimination of rotor currents from (3.64)–(3.72) gives the final dynamic model with general nonlinearities.

$$\frac{1}{\omega_s}\frac{d\psi_d}{dt} = R_s I_d + \frac{\omega}{\omega_s}\psi_q + V_d \tag{3.191}$$

$$\frac{1}{\omega_s}\frac{d\psi_q}{dt} = R_s I_q - \frac{\omega}{\omega_s}\psi_d + V_q \tag{3.192}$$

$$\frac{1}{\omega_s}\frac{d\psi_o}{dt} = R_s I_o + V_o \tag{3.193}$$

$$T'_{do}\frac{dE'_q}{dt} = -E'_q - (X_d - X'_d)[I_d - \frac{X'_d - X''_d}{(X'_d - X_{\ell s})^2}(\psi_{1d} + (X'_d - X_{\ell s})I_d$$
$$-E'_q + S^{(2)}_{1d}(Y_2))] - S^{(2)}_{fd}(Y_2) + E_{fd} \tag{3.194}$$

$$T''_{do}\frac{d\psi_{1d}}{dt} = -\psi_{1d} + E'_q - (X'_d - X_{\ell s})I_d - S^{(2)}_{1d}(Y_2) \tag{3.195}$$

$$T'_{qo}\frac{dE'_d}{dt} = -E'_d + (X_q - X'_q)[I_q - \frac{X'_q - X''_q}{(X'_q - X_{\ell s})^2}(\psi_{2q} + (X'_q - X_{\ell s})I_q$$
$$+E'_d + S^{(2)}_{2q}(Y_2))] + S^{(2)}_{1q}(Y_2) \tag{3.196}$$

$$T''_{qo}\frac{d\psi_{2q}}{dt} = -\psi_{2q} - E'_d - (X'_q - X_{\ell s})I_q - S^{(2)}_{2q}(Y_2) \tag{3.197}$$

$$\frac{d\delta}{dt} = \omega - \omega_s \tag{3.198}$$

$$\frac{2H}{\omega_s}\frac{d\omega}{dt} = T_M - (\psi_d I_q - \psi_q I_d) - T_{FW} \tag{3.199}$$

with the three algebraic equations,

$$\psi_d = -X''_d I_d + \frac{(X''_d - X_{\ell s})}{(X'_d - X_{\ell s})}E'_q + \frac{(X'_d - X''_d)}{(X'_d - X_{\ell s})}\psi_{1d}$$
$$-S^{(2)}_d(Y_2) \tag{3.200}$$

$$\psi_q = -X''_q I_q - \frac{(X''_q - X_{\ell s})}{(X'_q - X_{\ell s})}E'_d + \frac{(X'_q - X''_q)}{(X'_q - X_{\ell s})}\psi_{2q}$$
$$-S^{(2)}_q(Y_2) \tag{3.201}$$

$$\psi_o = -X_{\ell s}I_o - S^{(2)}_o(Y_2) \tag{3.202}$$

$$Y_2 = [I_d \ E'_q \ \psi_{1d} \ I_q \ E'_d \ \psi_{2q} \ I_o]^t \tag{3.203}$$

As in the last section, new speeds could be defined so that each dynamic state includes a time constant. It is important to note, however, that, as in the last section, terms on the right-hand side of the dynamic state model imply that these time constants do not necessarily completely identify the speed of response of each variable. This is even more evident with the addition of nonlinearities.

One purpose for beginning this section by returning to the *abc* variables was to trace the nonlinearities through the transformation and scaling process. This ensures that the resulting model with nonlinearities is, in some sense, consistent. This was partly motivated by the proliferation of different methods to account for saturation in the literature. For example, the literature talks about "X_{md}" saturating, or X_{md} being a function of the dynamic states. This could imply that many constants we have defined would change when saturation is considered. With the presentation given above, it is clear that all constants can be left unchanged, while the nonlinearities are included in a set of functions to be specified based on some design calculation or test procedure.

It is interesting to compare these general nonlinearity functions with other methods that have appeared in the literature [20, 22, 23, 26, 27, 35, 36] – [50]. Reference [37] discusses a typical representation that uses:

$$S_d^{(2)} = 0, \ S_{1d}^{(2)} = 0, \ S_q^{(2)} = 0, \ S_{2q}^{(2)} = 0, \ S_o^{(2)} = 0 \qquad (3.204)$$

and keeps $S_{fd}^{(2)}$ and $S_{1q}^{(2)}$ expressed as

$$S_{fd}^{(2)} = \frac{\psi_d''}{|\psi''|} S_G(|\psi''|) \qquad (3.205)$$

$$S_{1q}^{(2)} = \frac{\psi_q''(X_q - X_{\ell s})}{|\psi''|(X_d - X_{\ell s})} S_G(|\psi''|) \qquad (3.206)$$

where

$$|\psi''| \triangleq (\psi_d''^2 + \psi_q''^2)^{1/2} \qquad (3.207)$$

and

$$\psi_d'' \triangleq \left(\frac{X_d'' - X_{\ell s}}{X_d' - X_{\ell s}} \right) E_q' + \left(\frac{X_d' - X_d''}{X_d' - X_{\ell s}} \right) \psi_{1d} \qquad (3.208)$$

$$\psi_q'' \triangleq -\left(\frac{X_q'' - X_{\ell s}}{X_q' - X_{\ell s}} \right) E_d' + \left(\frac{X_q' - X_q''}{X_q' - X_{\ell s}} \right) \psi_{2q} \qquad (3.209)$$

The saturation function S_G should be correct under open-circuit conditions. For steady state with

$$I_d = I_q = I_o = I_{1q} = I_{2q} = 0$$

$$\psi_q = -E'_d = \psi''_q = -V_d = \psi_{1q} = \psi_{2q} = 0$$

$$\psi_d = E'_q = \psi''_d = V_q = \psi_{1d} = E_{fd} - S^{(2)}_{fd} \qquad (3.210)$$

the open-circuit terminal voltage is

$$V_{t_{oc}} = \sqrt{V_d^2 + V_q^2} = E'_q \qquad (3.211)$$

and the field current is

$$I_{fd} = \frac{E'_q + S^{(2)}_{fd}}{X_{md}} \qquad (3.212)$$

From the saturation representation of (3.205),

$$S^{(2)}_{fd} = S_G(V_{t_{oc}}) \qquad (3.213)$$

$$X_{md}I_{fd} = V_{t_{oc}} + S_G(V_{t_{oc}}) \qquad (3.214)$$

The function S_G can then be obtained from an open-circuit characteristic, as shown in Figure 3.3. While this illustrates the validity of the saturation function under open-circuit conditions, it does not totally support its use under load. In addition, it has been shown that this representation does not satisfy the assumption of a conservative coupling field [51].

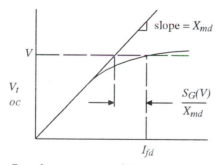

Figure 3.3: *Synchronous machine open-circuit characteristic*

3.6 Single-Machine Steady State

To introduce steady state, we assume constant states and look at the algebraic equations resulting from the dynamic model. We will analyze the system under the condition of a linear magnetic circuit. Thus, beginning with (3.148)–(3.159), we observe that, for constant states, we must have constant speed ω and constant angle δ, thus requiring $\omega = \omega_s$ and, therefore,

$$V_d = -R_s I_d - \psi_q \tag{3.215}$$

$$V_q = -R_s I_q + \psi_d \tag{3.216}$$

Assuming a balanced three-phase operation, all of the "zero" variables and damper winding currents are zero. The fact that damper-winding currents are zero can be seen by recalling that the right-hand sides of (3.152)–(3.154) are actually scaled damper-winding currents. Using these to simplify (3.151), (3.153), (3.157), and (3.158), the other algebraic equations to be solved are

$$0 = -E_q' - (X_d - X_d')I_d + E_{fd} \tag{3.217}$$

$$0 = -\psi_{1d} + E_q' - (X_d' - X_{\ell s})I_d \tag{3.218}$$

$$0 = -E_d' + (X_q - X_q')I_q \tag{3.219}$$

$$0 = -\psi_{2q} - E_d' - (X_q' - X_{\ell s})I_q \tag{3.220}$$

$$0 = T_M - (\psi_d I_q - \psi_q I_d) - T_{FW} \tag{3.221}$$

$$\psi_d = E_q' - X_d' I_d \tag{3.222}$$

$$\psi_q = -E_d' - X_q' I_q \tag{3.223}$$

Except for (3.221), these are all linear equations that can easily be solved for various steady-state representations. Substituting for ψ_d and ψ_q in (3.215) and (3.216) gives

$$V_d = -R_s I_d + E_d' + X_q' I_q \tag{3.224}$$

$$V_q = -R_s I_q + E_q' - X_d' I_d \tag{3.225}$$

These two real algebraic equations can be written as one complex equation of the form

$$(V_d + jV_q)e^{j(\delta - \pi/2)} = -(R_s + jX_q)(I_d + jI_q)e^{j(\delta - \pi/2)} + \overline{E} \tag{3.226}$$

where

$$
\begin{aligned}
\overline{E} &= [(E_d' - (X_q - X_d')I_q) + j(E_q' + (X_q - X_d')I_d)]e^{j(\delta - \pi/2)} \\
&= j[(X_q - X_d')I_d + E_q']e^{j(\delta - \pi/2)}
\end{aligned}
\tag{3.227}
$$

Clearly, many alternative complex equations can be written from (3.224) and (3.225), depending on what is included in the "internal" voltage $\overline{E}$. For balanced symmetrical sinusoidal steady-state abc voltages and currents, the quantities $(V_d + jV_q)e^{j(\delta - \pi/2)}$ and $(I_d + jI_q)e^{j(\delta - \pi/2)}$ are the per-unit RMS phasors for a phase voltage and current (see (3.73)–(3.91)). This gives considerable physical significance to the circuit form of (3.226) shown in Figure 3.4. The internal voltage $\overline{E}$ can be further simplified, using (3.217), as

$$
\begin{aligned}
\overline{E} &= j[(X_q - X_d)I_d + E_{fd}]e^{j(\delta - \pi/2)} \\
&= [(X_q - X_d)I_d + E_{fd}]e^{j\delta}
\end{aligned}
\tag{3.228}
$$

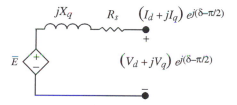
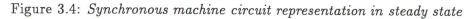

Figure 3.4: *Synchronous machine circuit representation in steady state*

An important observation is

$$
\delta = \text{angle on } \overline{E}
\tag{3.229}
$$

Also from (3.142) and (3.143),

$$
I_{fd} = E_{fd}/X_{md}
\tag{3.230}
$$

Several other points are worth noting. First, the open-circuit (or zero stator current) terminal voltage is

$$
(V_d + jV_q)e^{j(\delta - \pi/2)} \mid_{I_d = I_q = 0} = E_{fd}e^{j\delta}
\tag{3.231}
$$

Therefore, for $E_{fd} = 1$, the open-circuit terminal voltage is 1, and field current is $1/X_{md}$. Also,

$$
V_d \mid_{I_d = I_q = 0} = E_d' \mid_{I_d = I_q = 0} = -\psi_q \mid_{I_d = I_q = 0} = 0
\tag{3.232}
$$

$$
V_q \mid_{I_d = I_q = 0} = E_q' \mid_{I_d = I_q = 0} = \psi_d \mid_{I_d = I_q = 0} = E_{fd}
\tag{3.233}
$$

The electrical torque is

$$T_{ELEC} = \psi_d I_q - \psi_q I_d = V_d I_d + V_q I_q + R_s(I_d^2 + I_q^2) \qquad (3.234)$$

This torque is precisely the "real power" delivered by the controlled source of Figure 3.4. That is, for $\overline{I} = (I_d + jI_q)e^{j(\delta - \pi/2)}$,

$$T_{ELEC} = T_M = \text{Real}[\overline{E}\overline{I}^*] \qquad (3.235)$$

We can then conclude that the electrical torque from the shaft is equal to the power delivered by the controlled source. In steady state, the electrical torque from the shaft equals T_M when $T_{FW} = 0$. From the circuit with $\overline{V} = (V_d + jV_q)e^{j(\delta - \pi/2)}$,

$$T_{ELEC} = \text{Real}\left[\overline{E}\left(\frac{\overline{E} - \overline{V}}{R_s + jX_q}\right)^*\right] \qquad (3.236)$$

For zero stator resistance and round rotor,

$$T_{ELEC}\big|_{\substack{R_s=0 \\ X_d=X_q}} = \text{Real}\left[E_{fd}e^{j\delta}\left(\frac{E_{fd}e^{-j\delta} - \overline{V}^*}{-jX_d}\right)\right] \qquad (3.237)$$

or

$$T_{ELEC}\big|_{\substack{R_s=0 \\ X_d=X_q}} = \frac{E_{fd}V}{X_d}\sin\delta_T \qquad (3.238)$$

where the angle δ_T is called the torque angle,

$$\delta_T \triangleq \delta - \theta \qquad (3.239)$$

with

$$\overline{V} = Ve^{j\theta} \qquad (3.240)$$

Under these conditions, and this definition of the torque angle, $\delta_T < 0$ for a motor and $\delta_T > 0$ for a generator.

Example 3.1

Consider a synchronous machine (without saturation) serving a load with

$$\overline{V} = 1\angle 10° \text{ pu} \quad \overline{I} = 0.5\angle -20° \text{ pu}$$

It has $X_d = 1.2$, $X_q = 1.0$, $X_{md} = 1.1$, $X'_d = 0.232$, $R_s = 0$ (all in pu). Find δ, δ_T, I_d, I_q, V_d, V_q, ψ_d, ψ_q, E'_q, E_{fd}, I_{fd} (all in pu except angles in degrees).

Solution:

$$
\begin{aligned}
E'_q &= -(1.2 - 0.232)I_d + E_{fd} \\
(0.768I_d + E'_q)\angle\delta &= 1\angle 90 \times 0.5\angle -20° + 1\angle 10° \\
&= 1.323\angle 29.1°
\end{aligned}
$$

so

$$
\delta = 29.1° \quad \delta_T = 29.1° - 10° = 19.1°
$$
$$
I_d + jI_q = 0.5\angle -20° - 29.1° + 90° = 0.5\angle 40.9°
$$
$$
I_d = 0.378 \quad I_q = 0.327
$$
$$
V_d + jV_q = 1\angle 10° - 29.1° + 90° = 1\angle 70.9°
$$
$$
V_d = 0.327 \quad V_q = 0.945
$$
$$
\psi_d = V_q + 0I_q = 0.945
$$
$$
\psi_q = -V_d - 0I_d = -0.327
$$

To find E'_q and E_{fd}, return to $|\,\overline{E}\,|$

$$
0.768 \times 0.378 + E'_q = 1.323
$$
$$
E'_q = 1.033
$$
$$
1.033 = -(1.2 - 0.232) \times 0.378 + E_{fd}
$$
$$
E_{fd} = 1.399
$$

To find I_{fd}, it is easy to show that

$$
I_{fd} = \frac{E_{fd}}{X_{md}} = \frac{1.399}{1.1} = 1.27
$$

These solutions can be checked by noting that, in scaled per unit, T_{ELEC} is equal to P_{OUT}.

$$
\begin{aligned}
T_{ELEC} &= \psi_d I_q - \psi_q I_d = 0.4326 \\
P_{OUT} &= \text{Real}\,(\overline{V}\,\overline{I}^*) = \text{Real}\,(0.5\angle + 30°) = 0.433
\end{aligned}
$$

Also,

$$
\begin{aligned}
Q_{OUT} &= \text{Imag}(\overline{V}\,\overline{I}^*) = \text{Imag}(0.5\angle 30^o) = 0.25 \\
&= \text{Imag}((V_d + jV_q)e^{-j(\delta-\pi/2)}(I_d - jI_q)e^{j(\delta-\pi/2)}) \\
&= \text{Imag}((V_d + jV_q)(I_d - jI_q)) \\
&= \psi_d I_d + \psi_q I_q = 0.25
\end{aligned}
$$

□

The steady-state analysis of a given problem involves certain constraints. For example, depending on what is specified, the solution of the steady-state equations may be very difficult to solve. The solution of steady-state in multimachine power systems is usually called load flow, and is discussed in later chapters. The extension of this steady-state analysis to include saturation is left as an exercise.

3.7   Operational Impedances and Test Data

The synchronous machine model derived in this chapter was based on the initial assumption of three stator windings, one field winding, and three damper windings ($1d$, $1q$, $2q$). In addition, the machine reactances and time constants were defined in terms of this machine structure. This is consistent with [20] and many other references. It was noted earlier, however, that many of the machine reactances and time constants have been defined through physical tests or design parameters rather than a presupposed physical structure and model. Regardless of the definition of constants, a given model contains quantities that must be replaced by numbers in a specific simulation. Since designers use considerably more detailed modeling, and physical tests are model independent, there could be at least three different ways to arrive at a value for a constant denoted by the symbols used in the model of this chapter. For example, a physical test can be used to compute a value of T_{do}'' if T_{do}'' is defined through the outcome of a test. A designer can compute a value of T_{do}'' from physical parameters such that the value would approximate the test value. The definition of T_{do}'' in this chapter was not based on any test and could, therefore, be different from that furnished by a manufacturer. For this reason, it is important to always verify the definitions of all constants to ensure that the numerical value is a good approximation of the constant used in the model.

The concept of operational impedance was introduced as a means for relating test data to model constants. The concept is based on the response of a machine to known test voltages. These test voltages may be either dc or sinusoidal ac of variable frequency. The stator equations in the transformed and scaled variables can be written in the Laplace domain from (3.64)–(3.72) with constant speed ($\omega = \omega_{ss}$) as

$$\overline{V}_d = -R_s \overline{I}_d - \frac{\omega_{ss}}{\omega_s}\overline{\psi}_q + \frac{s}{\omega_s}\overline{\psi}_d \qquad (3.241)$$

$$\overline{V}_q = -R_s \overline{I}_q + \frac{\omega_{ss}}{\omega_s}\overline{\psi}_d + \frac{s}{\omega_s}\overline{\psi}_q \qquad (3.242)$$

$$\overline{V}_o = -R_s \overline{I}_o + \frac{s}{\omega_s}\overline{\psi}_o \qquad (3.243)$$

where s is the Laplace domain operator, which, in sinusoidal steady state with frequency ω_o in radians/sec, is

$$s = j\omega_o \qquad (3.244)$$

If we make the assumption that the magnetic circuit has a linear flux linkage/current relationship that satisfies assumptions (3) and (4) of Section 3.3, we can propose that we have the Laplace domain relationship for any number of rotor-windings (or equivalent windings that represent solid iron rotor effects). When scaled, these relationships could be solved for $\overline{\psi}_d$, $\overline{\psi}_q$, and $\overline{\psi}_o$ as functions of $\overline{I}_d, \overline{I}_q, \overline{I}_o$, all rotor winding voltages and the operator s. For balanced, symmetric windings and all-rotor winding voltages zero except for V_{fd}, the result would be

$$\overline{\psi}_d = -X_{dop}(s)\overline{I}_d + \overline{G}_{op}(s)\overline{V}_{fd} \qquad (3.245)$$

$$\overline{\psi}_q = -X_{qop}(s)\overline{I}_q \qquad (3.246)$$

$$\overline{\psi}_o = -X_{oop}\overline{I}_o \qquad (3.247)$$

To see how this could be done for a specific model, consider the three-damper-winding model of the previous sections. The scaled Kirchhoff equations are given as (3.67)–(3.70), and the scaled linear magnetic circuit equations are given as (3.122)–(3.128). From these equations in the Laplace domain with operator s and $V_{1d} = V_{1q} = V_{2q} = 0$,

$$\frac{s}{\omega_s}[X_{md}(-\overline{I}_d) + X_{fd}\overline{I}_{fd} + c_d X_{md}\overline{I}_{1d}] = -R_{fd}\overline{I}_{fd} + \overline{V}_{fd} \qquad (3.248)$$

$$\frac{s}{\omega_s}[X_{md}(-\overline{I}_d) + c_d X_{md}\overline{I}_{fd} + X_{1d}\overline{I}_{1d}] = -R_{1d}\overline{I}_{1d} \tag{3.249}$$

$$\frac{s}{\omega_s}[X_{mq}(-\overline{I}_q) + X_{1q}\overline{I}_q + c_q X_{mq}\overline{I}_{2q}] = -R_{1q}\overline{I}_{1q} \tag{3.250}$$

$$\frac{s}{\omega_s}[X_{mq}(-\overline{I}_q) + c_q X_{mq}\overline{I}_{1q} + X_{2q}\overline{I}_{2q}] = -R_{2q}\overline{I}_{2q} \tag{3.251}$$

$$\overline{\psi}_o = X_{\ell s}(-\overline{I}_o) \tag{3.252}$$

The two d equations can be solved for $\overline{I}_{fd}$ and $\overline{I}_{1d}$ as functions of s times $\overline{I}_d$ and $\overline{V}_{fd}$. The two q axis equations can be solved for $\overline{I}_{1q}$ and $\overline{I}_{2q}$ as functions of s times $\overline{I}_q$. When substituted into (3.122) and (3.125), this would produce the following operational functions for this given model:

$$\overline{X}_{dop}(s) = X_d$$
$$- \left[\frac{\frac{s}{\omega_s}X_{md}^2(R_{1d} + \frac{s}{\omega_s}X_{1d} - 2\frac{s}{\omega_s}c_d X_{md} + R_{fd} + \frac{s}{\omega_s}X_{fd})}{(R_{fd} + \frac{s}{\omega_s}X_{fd})(R_{1d} + \frac{s}{\omega_s}X_{1d}) - (\frac{s}{\omega_s}c_d X_{md})^2} \right] \tag{3.253}$$

$$\overline{G}_{op}(s) = \left[\frac{X_{md}(R_{1d} + \frac{s}{\omega_s}X_{1d} - \frac{s}{\omega_s}c_d X_{md})}{(R_{fd} + \frac{s}{\omega_s}X_{fd})(R_{1d} + \frac{s}{\omega_s}X_{1d}) - (\frac{s}{\omega_s}c_d X_{md})^2} \right] \tag{3.254}$$

$$\overline{X}_{qop}(s) = X_q$$
$$- \frac{\frac{s}{\omega_s}X_{mq}^2(R_{2q} + \frac{s}{\omega_s}X_{2q} - 2\frac{s}{\omega_s}c_q X_{mq} + R_{1q} + \frac{s}{\omega_s}X_{1q})}{(R_{2q} + \frac{s}{\omega_s}X_{2q})(R_{1q} + \frac{s}{\omega_s}X_{1q}) - (\frac{s}{\omega_s}c_q X_{mq})^2} \tag{3.255}$$

$$\overline{X}_{oop}(s) = X_{\ell s} \tag{3.256}$$

Note that for $s = 0$ in this model

$$\overline{X}_{dop}(0) = X_d \tag{3.257}$$

$$\overline{G}_{op}(0) = \frac{X_{md}}{R_{fd}} \tag{3.258}$$

$$\overline{X}_{qop}(0) = X_q \tag{3.259}$$

and for $s = \infty$ in this model with $c_d = c_q = 1$

$$\overline{X}_{dop}(\infty) = X_d'' \tag{3.260}$$

$$\overline{X}_{qop}(\infty) = X_q'' \tag{3.261}$$

For this model, it is also possible to rewrite (3.253)–(3.256) as a ratio of polynomials in s that can be factored to give a time constant representation.

The purpose for introducing this concept of operational functions is to show one possible way in which a set of parameters may be obtained from a machine test. At standstill, the Laplace domain equations (3.241)–(3.243) and (3.245)–(3.247) are

$$\overline{V}_d = -\left(R_s + \frac{s}{\omega_s}X_{dop}(s)\right)\overline{I}_d + \frac{s}{\omega_s}\overline{G}_{op}(s)\overline{V}_{fd} \qquad (3.262)$$

$$\overline{V}_q = -\left(R_s + \frac{s}{\omega_s}X_{qop}(s)\right)\overline{I}_q \qquad (3.263)$$

$$\overline{V}_o = -\left(R_s + \frac{s}{\omega_s}X_{oop}(s)\right)\overline{I}_o \qquad (3.264)$$

To see how these can be used with a test, consider the schematic of Figure 3.1 introduced earlier. With all the *abc* dot ends connected together to form a neutral point, the three *abc x* ends form the stator terminals. If a scaled voltage V_{test} is applied across *bc*, with the *a* terminal open, the scaled series $I_c(-I_b)$ current establishes an axis that is 90° ahead of the original *a*-axis, as shown in Figure 3.5.

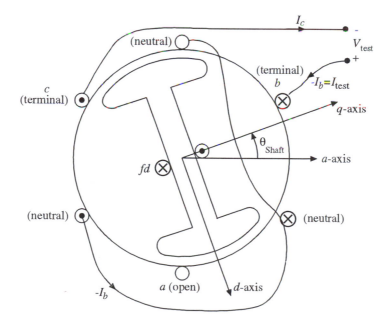

Figure 3.5: *Standstill test schematic*

With this symmetry and $\theta_{\text{shaft}} = \frac{2\pi}{P}$ mech. rad (found by observing the field voltage as the rotor is turned), the scaled voltage V_a will be zero even for nonzero $I_c = -I_b$ and I_{fd}, since its axis is perpendicular to both the b- and c-axis and the field winding d-axis. Also, the unscaled test voltage is

$$v_{\text{test}} = v_b - v_c \tag{3.265}$$

and, by symmetry with $\theta_{\text{shaft}} = \frac{2\pi}{P}$ mechanical radians,

$$v_b = -v_c \tag{3.266}$$

The unscaled transformed voltages are

$$v_d = \frac{\sqrt{3}}{2}v_b - \frac{\sqrt{3}}{2}v_c = \frac{\sqrt{3}}{2}v_{\text{test}} \tag{3.267}$$

$$v_q = v_o = 0 \tag{3.268}$$

and the unscaled transformed currents are

$$i_d = \frac{\sqrt{3}}{2}i_b - \frac{\sqrt{3}}{2}i_c = \sqrt{3}i_b = \sqrt{3}i_{\text{test}} \tag{3.269}$$

$$i_q = i_o = 0 \tag{3.270}$$

For a test set of voltage and current,

$$v_{\text{test}} = \sqrt{3}\sqrt{2}V_{to}\cos(\omega_o t + \theta_o) \tag{3.271}$$

$$i_{\text{test}} = \sqrt{2}I_{to}\cos(\omega_o t + \phi_o) \tag{3.272}$$

with scaled RMS cosine reference phasors defined as

$$\overline{V}_{\substack{\text{test}\\pu}} \overset{\Delta}{=} \frac{\sqrt{3}V_{to}}{V_{BABC}}e^{j\theta_o} \tag{3.273}$$

$$\overline{I}_{\substack{\text{test}\\pu}} \overset{\Delta}{=} \frac{I_{to}}{I_{BABC}}e^{j\phi_o} \tag{3.274}$$

the scaled quantities V_d and I_d are

$$V_d = \frac{v_d}{V_{BDQ}} = \frac{1}{\sqrt{2}V_{BABC}}\frac{\sqrt{3}}{2}\sqrt{3}\sqrt{2}V_{to}\cos(\omega_o t + \theta_o) \tag{3.275}$$

$$I_d = \frac{-i_d}{I_{BDQ}} = \frac{-1}{\sqrt{2}I_{BABC}}\sqrt{3}\sqrt{2}I_{to}\cos(\omega_o t + \phi_o) \tag{3.276}$$

or

$$V_d = \frac{\sqrt{3}}{2}|\overline{V}_{\substack{\text{test}\\pu}}|\cos(\omega_o t + \theta_o) \tag{3.277}$$

$$I_d = -\sqrt{3}|\overline{I}_{\substack{\text{test}\\pu}}|\cos(\omega_o t + \phi_o) \tag{3.278}$$

$$(\overline{V}_d = \frac{\sqrt{3}}{2}|\overline{V}_{\substack{\text{test}\\pu}}|e^{j\theta_o}, \overline{I}_d = -\sqrt{3}|\overline{I}_{\substack{\text{test}\\pu}}|e^{j\phi_o})$$

For $s = j\omega_o$, the ratio of test voltage to current is

$$\overline{Z}_o(j\omega_o) \triangleq \frac{\overline{V}_{\substack{\text{test}\\pu}}}{\overline{I}_{\substack{\text{test}\\pu}}} = -2\frac{\overline{V}_d}{\overline{I}_d} \tag{3.279}$$

which can be written as a real plus imaginary function

$$\overline{Z}_o(j\omega_o) = R_o(\omega_o) + jX_o(\omega_o) \tag{3.280}$$

Therefore, from (3.262), with $\overline{V}_{fd} = 0$ and $s = j\omega_o$,

$$\left(R_s + j\frac{\omega_o}{\omega_s}\overline{X}_{dop}(j\omega_o)\right) = \frac{1}{2}\overline{Z}_o(j\omega_o) \tag{3.281}$$

Clearly R_s is one-half the ratio of a set of dc test voltage and current. After R_s is determined, the operational impedance for any frequency ω_o is

$$\overline{X}_{dop}(j\omega_o) = -j\frac{\omega_s}{\omega_o}\left(\frac{\overline{V}_{\substack{\text{test}\\pu}}}{2\overline{I}_{\substack{\text{test}\\pu}}} - R_s\right) \tag{3.282}$$

A typical test result for a salient pole machine is shown in Figure 3.6 [20, 52] ($\overline{X}_{dop}(j\omega_o)$ is a complex quantity). The plot typically has three levels with two somewhat distinct breakpoints, as shown. For round rotor machines, the plot is more uniform, decreasing with a major breakpoint at 0.01 Hz. A similar test with different rotor positions could be used to compute $\overline{X}_{qop}(j\omega_o)$. These tests, as well as others, are described in considerable detail in [20, 52, 53]. Reference [53] also includes a discussion on curve-fitting procedures to compute "best-fit" model parameters from the frequency response test data. For example, the operational impedance $\overline{X}_{dop}(s)$ for the model used earlier is given by (3.253). If a plot similar to Figure 3.6 (together with a plot of the angle on $\overline{X}_{dop}(s)$) is available, the model parameters can be computed so that (3.253) is some "best-fit" to Figure 3.6. It is possible to make the fit better for different frequency ranges. Thus, the data to be used for a given model can be adjusted to make the model more accurate in certain frequency ranges. The only way to make the model more accurate in all frequency ranges is to increase the dynamic order of the model by adding more "damper" windings in an attempt to reach a better overall fit of $\overline{X}_{dop}(s)$ to the standstill frequency response. Similarly it is clear that the key to model reduction is to properly eliminate dynamic states while still preserving some phenomena (or frequency range response) of interest. This is discussed extensively in later chapters.

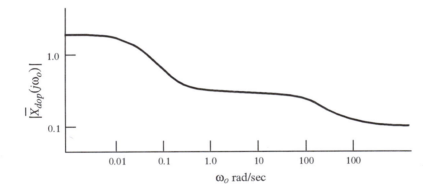

Figure 3.6: *Salient-pole machine operational impedance*

In summary, it is important to note several important points about synchronous machine dynamic modeling. First, the literature abounds with various notational conventions and definitions. The notations and conventions used in this text are as standard as possible given the proliferation of models. It basically follows that of [20]–[29]. Second, it is important to repeat that standard symbols have been defined through the model proposed in this text (and [20]–[29]) with a few noted exceptions. The procedures followed by the industry often define constants through well-defined tests. In some cases, these definitions do not coincide precisely with the model definitions in this text. Thus, when using data obtained from manufacturers, it is important to clarify which definition of symbols is being used. If frequency response data are given, the model parameters can be computed using curve-fitting techniques.

3.8   Problems

3.1 The text uses T_{qdo} to transform abc variables into dqo variables. Consider the following alternative transformation matrix:

$$P_{dqo} = \sqrt{\frac{2}{3}} \begin{bmatrix} \cos \frac{P}{2}\theta_{\text{shaft}} & \cos \left(\frac{P}{2}\theta_{\text{shaft}} - \frac{2\pi}{3}\right) & \cos \left(\frac{P}{2}\theta_{\text{shaft}} + \frac{2\pi}{3}\right) \\ \sin \frac{P}{2}\theta_{\text{shaft}} & \sin \left(\frac{P}{2}\theta_{\text{shaft}} - \frac{2\pi}{3}\right) & \sin \left(\frac{P}{2}\theta_{\text{shaft}} + \frac{2\pi}{3}\right) \\ \frac{1}{\sqrt{2}} & \frac{1}{\sqrt{2}} & \frac{1}{\sqrt{2}} \end{bmatrix}$$

Show that $P_{dqo}^t = P_{dqo}^{-1}$ (P_{dqo} is orthogonal).

3.2 Given the following model

$$v = 10i + \frac{d\lambda}{dt}, \ \lambda = 0.05i$$

scale $v, i,$ and λ as follows:

$$V = \frac{v}{V_B}, \ I = \frac{i}{I_B}, \ \psi = \frac{\lambda}{\Lambda_B}$$

to get

$$V = RI + \frac{1}{\omega_B}\frac{d\psi}{dt}, \ \psi = XI$$

Find R and X if $V_B = 10,000$ volts, $S_B = 5 \times 10^6 \ VA$, $\omega_B = 2\pi 60$ rad/sec, and $I_B = S_B/V_B$, $\Lambda_B = V_B/\omega_B$.

3.3 Using the P_{dqo} of Problem 3.1 with

$$v_a = \sqrt{2}V \cos(\omega_s t + \theta)$$
$$v_b = \sqrt{2}V \cos(\omega_s t + \theta - \frac{2\pi}{3})$$
$$v_c = \sqrt{2}V \cos(\omega_s t + \theta + \frac{2\pi}{3})$$

and $\hat{\delta} \triangleq \frac{P}{2}\theta_{\text{shaft}} - \omega_s t - \frac{\pi}{2}$, express the phasor $\overline{V} = Ve^{j\theta}$ in terms of $v_d, v_q,$ and $\hat{\delta}$ that you get from using P_{dqo} to transform v_a, v_b, v_c into v_d, v_q, v_o.

3.4 Neglect saturation and derive an expression for $\overline{E}'$ in the following alternative steady-state equivalent circuit:

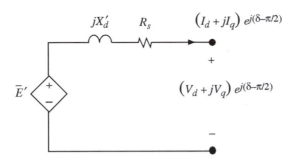

Write $\overline{E}'$ as a function of E_q', I_d, I_q, δ.

3.5 Given the magnetization curve shown in pu, compute X_{md} and plot $S_G(\psi)$ for

$$\psi = X_{md}I - S_G(\psi)$$

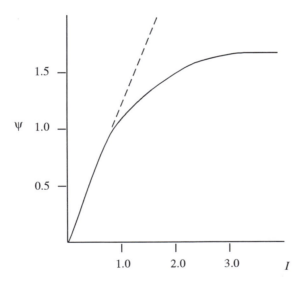

3.6 Repeat the derivation of the single-machine steady-state equivalent circuit using the following saturation functions:

$$S_d^{(2)} = S_{1d}^{(2)} = S_q^{(2)} = S_{1q}^{(2)} = S_{2q}^{(2)} = S_o^{(2)} = 0$$
$$S_{fd}^{(2)} = S_G(E_q')$$

where S_G is obtained from the open-circuit characteristic as given in the text.

3.7 From Ref. [51], show that the saturation model of Problem 3.6 does satisfy all conditions for a conservative coupling field.

3.8 Given (3.122)–(3.124) and (3.125)–(3.127), together with (3.64)–(3.70), find a circuit representation for these mathematical models. (Hint: Split X_d, X_{fd}, X_{1d}, etc. into leakage plus magnetizing.)

3.9 Given the following nonlinear magnetic circuit model for a synchronous machine:

$$
\begin{aligned}
\psi_d &= X_d(-I_d) + X_{md}I_{fd} - S_d(\psi_d, \psi_{fd}) \\
\psi_{fd} &= X_{md}(-I_d) + X_{fd}I_{fd} - S_{fd}(\psi_d, \psi_{fd}) \\
\psi_q &= X_q(-I_q) + X_{mq}I_{1q} - S_q(\psi_q, \psi_{1q}) \\
\psi_{1q} &= X_{mq}(-I_q) + X_{1q}I_{1q} - S_{1q}(\psi_q, \psi_{1q})
\end{aligned}
$$

(a) Find the constraints on the saturation functions S_d, S_{fd}, S_q, S_{1q} such that the overall model does not violate the assumption of a conservative coupling field.

(b) If the other steady-state equations are

$$
\begin{aligned}
V_d = -R_s I_d - \psi_q \qquad V_{fd} = R_{fd}I_{fd} \\
V_q = -R_s I_q + \psi_d \qquad 0 = R_{1q}I_{1q}
\end{aligned}
$$

find an expression for $\overline{E}$, where $\overline{E}$ is the voltage "behind" $R_s + jX_q$ in a circuit that has a terminal voltage

$$
\overline{V} = (V_d + jV_q)e^{j(\delta - \pi/2)}
$$

Chapter 4

SYNCHRONOUS MACHINE CONTROL MODELS

4.1 Voltage and Speed Control Overview

The primary objective of an electrical power system is to maintain balanced sinusoidal voltages with virtually constant magnitude and frequency. In the synchronous machine models of the last chapter, the terminal constraints (relationships between V_d, I_d, V_q, I_q, V_o, and I_o) were not specified. These will be discussed in the next chapter. In addition, the two quantities E_{fd} and T_M were left as inputs to be specified. E_{fd} is the scaled field voltage, which, if set equal to 1.0 pu, gives 1.0 pu open-circuit terminal voltage. T_M is the scaled mechanical torque to the shaft. If it is specified as a constant, the machine terminal constraints will determine the steady-state speed. Specifying E_{fd} and T_M to be constants in the model means that the machine does not have voltage or speed (and, hence, frequency) control. If a synchronous machine is to be useful for a wide range of operating conditions, it should be capable of participating in the attempt to maintain constant voltage and frequency. This means that E_{fd} and T_M should be systematically adjusted to accommodate any change in terminal constraints. The physical device that provides the value of E_{fd} is called the exciter. The physical device that provides the value of T_M is called the prime mover. This chapter is devoted to basic mathematical models of these components and their associated control systems.

4.2   Exciter Models

One primary reason for using three-phase generators is the constant electrical torque developed in steady state by the interaction of the magnetic fields produced by the armature ac currents with the field dc current. Furthermore, for balanced three-phase machines, a dc current can be produced in the field winding by a dc voltage source. In steady state, adjustment of the field voltage changes the field current and, therefore, the terminal voltage. Perhaps the simplest scheme for voltage control would be a battery with a rheostat adjusted voltage divider connected to the field winding through slip rings. Manual adjustment of the rheostat could be used to continuously react to changes in operating conditions to maintain a voltage magnitude at some point. Since large amounts of power are normally required for the field excitation, the control device is usually not a battery, and is referred to as the main exciter. This main exciter may be either a dc generator driven off the main shaft (with brushes and slip rings), an inverted ac generator driven off the main shaft (brushless with rotating diodes), or a static device such as an ac-to-dc converter fed from the synchronous machine terminals or auxiliary power (with slip rings). The main exciter may have a pilot exciter that provides the means for changing the output of the main exciter. In any case, E_{fd} normally is not manipulated directly, but is changed through the actuation of the exciter or pilot exciter.

Consider first the model for rotating dc exciters. One circuit for a separately excited dc generator is shown in Figure 4.1 [21].

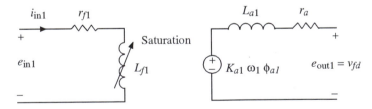

Figure 4.1: *Separately excited dc machine circuit*

Its output is the unscaled synchronous machine field voltage v_{fd}. For small r_{a1} and L_{a1}, this circuit has the dynamic model

$$e_{in1} = i_{in1}r_{f1} + \frac{d\lambda_{f1}}{dt} \tag{4.1}$$

$$v_{fd} = K_{a1}\omega_1\phi_{a1} \tag{4.2}$$

with the exciter field flux linkage related to field flux ϕ_{f1} by

$$\lambda_{f1} = N_{f1}\phi_{f1} \tag{4.3}$$

Assuming a constant percent leakage (coefficient of dispersion σ_1), the armature flux is

$$\phi_{a1} = \frac{1}{\sigma_1}\phi_{f1} \tag{4.4}$$

Assuming constant exciter shaft speed ω_1,

$$\lambda_{f1} = \frac{N_{f1}\sigma_1}{K_{a1}\omega_1}v_{fd} \tag{4.5}$$

Now, the relationship between v_{fd} and i_{in1} is nonlinear due to saturation of the exciter iron, as shown in Figure 4.2.

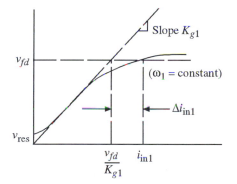

Figure 4.2: *Exciter saturation curve*

Without armature resistance or inductance, this curve is valid for open-circuit or loaded conditions. The slope of the unsaturated curve (air-gap line) is [55]

$$K_{g1} = \frac{K_{a1}\omega_1}{N_{f1}\sigma_1}L_{f1us} \tag{4.6}$$

where L_{f1us} is referred to as the "unsaturated" field inductance. The saturation can be accounted for by a saturation function f_{sat} defined through Figure 4.2 as

$$f_{sat}(v_{fd}) = \Delta i_{in1}/v_{fd} \tag{4.7}$$

In terms of these quantities,

$$\lambda_{f1} = \frac{L_{f1us}}{K_{g1}}v_{fd} \tag{4.8}$$

and

$$i_{in1} = \frac{v_{fd}}{K_{g1}} + f_{sat}(v_{fd})v_{fd} \tag{4.9}$$

With these assumptions, the unscaled exciter dynamic model is

$$e_{in1} = \frac{r_{f1}}{K_{g1}}v_{fd} + r_{f1}f_{sat}(v_{fd})v_{fd} + \frac{L_{f1us}}{K_{g1}}\frac{dv_{fd}}{dt} \tag{4.10}$$

This equation must now be scaled for use with the previously scaled synchronous model. Since the armature terminals of the exciter are connected directly to the terminals of the synchronous machine field winding, we must scale v_{fd} as before and use the same system power base. Thus, using (3.54), (3.59), (3.113), and (3.139), we define

$$V_R \triangleq \frac{X_{md}e_{in1}}{R_{fd}V_{BFD}} \tag{4.11}$$

$$K_{E_{sep}} \triangleq \frac{r_{f1}}{K_{g1}} \tag{4.12}$$

$$T_E \triangleq \frac{L_{f1us}}{K_{g1}} \tag{4.13}$$

$$S_E(E_{fd}) \triangleq r_{f1}f_{sat}\left(\frac{V_{BFD}R_{fd}}{X_{md}}E_{fd}\right) \tag{4.14}$$

With these assumptions and definitions, the scaled model of a separately excited dc generator main exciter is

$$T_E\frac{dE_{fd}}{dt} = -\left(K_{E_{sep}} + S_E(E_{fd})\right)E_{fd} + V_R \tag{4.15}$$

The input V_R is normally the scaled output of the amplifier (or pilot exciter), which is applied to the field of the separately excited main exciter.

When the main dc generator is self-excited, this amplifier voltage appears in series with the exciter field, as shown in Figure 4.3. This field circuit has the dynamic equation

$$e_{in1} = r_{f1}i_{in1} + \frac{d\lambda_{f1}}{dt} - v_{fd} \tag{4.16}$$

with the same assumptions as above, and the new constant $K_{E_{self}}$ defined as

$$K_{E_{self}} \triangleq K_{E_{sep}} - 1 \tag{4.17}$$

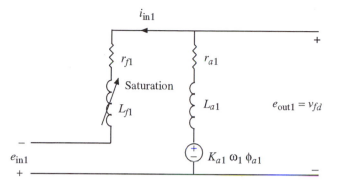

Figure 4.3: *Self-excited dc generator circuit*

The scaled model of a self-excited dc generator main exciter is, then,

$$T_E \frac{dE_{fd}}{dt} = -(K_{E_{\text{self}}} + S_E(E_{fd}))E_{fd} + V_R \qquad (4.18)$$

For typical machines, $K_{E_{\text{self}}}$ is a small negative number. To allow voltage buildup, it would be necessary to specify V_R to include the residual voltage that exists at zero $i_{\text{in}1}$, as shown in Figure 4.2. Also, if $e_{\text{in}1}$ is replaced by a rheostat whose variable resistance is included in r_{f1}, $K_{E_{\text{self}}}$ and S_E would be functions of both the actual field resistance and the rheostat resistance. This rheostat can be set to produce the required terminal voltage. In steady state with $V_R = V_{\text{res}}$ this requires

$$K_{E_{\text{self}}} = \left(\frac{V_{\text{res}}}{E_{fd}}\right) - S_E(E_{fd}) \qquad (4.19)$$

That is, since r_{f1} includes the rheostat setting, it can be adjusted for a given steady-state condition. This would make r_{f1}, rather than V_R, a control variable.

In steady state, the exciter equation (written simply with K_E) is

$$0 = -(K_E + S_E(E_{fd}))E_{fd} + V_R \qquad (4.20)$$

The input V_R usually has a maximum $V_{R_{\text{max}}}$, which produces maximum (ceiling) excitation voltage $E_{fd_{\text{max}}}$. Since S_E is a function of this excitation level, these quantities must satisfy

$$0 = -(K_E + S_{E_{\text{max}}})E_{fd_{\text{max}}} + V_{R_{\text{max}}} \qquad (4.21)$$

When specifying exciter data, it is common to specify the saturation function through the number $S_{E_{\max}}$ and the number $S_{E0.75_{\max}}$, which is the saturation level when E_{fd} is 0.75 $E_{fd_{\max}}$. A saturation function is then fitted to these two points. One typical function is

$$S_E(E_{fd}) = A_x e^{B_x E_{fd}} \tag{4.22}$$

When evaluated at two points, this function gives

$$S_{E_{\max}} = A_x e^{B_x E_{fd_{\max}}} \tag{4.23}$$

$$S_{E0.75_{\max}} = A_x e^{B_x \frac{3}{4} E_{fd_{\max}}} \tag{4.24}$$

For given values of K_E, $V_{R_{\max}}$, $S_{E_{\max}}$, and $S_{E0.75_{\max}}$, the constants A_x and B_x can be computed. This is illustrated in the following example.

Example 4.1

Given: $K_E = 1.0$, $V_{R_{\max}} = 7.3$, $S_{E_{\max}} = 0.86$, $S_{E0.75_{\max}} = 0.50$ (all in pu)

Find: A_x and B_x

Solution: From (4.20),

$$0 = -(1 + 0.86)E_{fd_{\max}} + 7.3$$
$$0.86 = A_x e^{B_x E_{fd_{\max}}}$$
$$0.50 = A_x e^{B_x \frac{3}{4} E_{fd_{\max}}}$$

Solving these three equations gives

$$E_{fd_{\max}} = 3.925$$
$$A_x = 0.09826$$
$$B_x = 0.5527$$

$\square$

References [55, 56] and [75] give additional information on these and other exciter models.

4.3   Voltage Regulator Models

The exciter provides the mechanism for controlling the synchronous machine terminal voltage magnitude. In order to automatically control terminal voltage, a transducer signal must be compared to a reference voltage and amplified to produce the exciter input V_R. The amplifier can be a pilot exciter (another dc generator) or a solid-state amplifier. In either case, the amplifier is often modeled as in the last section with a limiter replacing the saturation function.

$$T_A \frac{dV_R}{dt} = -V_R + K_A V_{\text{in}} \tag{4.25}$$

$$V_R^{\text{min}} \leq V_R \leq V_R^{\text{max}} \tag{4.26}$$

where V_{in} is the amplifier input, T_A is the amplifier time constant, and K_A is the amplifier gain. The V_R limit can be multivalued to allow a higher limit during transients. The steady-state limit would be lower to reflect thermal constraints on the exciter and synchronous machine field winding. Recall that V_R is the scaled input to the main exciter. This voltage may be anything between zero and its limits if the main exciter is self-excited, but must be nonzero if the exciter is separately excited. We have assumed that the amplifier data have been scaled according to our given per-unit system.

If the voltage V_{in} is simply the error voltage produced by the difference between a reference voltage and a conditioned potential transformer connected to the synchronous machine terminals, the closed-loop control system can exhibit instabilities. This can be seen by noting that the self-excited dc exciter can have a negative K_E such that its open-loop eigenvalue is positive for small saturation S_E. Even without this potential instability, there is always a need to shape the regulator response to achieve desirable dynamic performance. In many standard excitation systems, this is accomplished through a stabilizing transformer whose input is connected to the output of the exciter and whose output voltage is subtracted from the amplifier input. A scaled circuit showing the transformer output as V_F is shown in Figure 4.4.

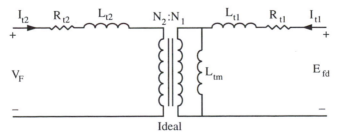

Figure 4.4: *Stabilizing transformer circuit*

If I_{t2} is initially zero and L_{t2} is very large, then I_{t2} must remain near zero.

With this assumption, an approximate dynamic model for this circuit is

$$E_{fd} = R_{t1}I_{t1} + (L_{t1} + L_{tm})\frac{dI_{t1}}{dt} \tag{4.27}$$

$$V_F = \frac{N_2}{N_1}L_{tm}\frac{dI_{t1}}{dt} \tag{4.28}$$

where V_F is a scaled output of the stabilizing transformer. Differentiating V_F and E_{fd} gives

$$\frac{dV_F}{dt} = \frac{N_2}{N_1}L_{tm}\left(\frac{1}{(L_{t1} + L_{tm})}\left(\frac{dE_{fd}}{dt} - \frac{R_{t1}}{L_{tm}}\frac{N_2}{N_1}V_F\right)\right) \tag{4.29}$$

Using (4.18) with general K_E and defining

$$T_F \triangleq \frac{L_{t1} + L_{tm}}{R_{t1}} \tag{4.30}$$

$$K_F \triangleq \frac{N_1}{N_2}\frac{L_{tm}}{R_{t1}} \tag{4.31}$$

the dynamic model of the stabilizing transformer can be written as

$$T_F\frac{dV_F}{dt} = -V_F + K_F\left(-\frac{K_E + S_E(E_{fd})}{T_E}E_{fd} + \frac{V_R}{T_E}\right) \tag{4.32}$$

Another form of this model is often used by defining

$$R_f \triangleq \frac{K_F}{T_F}E_{fd} - V_F \tag{4.33}$$

With R_f (called rate feedback) as the dynamic state,

$$T_F\frac{dR_f}{dt} = -R_f + \frac{K_F}{T_F}E_{fd} \tag{4.34}$$

This form will be used throughout the remainder of the text.

If the amplifier input V_{in} contained only the reference voltage minus the terminal voltage minus V_F, the voltage regulator could still have regulation and stability problems. There can be regulation problems when two or more synchronous machines are connected in parallel and each machine has an exciter plus voltage regulator. Since the synchronous machine field current has a major role in determining the reactive power output of the machine, parallel operation requires that the field currents be adjusted properly so that the machines share reactive power. This is accomplished through the addition of a load compensation circuit in the regulator, which makes the parallel operation appear as if there are two different terminal voltages even though both machines are paralleled to the same bus. This can be done by including stator current in the regulator input. To see how this can be modeled, consider the scaled terminal voltages V_a, V_b, and V_c found by transforming and rescaling V_d, V_q, and V_o. Using (3.14) with $V_o = 0$ and (3.39)

$$V_a = \sqrt{2}(V_d \sin(\omega_s t + \delta) + V_q \cos(\omega_s t + \delta)) \tag{4.35}$$

$$V_b = \sqrt{2}\left(V_d \sin\left(\omega_s t + \delta - \frac{2\pi}{3}\right) + V_q \cos\left(\omega_s t + \delta - \frac{2\pi}{3}\right)\right) \tag{4.36}$$

$$V_c = \sqrt{2}\left(V_d \sin\left(\omega_s t + \delta + \frac{2\pi}{3}\right) + V_q \cos\left(\omega_s t + \delta + \frac{2\pi}{3}\right)\right) \tag{4.37}$$

The stator line currents I_a, I_b, and I_c are related in the same way with V_d and V_q replaced by I_d and I_q. These expressions are valid for transients as well as steady state, and can be written alternatively as

$$V_a = \sqrt{2}\sqrt{V_d^2 + V_q^2} \cos\left(\omega_s t + \delta - \frac{\pi}{2} + \tan^{-1}\frac{V_q}{V_d}\right) \tag{4.38}$$

$$V_b = \sqrt{2}\sqrt{V_d^2 + V_q^2} \cos\left(\omega_s t + \delta - \frac{\pi}{2} + \tan^{-1}\frac{V_q}{V_d} - \frac{2\pi}{3}\right) \tag{4.39}$$

$$V_c = \sqrt{2}\sqrt{V_d^2 + V_q^2} \cos\left(\omega_s t + \delta - \frac{\pi}{2} + \tan^{-1}\frac{V_q}{V_d} + \frac{2\pi}{3}\right) \tag{4.40}$$

and similarly for I_a, I_b, and I_c. To see how load compensation can be performed, consider the case of an overexcited synchronous machine (serving inductive load), with the steady-state phasor diagram of Figure 4.5. Suppose that the sensed voltage is line-to-line RMS voltage $\overline{V}_{ac}$; then the uncompensated voltage is defined as

$$Vt_{\text{uncomp}} \triangleq \frac{1}{\sqrt{3}}|\overline{V}_{ac}| \tag{4.41}$$

Consider the compensated voltage defined as

$$Vt_{\text{comp}} \triangleq \frac{1}{\sqrt{3}}|\overline{V}_{ac} - \propto_{\text{comp}} \overline{I}_b| \tag{4.42}$$

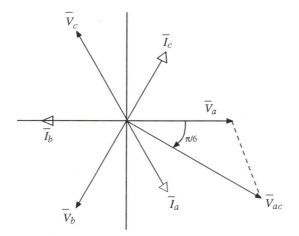

Figure 4.5: *Steady-state overexcited phasor diagram*

where α_{comp} is some positive compensation constant. The compensated terminal voltage will differ from the uncompensated terminal voltage by an amount proportional to α_{comp} and the phase shift of $\overline{I}_b$. With $\overline{I}_b$ in the position shown in the phasor diagram, and if the reference voltage is near the uncompensated value of V_t, the error signal from the compensated V_t will tell the regulator to lower the terminal voltage (the reverse of what you might expect!). When the machine is underexcited (serving capacitive load), the current $\overline{I}_b$ will lead $\overline{V}_b$, and the compensated voltage will be less than the uncompensated voltage. This voltage, when compared to a reference voltage near the uncompensated voltage, will tell the regulator to raise the terminal voltage (again, the reverse of what you might expect). When two generators are operating in parallel, if the field current on one generator becomes excessive and causes "circulating" reactive power, this reactive power will be an inductive load to the excessively excited machine and a capacitive load to the other. The compensation circuits then will cause the excessive excitation to be reduced and the other to be raised, thus balancing the reactive power loading. This type of load compensation is called parallel droop compensation. Additional types of compensation that do not result in a drop in voltage under inductive load are also available.

In terms of the *dq* components of $\overline{V}_a$, $\overline{V}_c$, and $\overline{I}_b$, we can define the compensated and uncompensated voltages as

$$V_{t\text{comp}} \triangleq \frac{1}{\sqrt{3}} \left[\left(\frac{3}{2}V_d + \frac{\sqrt{3}}{2}V_q - \propto_{\text{comp}} I_d \right)^2 \right.$$

$$\left. + \left(\frac{3}{2}V_q - \frac{\sqrt{3}}{2}V_d - \propto_{\text{comp}} I_q \right)^2 \right]^{1/2} \quad (4.43)$$

This gives

$$V_{t\text{uncomp}} = V_{t\text{comp}}|_{\propto\text{comp}=0} = \sqrt{V_d^2 + V_q^2} \quad (4.44)$$

Example 4.2

$$\overline{V}_a = 1\angle 0, \overline{V}_b = 1\angle -120°, \overline{V}_c = 1\angle 120°$$

$$\overline{I}_b = 0.5\angle 180° \text{ (all pu)}$$

This is a power factor $= \cos 60°$ (lagging power factor load)

$$V_{t\text{uncomp}} = \frac{1}{\sqrt{3}}\sqrt{3} = 1.0$$

$$V_{t\text{comp}} = \frac{1}{\sqrt{3}} | \sqrt{3}\angle -30° - \propto_{\text{comp}} 0.5\angle 180 |$$

For $\propto_{\text{comp}} = 0.1$

$$V_{t\text{comp}} = 1.025$$

If the reference voltage is 1.0, the voltage regulator will attempt to *lower* the voltage by lowering E_{fd}.

□

Example 4.3

Let us use the same voltages, $\bar{I}_b = 0.5\angle -90°$ (pu). This is the power factor $\cos 30°$ (leading).

$$V_{t\text{uncomp}} = \frac{1}{\sqrt{3}}\sqrt{3} = 1.0$$

$$V_{t\text{comp}} = \frac{1}{\sqrt{3}} | \sqrt{3}\angle -30° - 0.05\angle -90° |$$
$$= 0.986$$

If the reference voltage is 1.0, the voltage regulator will attempt to *raise* the voltage by raising E_{fd}. This compensation will make two parallel generators share the total reactive power output.

$\square$

Realizing that the sensed voltage may be either compensated or uncompensated, we simply drop the subscript and write the final general expression for the amplifier input voltage using the stabilizer feedback variable R_f, rather than V_F, as

$$V_{\text{in}} = V_{\text{ref}} - V_t + R_f - \frac{K_F}{T_F}E_{fd} \qquad (4.45)$$

In this model, we have not included any dynamics for the transducer, which could be a potential transformer, filters, and smoothers. There are many other fine details about excitation systems that may be important for some simulations. In keeping with the philosophy of this text as one on fundamental dynamic modeling, we conclude this section with a summary of a fundamental model of an excitation system:

$$T_E \frac{dE_{fd}}{dt} = -(K_E + S_E(E_{fd}))E_{fd} + V_R \qquad (4.46)$$

$$T_F \frac{dR_f}{dt} = -R_f + \frac{K_F}{T_F} E_{fd} \qquad (4.47)$$

$$T_A \frac{dV_R}{dt} = -V_R + K_A R_f - \frac{K_A K_F}{T_F} E_{fd}$$

$$+ K_A(V_{\text{ref}} - V_t) \qquad (4.48)$$

$$V_R^{\text{min}} \le V_R \le V_R^{\text{max}} \qquad (4.49)$$

where K_E may be the previously defined separate or self-excited constant, and V_t may be either of the previously defined compensated or uncompensated terminal voltages. To be complete, the model should include an expression for S_E and an algebraic equation for V_t.

4.4   Turbine Models

The frequency of the ac voltage at the terminals of a synchronous machine is determined by its shaft speed and the number of magnetic poles of the machine. The steady-state speed of a synchronous machine is determined by the speed of the prime mover that drives its shaft. Typical prime movers are diesel engines, gasoline engines, steam turbines, hydroturbines (water wheels), and gas turbines. The prime mover output affects the input T_M in the model of the last chapter. This section presents basic models for the hydroturbine and steam turbine.

Hydroturbines

Hydropower plants have essentially five major components. These are the storage reservoir, intake tunnel, surge tank, penstock, and water (hydro) turbine. Precise nonlinear models of these components are not typically used in power system dynamic analysis. Alternatively, approximate linear models are used to capture the fundamental characteristics of the plant and its impact on the electrical system. Thus, the following models should not be used for studies where large changes in turbine power are expected. Since the turbine torque is the primary variable of interest, most models are made as simple as possible while still preserving the turbine torque and speed control characteristics. The power to the water turbine depends on the position of the gate valve at the bottom of the penstock. The power is derived from the water pressure that arises due to the water head (elevation). The penstock is the water channel from the intake tunnel of the elevated reservoir down to the gate valves and turbine. The gate valve position then corresponds to a certain level of power P_{HV} at rated speed. Using scaled parameters consistent with the last chapter, a simplified model of hydroturbine small-change dynamics is [57]

$$\Delta FR = A_{11}\Delta TH + A_{12}\Delta\omega_{HT} + A_{13}\Delta P_{HV} \qquad (4.50)$$

$$\Delta T_{HT} = A_{21}\Delta TH + A_{22}\Delta\omega_{HT} + A_{23}\Delta P_{HV} \qquad (4.51)$$

where ΔFR is the change in water flow rate, ΔTH is the change in turbine head, $\Delta\omega_{HT}$ is the change in hydroturbine speed, and ΔT_{HT} is the change in hydroturbine output torque, all scaled consistently with the last chapter. One common model considers only two dynamic phenomena, the rate of change of flow deviation as [57]

$$\frac{d\Delta FR}{dt} = -\frac{\Delta TH}{T_w} \qquad (4.52)$$

and Newton's law for the turbine mass with scaled inertia constant H_{HT}

$$\frac{d\Delta\delta_{HT}}{dt} = \Delta\omega_{HT} \qquad (4.53)$$

$$\frac{2H_{HT}}{\omega_s}\frac{d\Delta\omega_{HT}}{dt} = \Delta T_{HT} - \Delta T_M \qquad (4.54)$$

where T_w is the water starting time in seconds, and T_M is as previously defined for the synchronous machine shaft dynamics. The last equation allows for the case where the shaft connection between the hydroturbine output and the synchronous machine input is considered a stiff spring rather than a rigid connection. In this case, the change in torque into the synchronous machine would be

$$\Delta T_M = -K_{HM}(\Delta\delta - \Delta\delta_{HT}) \qquad (4.55)$$

where K_{HM} represents the stiffness of the coupling between the turbine and the synchronous machine, and δ is the machine angle, as previously defined. While it is clearly possible to keep all terms, it is customary to assume that $A_{12}\Delta\omega_{HT}$ and $A_{22}\Delta\omega_{HT}$ are small compared to other terms.

For operation near an equilibrium point (denoted as superscript o) with

$$T^o_{HT} = A_{23}P^o_{HV} = T^o_M = -K_{HM}(\delta^o - \delta^o_{HT}) \qquad (4.56)$$

and

$$\omega^o_{HT} = \omega_s \qquad (4.57)$$

the actual variables are

$$T_{HT} = T^o_{HT} + \Delta T_{HT} \qquad (4.58)$$

$$T_M = T^o_M + \Delta T_M \qquad (4.59)$$

$$P_{HV} = P^o_{HV} + \Delta P_{HV} \qquad (4.60)$$

$$\delta = \delta^o + \Delta\delta \qquad (4.61)$$

$$\delta_{HT} = \delta^o_{HT} + \Delta\delta_{HT} \qquad (4.62)$$

$$\omega_{HT} = \omega^o_{HT} + \Delta\omega_{HT} \qquad (4.63)$$

The hydroturbine model written in actual per-unit torques (but valid only for small changes about an equilibrium point) is written with T_{HT} as a state:

$$T_w \frac{dT_{HT}}{dt} = -\frac{1}{A_{11}}T_{HT} + \frac{A_{23}}{A_{11}}P_{HV}$$
$$+ T_w\left(A_{23} - \frac{A_{13}A_{21}}{A_{11}}\right)\frac{dP_{HV}}{dt} \qquad (4.64)$$

$$\frac{d\delta_{HT}}{dt} = \omega_{HT} - \omega_s \qquad (4.65)$$

$$\frac{2H_{HT}}{\omega_s}\frac{d\omega_{HT}}{dt} = T_{HT} - T_M \qquad (4.66)$$

$$T_M = -K_{HM}(\delta - \delta_{HT}) \qquad (4.67)$$

When used with the synchronous machine model, the above T_M expression must be used in the speed equation for the synchronous machine. Thus, the synchronous machine and turbine are coupled through the angles δ and δ_{HT}. The hydrovalve power P_{HV} will become a dynamic state when the hydro-governor is added. For a rigid connection, T_{HT} becomes T_M and the turbine inertia is simply added to the synchronous machine inertia H, giving only

$$T_w \frac{dT_M}{dt} = -\frac{1}{A_{11}}T_M + \frac{A_{23}}{A_{11}}P_{HV} + T_w\left(A_{23} - \frac{A_{13}A_{21}}{A_{11}}\right)\frac{dP_{HV}}{dt} \qquad (4.68)$$

and T_M remains in the synchronous machine speed equation. For an ideal lossless hydroturbine at full load [57]

$$A_{11} = 0.5, \quad A_{21} = 1.5, \quad A_{13} = 1.0, \quad A_{23} = 1.0 \qquad (4.69)$$

and T_w ranges between 0.5 and 5 seconds. Additional hydroturbine models are discussed in references [57]–[59].

Steam turbines

Steam plants consist of a fuel supply to a steam boiler that supplies a steam chest. The steam chest contains pressurized steam that enters a high-pressure (HP) turbine through a steam valve. As in the hydroturbine, the power into the high-pressure turbine is proportional to the valve opening. A nonreheat system would then terminate in the condenser and cooling systems, with the HP turbine shaft connected to the synchronous machine. It is common to include additional stages, such as the intermediate (IP) and low (LP) pressure turbines. The steam is reheated upon leaving the high-pressure turbine, and either reheated or simply crossed over between the IP and LP turbines. The dynamics that are normally represented are the steam chest delay, the reheat delay, or the crossover piping delay. In a tandem connection, all stages are on the same shaft. In a cross-compound system, the different stages may be connected on different shafts. These two shafts then supply the torque for two generators. In this analysis, we model the steam chest dynamics, the single reheat dynamics, and the mass dynamics for a two-stage (HP and LP) turbine tandem mounted. In this model, we are interested in the effect of the steam valve position (power P_{SV}) on the synchronous machine torque T_M. The incremental steam chest dynamic model is a simple linear single time constant with unity gain, written in scaled variables consistent with the last chapter as

$$T_{CH}\frac{d\Delta P_{CH}}{dt} = -\Delta P_{CH} + \Delta P_{SV} \qquad (4.70)$$

where ΔP_{CH} is the change in output power of the steam chest. This output is either converted into torque on the HP turbine or passed on to the reheat cycle. Let the fraction that is converted into torque be

$$\Delta T_{HP} = K_{HP}\Delta P_{CH} \qquad (4.71)$$

and the fraction passed on to the reheater be $(1-K_{HP})\Delta P_{CH}$. The dynamics of the HP turbine mass in incremental scaled variables are

$$\frac{d\Delta \delta_{HP}}{dt} = \Delta\omega_{HP} \qquad (4.72)$$

$$\frac{2H_{HP}}{\omega_s}\frac{d\Delta\omega_{HP}}{dt} = \Delta T_{HP} - \Delta T_{HL} \tag{4.73}$$

where ΔT_{HL} is the incremental change in torque transmitted through the shaft to the LP turbine. This is modeled as a stiff spring:

$$\Delta T_{HL} = -K_{HL}(\Delta\delta_{LP} - \Delta\delta_{HP}) \tag{4.74}$$

The reheat process has a time delay that can be modeled similarly as

$$T_{RH}\frac{d\Delta P_{RH}}{dt} = -\Delta P_{RH} + (1 - K_{HP})\Delta P_{CH} \tag{4.75}$$

where ΔP_{RH} is the change in output power of the reheater. Assuming that this output is totally converted into torque on the LP turbine,

$$\Delta T_{LP} = \Delta P_{RH} \tag{4.76}$$

The dynamics of the LP turbine mass in incremental scaled variables are

$$\frac{d\Delta\delta_{LP}}{dt} = \Delta\omega_{LP} \tag{4.77}$$

$$\frac{2H_{LP}}{\omega_s}\frac{d\Delta\omega_{LP}}{dt} = \Delta T_{HL} + \Delta T_{LP} - \Delta T_M \tag{4.78}$$

where the torque to the connection of the LP turbine to the synchronous machine is assumed to be transmitted through a stiff spring as

$$\Delta T_M = -K_{LM}(\Delta\delta - \Delta\delta_{LP}) \tag{4.79}$$

For operation near an equilibrium point (denoted by superscript o) with

$$P_{CH}^o = P_{SV}^o, \ T_{HP}^o = T_{HL}^o = K_{HP}P_{CH}^o = -K_{HL}(\delta_{LP}^o - \delta_{HP}^o),$$
$$T_{LP}^o = P_{RH}^o = (1 - K_{HP})P_{CH}^o, \ \omega_{LP}^o = \omega_s, \ \omega_{HP}^o = \omega_s,$$
$$T_M^o = P_{CH}^o = -K_{LM}(\delta^o - \delta_{LP}^o) \tag{4.80}$$

the actual variables are

$$P_{CH} = P_{CH}^o + \Delta P_{CH} \tag{4.81}$$

$$P_{SV} = P_{SV}^o + \Delta P_{SV} \tag{4.82}$$

$$\delta_{HP} = \delta_{HP}^o + \Delta\delta_{HP} \tag{4.83}$$

$$\omega_{HP} = \omega_{HP}^o + \Delta\omega_{HP} \tag{4.84}$$

$$P_{RH} = P_{RH}^o + \Delta P_{RH} \tag{4.85}$$

$$\delta_{LP} = \delta_{LP}^o + \Delta\delta_{LP} \tag{4.86}$$

$$\omega_{LP} = \omega_{LP}^o + \Delta\omega_{LP} \tag{4.87}$$

$$\delta = \delta^o + \Delta\delta \tag{4.88}$$

$$T_M = T_M^o + \Delta T_M \tag{4.89}$$

The steam turbine model written in actual per-unit torques (but valid only for small changes about an equilibrium point) is written as

$$T_{CH}\frac{dP_{CH}}{dt} = -P_{CH} + P_{SV} \tag{4.90}$$

$$\frac{d\delta_{HP}}{dt} = \omega_{HP} - \omega_s \tag{4.91}$$

$$\frac{2H_{HP}}{\omega_s}\frac{d\omega_{HP}}{dt} = K_{HP}P_{CH} + K_{HL}(\delta_{LP} - \delta_{HP}) \tag{4.92}$$

$$T_{RH}\frac{dP_{RH}}{dt} = -P_{RH} + (1 - K_{HP})P_{CH} \tag{4.93}$$

$$\frac{d\delta_{LP}}{dt} = \omega_{LP} - \omega_s \tag{4.94}$$

$$\frac{2H_{LP}}{\omega_s}\frac{d\omega_{LP}}{dt} = -K_{HL}(\delta_{LP} - \delta_{HP}) + P_{RH} + K_{LM}(\delta - \delta_{LP}) \tag{4.95}$$

and T_M in the synchronous machine speed equation must be replaced by

$$T_M = -K_{LM}(\delta - \delta_{LP}) \tag{4.96}$$

The steam valve position P_{SV} will become a dynamic state when the steam-governor equations are added.

For rigid shaft couplings

$$K_{HP}P_{CH} = T_M - P_{RH} \tag{4.97}$$

and the two turbine masses are added into the synchronous machine inertia to give the following steam turbine model with T_M as a dynamic state:

$$T_{RH}\frac{dT_M}{dt} = -T_M + (1 - \frac{K_{HP}T_{RH}}{T_{CH}})P_{CH}$$

$$+ \frac{K_{HP}T_{RH}}{T_{CH}}P_{SV} \qquad (4.98)$$

$$T_{CH}\frac{dP_{CH}}{dt} = -P_{CH} + P_{SV} \qquad (4.99)$$

and T_M remains as a state in the synchronous machine model. It is possible to add more reheat stages and additional details. It is also possible to further simplify.

For a nonreheat system, simply set $T_{RH} = 0$ in (4.98)–(4.99), and the following model is obtained:

$$T_{CH}\frac{dT_M}{dt} = -T_M + P_{SV} \qquad (4.100)$$

where again T_M remains a state in the synchronous machine model and P_{SV} will become a state when the governor is added. The above non-reheat model is often referred to as a **Type A** steam turbine model [59, 60].

4.5 Speed Governor Models

The prime mover provides the mechanism for controlling the synchronous machine speed and, hence, terminal voltage frequency. To automatically control speed (and therefore frequency), a device must sense either speed or frequency in such a way that comparison with a desired value can be used to create an error signal to take corrective action. In order to give a physical feeling to the governor process, we will derive the dynamics of what could be considered a crude (and yet practical) mechanical hydraulic governor. This illustration and the derivation were originally given by Elgerd in [61].

Figure 4.6 gives a simple schematic of a flyball speed sensor with ideal linkage to a hydraulic amplifier and piston for main valve control.
Suppose that the distance of points a and e from a fixed higher horizontal reference are related to the per-unit values of a power change setting P_C and value power P_{SV}, respectively. To see how the flyball functions for some

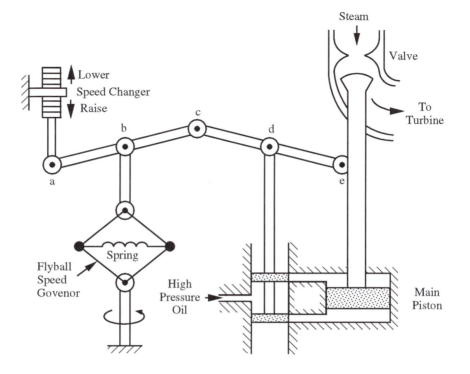

Figure 4.6: *Mechanical–hydraulic speed governor [61]*

fixed P_C^o, suppose that a load is removed from the generator such that an excess of power is being supplied to the turbine through the valve. This excess power will cause a change in generator speed $\Delta\omega$, which will increase the velocity of the flyballs and hence lower point b. Lowering point b results in a lowering of point c since they are assumed to be connected by a rigid rod. Lowering point c must either lower d (if e does not change) or raise e (if d does not change). If point d is lowered, the high-pressure fluid will enter the hydraulic servo through the lower channel and exert a force on the main piston to move up point e. Thus, in any case, lowering c results in a raising of e and a corresponding decrease in P_{SV}. The decrease in P_{SV} will eventually stop the increase in speed that initiated the movement of point b. To model this action, we analyze the linkages and note that any incremental change in the positions of points a, b, and c are related by

$$\Delta y_b = K_{ba}\Delta y_a + K_{bc}\Delta y_c \qquad (4.101)$$

Any incremental change in the position of points c, d, and e are related by

$$\Delta y_d = K_{dc}\Delta y_c + K_{de}\Delta y_e \qquad (4.102)$$

The position of point a is related to the scaled value of P_C so that

$$\Delta P_C = K_a \Delta y_a \qquad (4.103)$$

Neglecting the flyball dynamics, the position of point b changes in proportion to a change in electrical speed as

$$\Delta \omega = \omega_s K_b \Delta y_b \qquad (4.104)$$

A change in the position of point d affects the position of point e through the time delay associated with the fluid in the servo. We assume a linear dynamic response for this time delay:

$$\frac{d\Delta y_e}{dt} = -K_e \Delta y_d \qquad (4.105)$$

Substituting the linkage relations in terms of the power change setting and the speed change:

$$\begin{aligned}
\frac{d\Delta y_e}{dt} &= -K_e(K_{dc}\Delta y_c + K_{de}\Delta y_e) \\
&= -K_e K_{dc}\left(\frac{\Delta y_b - K_{ba}\Delta y_a}{K_{bc}}\right) - K_e K_{de}\Delta y_e \\
&= -\frac{K_e K_{dc}}{K_{bc}K_b}\frac{\Delta\omega}{\omega_s} + \frac{K_e K_{dc}K_{ba}}{K_{bc}K_a}\Delta P_C - K_e K_{de}\Delta y_e \qquad (4.106)
\end{aligned}$$

Using the proportionality between ΔP_{SV} and Δy_e as

$$\Delta y_e = \frac{K_{dc}K_{ba}}{K_{de}K_{bc}K_a}\Delta P_{SV} \qquad (4.107)$$

and defining

$$\mathrm{DROOP} \triangleq \frac{K_{ba}K_b}{K_a}\left(\frac{\omega_s}{2\pi}\right) \qquad (4.108)$$

$$T_{SV} \triangleq \frac{1}{K_e K_{de}} \qquad (4.109)$$

the incremental governor model is

$$T_{SV}\frac{d\Delta P_{SV}}{dt} = -\Delta P_{SV} + \Delta P_C - \left(\frac{\omega_s}{2\pi \text{droop}}\right)\frac{\Delta\omega}{\omega_s} \tag{4.110}$$

The quantity "droop" is a speed droop expressed in Hz/per-unit megawatts. Alternatively, we define a speed regulation quantity R_D as

$$R_D \triangleq \frac{2\pi \text{droop}}{\omega_s} \tag{4.111}$$

For operation near an equilibrium point (denoted by superscript o) with

$$P_C^o = P_{SV}^o + \frac{1}{R_D}\left(\frac{\omega^o}{\omega_s} - 1\right) \tag{4.112}$$

the actual per-unit variables are

$$P_{SV} = P_{SV}^o + \Delta P_{SV} \tag{4.113}$$

$$P_C = P_C^o + \Delta P_C \tag{4.114}$$

$$\omega = \omega^o + \Delta\omega \tag{4.115}$$

In these variables, the governor model including limits on the value position is

$$T_{SV}\frac{dP_{SV}}{dt} = -P_{SV} + P_C - \frac{1}{R_D}\left(\frac{\omega}{\omega_s} - 1\right) \tag{4.116}$$

$$0 \leq P_{SV} \leq P_{SV}^{\max} \tag{4.117}$$

In addition to the limit on the valve position P_{SV}, it may also be important to constrain the derivative of P_{SV} as rate limits. If this is done, the above model corresponds to a General Electric type EH [59].

This model is not valid for large changes, but to illustrate the significance of P_C and R_D, suppose that the machine is unloaded with $P_{SV} = P_c = 0$. If P_C is left at 0 and the machine is loaded to its rating ($P_{SV} = 1$), the full load speed would be $(1 - R_D)\omega_s$. So, if the speed regulation is set for 5% droop ($R_D = 0.05$), the change in speed between no load and full load would be 5% of the rated load. Thus, R_D can be written as

$$R_D = \frac{\% \text{ droop}}{100} \tag{4.118}$$

The quantity P_C is a control input that can be either a constant, or the output of an automatic generation control (AGC) scheme. To provide zero steady-state error in speed (and therefore frequency), an integral control is needed. In multimachine power systems, this load frequency control (LFC) is used together with economic dispatch to maintain frequency at minimum cost on an areawide and systemwide basis. In this case, P_C would be the output of a load reference motor, which is driven by an AGC signal based on a unit control error. While this control ideally would maintain rated frequency in steady state, the accumulated error during transients makes it necessary to have time corrections whenever the total accumulated time error passes a specified threshold [61]. These controls involve fuel and boiler dynamics that are often considered slow enough to be constants.

4.6 Problems

4.1 Using the steady-state exciter model of (4.20) and (4.22) with $K_E = 1.0$, $V_{R\max} = 8.0$, $S_{E\max} = 0.9$, $S_E 0.75\max = 0.5$ (all pu), find $E_{fd\max}$, A_x, and B_x.

4.2 Using the exciter model of (4.46) with $K_E = 1.0$, $S_E = 0$, and $T_E = 0.5$ sec, compute the response of E_{fd} for a constant input of $V_R = 1.0$. Use an initial value of $E_{fd} = 0$ (all pu).

4.3 Using the exciter model of (4.18) and (4.22) with $V_R = V_{\text{res}} = 0.05$, find $K_{E_{\text{self}}}$ so that $E_{fd} = 1.0$ when $A_x = 0.1$ and $B_x = 0.6$ (all pu).

4.4 Using the answer to Problem 4.3, compute the response of E_{fd} for $T_E = 0.5$ sec when it starts at zero. This requires the solution of a nonlinear differential equation.

4.5 Using the excitation system models of (4.46)–(4.49), construct a block diagram in the Laplace domain that shows the control system with inputs V_{ref} and V_t, and output E_{fd}.

4.6 Repeat Problem 4.5 using V_F as a dynamic state, rather than R_f.

4.7 Starting with the dynamic model of (4.46)–(4.49), derive the following dynamic model (a fast static exciter/regulator):

$$T\frac{dE_{fd}}{dt} = -E_{fd} + K(V_{\text{ref}} - V_t)$$

4.8 Using the turbine/governor model of (4.100) and (4.116), with

$$T_{SV} = 2 \text{ sec} \quad P_C = 0.7 \text{ pu} \quad R_D = 0.05 \text{ pu}$$
$$T_{CH} = 4 \text{ sec} \quad \omega_s = 2\pi \ 60r/s$$

(a) Find the steady-state values of P_{SV} and T_M if $\omega = 376.9$ r/s.

(b) Find the dynamic response of P_{SV} and T_M if ω changes at time zero to be 376.8 r/s.

Chapter 5

SINGLE-MACHINE DYNAMIC MODELS

5.1 Terminal Constraints

Throughout Chapters 3 and 4, the constraints on I_d, I_q, I_o and V_d, V_q, V_o have been left unspecified. Perhaps the simplest terminal constraint that could be specified is that of an ideal balanced three-phase resistive load (R_{load} in per unit). This terminal constraint is

$$V_d = I_d R_{\text{load}} \quad (R_{\text{load}} < \infty) \tag{5.1}$$

$$V_q = I_q R_{\text{load}} \quad (R_{\text{load}} < \infty) \tag{5.2}$$

$$V_o = I_o R_{\text{load}} \quad (R_{\text{load}} < \infty) \tag{5.3}$$

and

$$I_d = I_q = I_o = 0 \quad (R_{\text{load}} = \infty) \tag{5.4}$$

The most commonly used terminal constraint for a single machine is the notorious infinite bus. In most power engineering terminology, an infinite bus is an ideal sinusoidal voltage source with constant magnitude, frequency, and phase. In three-phase systems, this implies an ideal balanced symmetrical three-phase set such as

$$V_a = \sqrt{2} V_s \cos(\omega_s t + \theta_{vs}) \tag{5.5}$$

$$V_b = \sqrt{2} V_s \cos\left(\omega_s t + \theta_{vs} - \frac{2\pi}{3}\right) \tag{5.6}$$

$$V_c = \sqrt{2}V_s \cos\left(\omega_s t + \theta_{vs} + \frac{2\pi}{3}\right) \qquad (5.7)$$

This is a positive phase sequence (ABC) set written in per-unit so that $V_s = 1.0$ for rated voltage and $\omega_s = 2\pi f_s$ for rated frequency f_s. In many studies, θ_{vs} is arbitrarily selected as zero. It is useful to know how a synchronous machine dynamic model can be made into an infinite bus. We begin by using the dynamic models of (3.148)–(3.159), (4.46)–(4.49), (4.100,) and (4.116)–(4.117) with V_t defined through (4.43) and (4.44). As discussed earlier, the "*a* phase" voltage of the machine during transients and steady state is

$$V_a = \sqrt{2}\sqrt{V_d^2 + V_q^2}\cos\left(\omega_s t + \delta - \frac{\pi}{2} + \tan^{-1}\frac{V_q}{V_d}\right) \qquad (5.8)$$

To qualify as an infinite bus, we must have

$$\sqrt{V_d^2 + V_q^2} = V_\infty \ (a \text{ constant}) \qquad (5.9)$$

and

$$\delta - \frac{\pi}{2} + \tan^{-1}\frac{V_q}{V_d} = \theta_{v\infty} \ (a \text{ constant}) \qquad (5.10)$$

Clearly, we must find parameter values that result in constants V_d, V_q, and δ.

Considering the voltage magnitude first, there are two ways in which $\sqrt{V_d^2 + V_q^2}$ can be made a constant. The first involves an infinitely high-gain voltage regulator with an infinitely fast amplifier and exciter. This makes the field winding flux linkage infinitely fast so that V_t is constant for all disturbances. This method, however, does not constrain V_d and V_q to be individually constant, and thus there is no way to force $\theta_{v\infty}$ to be constant. The second way to force the terminal voltage magnitude to be constant is actually the opposite of the high-gain regulator approach. Rather than force the field winding to be infinitely fast, we force it to be infinitely slow by letting T'_{do} equal infinity. In addition, we let T'_{qo} go to infinity as well as the machine inertia H. To complete the infinite bus specifications, we let R_s, X'_d, X'_q, T''_{do}, and T''_{qo} be zero. To see the result, we write the model (3.148)–(3.159) with these parameters:

$$\frac{1}{\omega_s}\frac{d\psi_{d\infty}}{dt} = \frac{\omega_\infty}{\omega_s}\psi_{q\infty} + V_{d\infty} \qquad (5.11)$$

$$\frac{1}{\omega_s}\frac{d\psi_{q\infty}}{dt} = -\frac{\omega_\infty}{\omega_s}\psi_{d\infty} + V_{q\infty} \qquad (5.12)$$

$$\frac{1}{\omega_s}\frac{d\psi_{o\infty}}{dt} = V_{o\infty} \tag{5.13}$$

$$\frac{dE'_{q\infty}}{dt} = 0 \tag{5.14}$$

$$\frac{dE'_{d\infty}}{dt} = 0 \tag{5.15}$$

$$\frac{d\delta_\infty}{dt} = \omega_\infty - \omega_s \tag{5.16}$$

$$\frac{d\omega_\infty}{dt} = 0 \tag{5.17}$$

$$\psi_{d\infty} = E'_{q\infty} \tag{5.18}$$

$$\psi_{q\infty} = -E'_{d\infty} \tag{5.19}$$

For $\omega_\infty(o) = \omega_s$, this model requires

$$E'_{q\infty} = E'_{q\infty}(0) \tag{5.20}$$

$$E'_{d\infty} = E'_{d\infty}(0) \tag{5.21}$$

$$\delta_\infty = \delta_\infty(0) \tag{5.22}$$

$$\psi_{d\infty} = E'_{q\infty}(0) \tag{5.23}$$

$$\psi_{q\infty} = -E'_{d\infty}(0) \tag{5.24}$$

which then gives

$$V_{d\infty} = E'_{d\infty}(0) \tag{5.25}$$

$$V_{q\infty} = E'_{q\infty}(0) \tag{5.26}$$

which satisfies the requirements of an infinite bus.

Consider the model of a synchronous machine connected to an infinite bus through a balanced three-phase line in unscaled parameters:

$$v_a = -i_a r_e - \frac{d\lambda_{ea}}{dt} + \sqrt{2}v_s \cos(\omega_s t + \theta_{vs}) \tag{5.27}$$

$$v_b = -i_b r_e - \frac{d\lambda_{eb}}{dt} + \sqrt{2}v_s \cos\left(\omega_s t + \theta_{vs} - \frac{2\pi}{3}\right) \tag{5.28}$$

$$v_c = -i_c r_e - \frac{d\lambda_{ec}}{dt} + \sqrt{2}v_s \cos\left(\omega_s t + \theta_{vs} + \frac{2\pi}{3}\right) \tag{5.29}$$

with

$$
\begin{bmatrix} \lambda_{ea} \\ \lambda_{eb} \\ \lambda_{ec} \end{bmatrix} = \begin{bmatrix} L_{es} & L_{em} & L_{em} \\ L_{em} & L_{es} & L_{em} \\ L_{em} & L_{em} & L_{es} \end{bmatrix} \begin{bmatrix} i_a \\ i_b \\ i_c \end{bmatrix} \tag{5.30}
$$

Transformation and scaling consistent with Chapter 3 give the following terminal constraints in per-unit:

$$
V_d = R_e I_d + \frac{\omega}{\omega_s}\psi_{eq} - \frac{1}{\omega_s}\frac{d\psi_{ed}}{dt} + V_s\sin(\delta - \theta_{vs}) \tag{5.31}
$$

$$
V_q = R_e I_q - \frac{\omega}{\omega_s}\psi_{ed} - \frac{1}{\omega_s}\frac{d\psi_{eq}}{dt} + V_s\cos(\delta - \theta_{vs}) \tag{5.32}
$$

$$
V_o = R_e I_o - \frac{1}{\omega_s}\frac{d\psi_{eo}}{dt} \tag{5.33}
$$

with

$$
\psi_{ed} = X_{ep}(-I_d) \tag{5.34}
$$

$$
\psi_{eq} = X_{ep}(-I_q) \tag{5.35}
$$

$$
\psi_{eo} = X_{eo}(-I_o) \tag{5.36}
$$

Note that V_s is the *RMS* per-unit infinite bus voltage, and all quantities are scaled on the machine ratings. It is customary to set θ_{vs} equal to zero, since one angle in any system can always be arbitrarily selected as a reference for all other angles. We will leave this angle as θ_{vs} for now.

While other terminal constraints can be specified, the infinite bus is the most widely used for single-machine analysis, partly because it has been traditional to study a single generator with the entire remaining network as a Thevenin equivalent impedance and voltage source. Such equivalents are clearly valid only for some steady-state conditions. Other uses of infinite bus models for machines have arisen recently as mechanisms for avoiding the problems associated with a reference angle and steady-state speed. Clearly, with an infinite bus in a system, the steady-state speed for synchronous machines must be ω_s, and the reference angle is conveniently specified. Because of their wide use, the remainder of this chapter is devoted to the analysis of single-machine infinite bus systems. It is also useful for illustrating several concepts of time scales in synchronous machines that will help in the extension to multimachine systems.

While the following sections are written to follow naturally, the Appendix gives an introduction to integral manifolds and singular perturbation, and provides the basic fundamentals used in the following sections to develop reduced-order models.

5.2 The Multi-Time-Scale Model

In this section, we study the special case of a single machine connected to an infinite bus. In particular, we use the synchronous machine model of (3.148)–(3.159), the exciter/AVR model of (4.46)–(4.49) without load compensation, the turbine/governor model of (4.100) and (4.116)–(4.117), and the terminal constraint of (5.31)–(5.33) and (5.34)–(5.36). Before combining these, we introduce the following scaled speed and time constant

$$\omega_t \triangleq T_s(\omega - \omega_s) \tag{5.37}$$

$$T_s \triangleq \sqrt{\frac{2H}{\omega_s}} \tag{5.38}$$

and the parameter

$$\epsilon \triangleq \frac{1}{\omega_s} \tag{5.39}$$

where we assume that H is large enough so that

$$\epsilon << T_s \tag{5.40}$$

We also combine the machine and line flux linkages and parameters as follows:

$$\psi_{de} \triangleq \psi_d + \psi_{ed}, \quad \psi_{qe} \triangleq \psi_q + \psi_{eq}, \quad \psi_{oe} \triangleq \psi_o + \psi_{eo},$$

$$X_{de} \triangleq X_d + X_{ep}, \quad X_{qe} \triangleq X_q + X_{ep}, \quad X'_{de} \triangleq X'_d + X_{ep},$$

$$X'_{qe} \triangleq X'_q + X_{ep}, \quad X''_{de} \triangleq X''_d + X_{ep}, \quad X''_{qe} \triangleq X''_q + X_{ep},$$

$$R_{se} \triangleq R_s + R_e \ , \quad X_{lse} \triangleq X_{ls} + X_{ep} \tag{5.41}$$

Substituting (5.31)–(5.33) into (3.148)–(3.159) and adding the other dynamic models give the following multi-time-scale model:

$$\epsilon \frac{d\psi_{de}}{dt} = R_{se}I_d + \left(1 + \frac{\epsilon}{T_s}\omega_t\right)\psi_{qe} + V_s \sin(\delta - \theta_{vs}) \tag{5.42}$$

$$\epsilon \frac{d\psi_{qe}}{dt} = R_{se}I_q - \left(1 + \frac{\epsilon}{T_s}\omega_t\right)\psi_{de} + V_s\cos(\delta - \theta_{vs}) \tag{5.43}$$

$$\epsilon \frac{d\psi_{oe}}{dt} = R_{se}I_o \tag{5.44}$$

$$T'_{do}\frac{dE'_q}{dt} = -E'_q - (X_d - X'_d)\left[I_d - \frac{X'_d - X''_d}{(X'_d - X_{\ell s})^2}(\psi_{1d} + (X'_d - X_{\ell s})I_d\right.$$

$$\left. -E'_q)\right] + E_{fd} \tag{5.45}$$

$$T''_{do}\frac{d\psi_{1d}}{dt} = -\psi_{1d} + E'_q - (X'_d - X_{\ell s})I_d \tag{5.46}$$

$$T'_{qo}\frac{dE'_d}{dt} = -E'_d + (X_q - X'_q)\left[I_q - \frac{X'_q - X''_q}{(X'_q - X_{\ell s})^2}(\psi_{2q} + (X'_q - X_{\ell s})I_q\right.$$

$$\left. +E'_d)\right] \tag{5.47}$$

$$T''_{qo}\frac{d\psi_{2q}}{dt} = -\psi_{2q} - E'_d - (X'_q - X_{\ell s})I_q \tag{5.48}$$

$$T_s\frac{d\delta}{dt} = \omega_t \tag{5.49}$$

$$T_s\frac{d\omega_t}{dt} = T_M - (\psi_{de}I_q - \psi_{qe}I_d) - T_{FW} \tag{5.50}$$

$$\psi_{de} = -X''_{de}I_d + \frac{(X''_d - X_{\ell s})}{(X'_d - X_{\ell s})}E'_q + \frac{(X'_d - X''_d)}{(X'_d - X_{\ell s})}\psi_{1d} \tag{5.51}$$

$$\psi_{qe} = -X''_{qe}I_q - \frac{(X''_q - X_{\ell s})}{(X'_q - X_{\ell s})}E'_d + \frac{(X'_q - X''_q)}{(X'_q - X_{\ell s})}\psi_{2q} \tag{5.52}$$

$$\psi_{oe} = -X_{oe}I_o \tag{5.53}$$

$$T_E\frac{dE_{fd}}{dt} = -(K_E + S_E(E_{fd}))E_{fd} + V_R \tag{5.54}$$

$$T_F\frac{dR_f}{dt} = -R_f + \frac{K_F}{T_F}E_{fd} \tag{5.55}$$

$$T_A\frac{dV_R}{dt} = -V_R + K_A R_f - \frac{K_A K_F}{T_F}E_{fd} + K_A(V_{\text{ref}} - V_t) \tag{5.56}$$

$$V_R^{\min} \leq V_R \leq V_R^{\max} \tag{5.57}$$

$$V_t = \sqrt{V_d^2 + V_q^2} \tag{5.58}$$

$$V_d = R_e I_d + \left(1 + \frac{\epsilon}{T_s}\omega_t\right)\psi_{eq} - \epsilon\frac{d\psi_{ed}}{dt} + V_s\sin(\delta - \theta_{vs}) \tag{5.59}$$

$$V_q = R_e I_q - \left(1 + \frac{\epsilon}{T_s}\omega_t\right)\psi_{ed} - \epsilon\frac{d\psi_{eq}}{dt} + V_s\cos(\delta - \theta_{vs}) \tag{5.60}$$

$$\psi_{ed} = -X_{ep}I_d \tag{5.61}$$

$$\psi_{eq} = -X_{ep}I_q \tag{5.62}$$

$$T_{CH}\frac{dT_M}{dt} = -T_M + P_{SV} \tag{5.63}$$

$$T_{SV}\frac{dP_{SV}}{dt} = -P_{SV} + P_C - \epsilon\frac{\omega_t}{R_D T_S} \tag{5.64}$$

$$0 \le P_{SV} \le P_{SV}^{\max} \tag{5.65}$$

This model could be put in closed form without algebraic equations by solving (5.51)–(5.53) and (5.61)–(5.62) for I_d, I_q, I_o, ψ_{ed}, and ψ_{eq} and substituting into the remaining differential equations, and by replacing the derivative terms in (5.59) and (5.60) with their respective functions, and substitution into (5.58) and (5.56).

This single-machine/infinite bus model has several classifications of dynamic time scales. The stator transients of ψ_{de} and ψ_{qe} are very fast relative to other dynamics. This shows up in the model as a small "time constant" ϵ multiplying their derivatives. The damper flux linkages ψ_{1d} and ψ_{2q} are also quite fast, since T_{do}'' and T_{qo}'' typically are quite small. The voltage regulator states E_{fd} and V_R tend to be fast because T_E and T_A are typically small. The field winding and rate feedback states E_q', R_f tend to be slow because T_{do}' and T_F can be relatively large. The turbine/governor states T_M, P_{SV} tend to be slow because T_{CH} and T_{SV} can be relatively large. This can be countered, however, by a small speed regulation R_D. The damper-winding flux linkage E_d' can be either fast or slow, depending on T_{qo}'. These classifications should be considered as general rule-of-thumb guidelines, and not absolute characteristics. The actual time-scale classification for a specific set of data can be quite different than that stated above, and can change between load levels.

The time-scale characteristics of a model are important since they determine the step size required in a time simulation. If a model contains fast dynamics, a small step size is needed. If the phenomenon of interest in a time simulation is known to be a predominantly average or slow response, it would be very helpful if the fast dynamics could be eliminated. They must be eliminated properly, so that their effect on the slower phenomena of interest is still preserved. Only in rare cases can fast dynamics be "eliminated" exactly. One such case is given in the next section.

5.3   Elimination of Stator/Network Transients

We begin this section by considering the special case of zero stator and network resistance. For this special (although common) assumption, (5.42)–(5.44) and (5.49) are

$$\epsilon \frac{d\psi_{de}}{dt} = \left(1 + \frac{\epsilon}{T_s}\omega_t\right)\psi_{qe} + V_s\sin(\delta - \theta_{vs}) \tag{5.66}$$

$$\epsilon \frac{d\psi_{qe}}{dt} = -\left(1 + \frac{\epsilon}{T_s}\omega_t\right)\psi_{de} + V_s\cos(\delta - \theta_{vs}) \tag{5.67}$$

$$\epsilon \frac{d\psi_{oe}}{dt} = 0 \tag{5.68}$$

$$T_s\frac{d\delta}{dt} = \omega_t \tag{5.69}$$

The first three differential equations have an explicit solution in terms of δ:

$$\psi_{de} = V_s\cos(\delta - \theta_{vs}) \tag{5.70}$$

$$\psi_{qe} = -V_s\sin(\delta - \theta_{vs}) \tag{5.71}$$

$$\psi_{oe} = 0 \tag{5.72}$$

This can be verified by substituting (5.70)–(5.72) into (5.66)–(5.69) and observing an exact identity. If the initial conditions on ψ_{de}, ψ_{qe}, ψ_{oe}, and δ satisfy (5.70)–(5.72), then ψ_{de}, ψ_{qe}, and ψ_{oe} are related to δ through (5.70)–(5.72) for all time. This makes (5.70)–(5.72) an integral manifold for ψ_{de}, ψ_{qe}, and ψ_{oe}. If the initial conditions do not satisfy (5.70)–(5.72) (the system starts off the manifold), there is still an exact explicit solution for ψ_{de}, ψ_{qe}, and ψ_{oe} as a function of δ and time t. This interesting fact about synchronous machine stator transients has been explained in considerable detail in [63] through [65]. The result is as follows. For any initial conditions ψ_{de}^o, ψ_{qe}^o, ψ_{oe}^o, δ^o, t_o and for any ϵ, the exact solutions of the differential equations (5.66)–(5.68) are

$$\psi_{de} = c_1\cos(\omega_s t + \delta - c_2) + V_s\cos(\delta - \theta_{vs}) \tag{5.73}$$

$$\psi_{qe} = -c_1 \sin(\omega_s t + \delta - c_2) - V_s \sin(\delta - \theta_{vs}) \qquad (5.74)$$

$$\psi_{oe} = c_3 \qquad (5.75)$$

with

$$c_1 = [(\psi_{de}^o - V_s \cos(\delta^o - \theta_{vs}))^2 + (\psi_{qe}^o + V_s \sin(\delta^o - \theta_{vs}))^2]^{\frac{1}{2}} \qquad (5.76)$$

$$c_2 = \omega_s t_o + \delta^o + \tan^{-1}\left(\frac{\psi_{qe}^o + V_s \sin(\delta^o - \theta_{vs})}{\psi_{de}^o - V_s \cos(\delta^o - \theta_{vs})}\right) \qquad (5.77)$$

$$c_3 = \psi_{oe}^o \qquad (5.78)$$

Clearly, if the initial conditions are on the manifold (satisfy (5.70)–(5.72)), this solution simplifies to the manifold itself ((5.70)–(5.72)). We emphasize that this interesting solution is valid for any ϵ, and requires only $R_{se} = 0$. This solution does, however, show the importance of R_{se}. If the initial condition is off the manifold, the exact solution of (5.73)–(5.75) reveals a sustained oscillation in ψ_{de} and ψ_{qe}, which means that the machine will never reach a constant speed equilibrium condition. Evidently, the resistance acts to damp out this oscillation in the actual model, keeping R_{se}.

When resistance R_{se} is not zero, an integral manifold for ψ_{de}, ψ_{qe}, and ψ_{oe} has been shown to exist only for the case in which ϵ is sufficiently small [63]–[65]. A first approximation of this integral manifold (keeping R_{se}) can be found by setting ϵ equal to zero on the left-hand side of (5.42)–(5.65) and solving the algebraic equations on the right-hand side for ψ_{de}, ψ_{qe} and ψ_{oe} as functions of all remaining dynamic states. As a matter of consistency, ϵ can also be set to zero on the right-hand side of (5.42)–(5.65). In either case, the resulting reduced-order model should approximate the exact model up to an "order ϵ" error [66, 67]. If the ϵ in (5.64) is set to zero, the turbine/governor dynamics would no longer depend on shaft speed. It could be argued that since R_D is usually also quite small, the ratio of ϵ/R_D should be kept and, thus, keep governor dynamics. For ϵ sufficiently small, this reduced-order model can be improved to virtually any order of accuracy (see Appendix).

It is customary to set ϵ equal to zero in (5.42)–(5.44) and (5.59)–(5.60) to give the following approximation of the exact stator transients integral manifold written together with the original algebraic equations for currents:

$$0 = R_{se}I_d + \psi_{qe} + V_s \sin(\delta - \theta_{vs}) \qquad (5.79)$$

$$0 = R_{se}I_q - \psi_{de} + V_s \cos(\delta - \theta_{vs}) \qquad (5.80)$$

$$0 = R_{se}I_o \qquad (5.81)$$

$$\psi_{de} = -X''_{de}I_d + \frac{(X''_d - X_{\ell s})}{(X'_d - X_{\ell s})}E'_q + \frac{(X'_d - X''_d)}{(X'_d - X_{\ell s})}\psi_{1d} \qquad (5.82)$$

$$\psi_{qe} = -X''_{qe}I_q - \frac{(X''_q - X_{\ell s})}{(X'_q - X_{\ell s})}E'_d + \frac{(X'_q - X''_q)}{(X'_q - X_{\ell s})}\psi_{2q} \qquad (5.83)$$

$$\psi_{oe} = -X_{oe}I_o \qquad (5.84)$$

These six algebraic equations clearly can be solved for the six variables ψ_{de}, ψ_{qe}, ψ_{oe}, I_d, I_q, and I_o as functions of δ, E'_q, E'_d, ψ_{1d}, and ψ_{2q}. If R_{se} is equal to zero, this approximation of the integral manifold becomes exact (compare (5.70)–(5.72) and (5.79)–(5.84)).

With the above results, the dynamic model has been reduced from a 14th-order model to an 11th-order model. The solution of (5.79)–(5.84) for ψ_{oe} and I_o is trivial, and ψ_{de}, ψ_{qe} can be eliminated, leaving only two equations to be solved for I_d and I_q as functions of δ, E'_q, E'_d, ψ_{1d}, and ψ_{2q}:

$$0 = R_{se}I_d - X''_{qe}I_q - \frac{(X''_q - X_{\ell s})}{(X'_q - X_{\ell s})}E'_d + \frac{(X'_q - X''_q)}{(X'_q - X_{\ell s})}\psi_{2q}$$

$$+V_s \sin(\delta - \theta_{vs}) \qquad (5.85)$$

$$0 = R_{se}I_q + X''_{de}I_d - \frac{(X''_d - X_{\ell s})}{(X'_d - X_{\ell s})}E'_q - \frac{(X'_d - X''_d)}{(X'_d - X_{\ell s})}\psi_{1d}$$

$$+V_s \cos(\delta - \theta_{vs}) \qquad (5.86)$$

These two real equations can be used to solve for I_d and I_q. They can also be used to make one complex equation and an interesting "dynamic" circuit. Adding (5.85) plus j times (5.86) and multiplying by $e^{j(\delta-\pi/2)}$ give the following "circuit" equation:

$$\left[\left[\frac{(X''_q - X_{\ell s})}{(X'_q - X_{\ell s})}E'_d - \frac{(X'_q - X''_q)}{(X'_q - X_{\ell s})}\psi_{2q} + (X''_q - X''_d)I_q\right]\right.$$

$$\left.+j\left[\frac{(X''_d - X_{\ell s})}{(X'_d - X_{\ell s})}E'_q + \frac{(X'_d - X''_d)}{(X'_d - X_{\ell s})}\psi_{1d}\right]\right]e^{j(\delta-\pi/2)}$$

$$= (R_s + jX''_d)(I_d + jI_q)e^{j(\delta-\pi/2)}$$

$$+(R_e + jX_{ep})(I_d + jI_q)e^{j(\delta-\pi/2)} + V_s e^{j\theta_{vs}} \qquad (5.87)$$

From (5.58)–(5.60) with $\epsilon = 0$,

$$V_t = \sqrt{V_d^2 + V_q^2} \qquad (5.88)$$

$$V_d = R_e I_d - X_{ep} I_q + V_s \sin(\delta - \theta_{vs}) \tag{5.89}$$

$$V_q = R_e I_q + X_{ep} I_d + V_s \cos(\delta - \theta_{vs}) \tag{5.90}$$

These last two equations can be written as one complex equation.

$$(V_d + jV_q)e^{j(\delta - \pi/2)} = (R_e + jX_{ep})(I_d + jI_q)e^{j(\delta - \pi/2)} + V_s e^{j\theta_{vs}} \tag{5.91}$$

Equations (5.87) and (5.91) can be expressed as a "dynamic equivalent circuit" with a controlled source voltage behind X''_d, as shown in Figure 5.1.

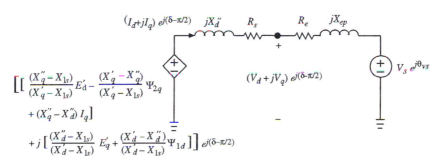

Figure 5.1: *Synchronous machine sub-transient dynamic circuit*

This circuit does not imply anything about steady state, but merely reflects the consequence of setting $\epsilon = 0$ in the original full dynamic model. Such a circuit is clearly not a unique representation of these algebraic equations. Many other similar circuits can be created by simply adding or subtracting reactances in the line and modifying the internal "source" accordingly [68].

An interesting and useful observation about this circuit is that the "real power from the internal source" is exactly equal to the electrical torque in the air gap (T_{ELEC}).

$$P_{\substack{\text{from} \\ \text{source}}} = \left[\frac{(X''_q - X_{ls})}{(X'_q - X_{ls})} E'_d - \frac{(X'_q - X''_q)}{(X'_q - X_{ls})} \psi_{2q} + (X''_q - X''_d) I_q \right] I_d$$
$$+ \left[\frac{(X''_d - X_{ls})}{(X'_d - X_{ls})} E'_q + \frac{(X'_d - X''_d)}{(X'_d - X_{ls})} \psi_{1d} \right] I_q$$

From (5.44) and (5.82) to (5.83):

$$T_{ELEC} = \left[-X''_{de} I_d + \frac{(X''_d - X_{ls})}{(X'_d - X_{ls})} E'_q + \frac{(X'_d - X''_d)}{(X'_d - X_{ls})} \psi_{1d} \right] I_q$$

$$- \left[-X''_{qe} I_q - \frac{(X''_q - X_{ls})}{(X'_q - X_{ls})} E'_d + \frac{(X'_q - X''_q)}{(X'_q - X_{ls})} \psi_{2q} \right] I_d$$

These are clearly equal, since the reactances X''_{de} and X''_{qe} each contain X_{ep}.

The final form of this model that has eliminated the stator/network transients by setting $\epsilon = 0$ on the left and right sides of (5.42)–(5.65) (except for (5.64), where we have chosen to keep the governor by arguing that R_D can be small) can be written in terms of the original variables as

$$T'_{do} \frac{dE'_q}{dt} = -E'_q - (X_d - X'_d)$$

$$\left[I_d - \frac{X'_d - X''_d}{(X'_d - X_{ls})^2} (\psi_{1d} + (X'_d - X_{ls}) I_d \right.$$

$$\left. - E'_q) \right] + E_{fd} \qquad (5.92)$$

$$T''_{do} \frac{d\psi_{1d}}{dt} = -\psi_{1d} + E'_q - (X'_d - X_{ls}) I_d \qquad (5.93)$$

$$T'_{qo} \frac{dE'_d}{dt} = -E'_d + (X_q - X'_q)$$

$$\left[I_q - \frac{X'_q - X''_q}{(X'_q - X_{ls})^2} (\psi_{2q} + (X'_q - X_{ls}) I_q + E'_d) \right] \quad (5.94)$$

$$T''_{qo} \frac{d\psi_{2q}}{dt} = -\psi_{2q} - E'_d - (X'_q - X_{ls}) I_q \qquad (5.95)$$

$$\frac{d\delta}{dt} = \omega - \omega_s \qquad (5.96)$$

$$\frac{2H}{\omega_s} \frac{d\omega}{dt} = T_M - \frac{(X''_d - X_{ls})}{(X'_d - X_{ls})} E'_q I_q - \frac{(X'_d - X''_d)}{(X'_d - X_{ls})} \psi_{1d} I_q$$

$$- \frac{(X''_q - X_{ls})}{(X'_q - X_{ls})} E'_d I_d + \frac{(X'_q - X''_q)}{(X'_q - X_{ls})} \psi_{2q} I_d$$

$$- (X''_q - X''_d) I_d I_q - T_{FW} \qquad (5.97)$$

$$T_E \frac{dE_{fd}}{dt} = -(K_E + S_E(E_{fd})) E_{fd} + V_R \qquad (5.98)$$

$$T_F \frac{dR_f}{dt} = -R_f + \frac{K_F}{T_F} E_{fd} \qquad (5.99)$$

$$T_A \frac{dV_R}{dt} = -V_R + K_A R_f - \frac{K_A K_F}{T_F} E_{fd} + K_A(V_{\text{ref}} - V_t) \quad (5.100)$$

$$T_{CH} \frac{dT_M}{dt} = -T_M + P_{SV} \quad (5.101)$$

$$T_{SV} \frac{dP_{SV}}{dt} = -P_{SV} + P_C - \frac{1}{R_D} \left(\frac{\omega}{\omega_s} - 1 \right) \quad (5.102)$$

with the limit constraints

$$V_R^{\text{min}} \leq V_R \leq V_R^{\text{max}} \quad (5.103)$$

$$0 \leq P_{SV} \leq P_{SV}^{\text{max}} \quad (5.104)$$

and the required algebraic equations, which come from the circuit of Figure 5.1, or the solution of the following for I_d, I_q:

$$0 = (R_s + R_e)I_d - (X_q'' + X_{ep})I_q$$
$$- \frac{(X_q'' - X_{\ell s})}{(X_q' - X_{\ell s})} E_d' + \frac{(X_q' - X_q'')}{(X_q' - X_{\ell s})} \psi_{2q} + V_s \sin(\delta - \theta_{vs}) \quad (5.105)$$

$$0 = (R_s + R_e)I_q + (X_d'' + X_{ep})I_d$$
$$- \frac{(X_d'' - X_{\ell s})}{(X_d' - X_{\ell s})} E_q' - \frac{(X_d' - X_d'')}{(X_d' - X_{\ell s})} \psi_{1d} + V_s \cos(\delta - \theta_{vs}) \quad (5.106)$$

and then substitution into the following equations for V_d, V_q:

$$V_d = R_e I_d - X_{ep} I_q + V_s \sin(\delta - \theta_{vs}) \quad (5.107)$$

$$V_q = R_e I_q + X_{ep} I_d + V_s \cos(\delta - \theta_{vs}) \quad (5.108)$$

and finally

$$V_t = \sqrt{V_d^2 + V_q^2} \quad (5.109)$$

The quantities V_{ref} and P_C remain as inputs.

5.4 The Two-Axis Model

The reduced-order model obtained in the last section still contains the damper-winding dynamics ψ_{1d} and ψ_{2q}. If T_{do}'' and T_{qo}'' are sufficiently small, there is an integral manifold for these dynamic states. A first approximation of the fast damper-winding integral manifold is found by setting T_{do}'' and T_{qo}'' equal to zero in (5.92)–(5.109) to obtain

$$0 = -\psi_{1d} + E_q' - (X_d' - X_{\ell s})I_d \tag{5.110}$$

$$0 = -\psi_{2q} - E_d' - (X_q' - X_{\ell s})I_q \tag{5.111}$$

When used to eliminate ψ_{1d} and ψ_{2q} in (5.107), the equations for I_d and I_q become

$$0 = (R_s + R_e)I_d - (X_q' + X_{ep})I_q - E_d' + V_s \sin(\delta - \theta_{vs}) \tag{5.112}$$

$$0 = (R_s + R_e)I_q + (X_d' + X_{ep})I_d - E_q' + V_s \cos(\delta - \theta_{vs}) \tag{5.113}$$

As in the last section, these two real equations can be written as one complex equation:

$$[E_d' + (X_q' - X_d')I_q + jE_q']e^{j(\delta - \frac{\pi}{2})}$$
$$= (R_s + jX_d')(I_d + jI_q)e^{j(\delta - \frac{\pi}{2})}$$
$$+ (R_e + jX_{ep})(I_d + jI_q)e^{j(\delta - \frac{\pi}{2})}$$
$$+ V_s e^{j\theta_{vs}} \tag{5.114}$$

The equations for V_d and V_q remain the same as in the last section so that the circuit of Figure 5.2 can be constructed to reflect the algebraic constraints for this two-axis model. It is easy to verify that, as in the last section, the "real power from" the internal source is exactly equal to the electrical torque across the air gap (T_{ELEC}) for this model.

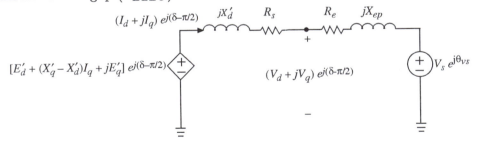

Figure 5.2: *Synchronous machine two-axis dynamic circuit*

The final form of this two-axis model, which has eliminated the stator/network and fast damper winding dynamics, is obtained by substituting

(5.110) and (5.111) into (5.92)–(5.109) to eliminate ψ_{1d} and ψ_{2q}:

$$T'_{do}\frac{dE'_q}{dt} = -E'_q - (X_d - X'_d)I_d + E_{fd} \tag{5.115}$$

$$T'_{qo}\frac{dE'_d}{dt} = -E'_d + (X_q - X'_q)I_q \tag{5.116}$$

$$\frac{d\delta}{dt} = \omega - \omega_s \tag{5.117}$$

$$\frac{2H}{\omega_s}\frac{d\omega}{dt} = T_M - E'_d I_d - E'_q I_q - (X'_q - X'_d)I_d I_q - T_{FW} \tag{5.118}$$

$$T_E\frac{dE_{fd}}{dt} = -(K_E + S_E(E_{fd}))E_{fd} + V_R \tag{5.119}$$

$$T_F\frac{dR_f}{dt} = -R_f + \frac{K_F}{T_F}E_{fd} \tag{5.120}$$

$$T_A\frac{dV_R}{dt} = -V_R + K_A R_F - \frac{K_A K_F}{T_F}E_{fd} + K_A(V_{ref} - V_t) \tag{5.121}$$

$$T_{CH}\frac{dT_M}{dt} = -T_M + P_{SV} \tag{5.122}$$

$$T_{SV}\frac{dP_{SV}}{dt} = -P_{SV} + P_C - \frac{1}{R_D}\left(\frac{\omega}{\omega_s} - 1\right) \tag{5.123}$$

with the limit constraints

$$V_R^{\min} \leq V_R \leq V_R^{\max} \tag{5.124}$$

$$0 \leq P_{SV} \leq P_{SV}^{\max} \tag{5.125}$$

and the required algebraic equations, which come from the circuit of Figure 5.2, or the solution of the following for I_d, I_q:

$$0 = (R_s + R_e)I_d - (X'_q + X_{ep})I_q - E'_d + V_s \sin(\delta - \theta_{vs}) \tag{5.126}$$

$$0 = (R_s + R_e)I_q + (X'_d + X_{ep})I_d - E'_q + V_s \cos(\delta - \theta_{vs}) \tag{5.127}$$

and then substitution into the following equations for V_d and V_q:

$$V_d = R_e I_d - X_{ep}I_q + V_s \sin(\delta - \theta_{vs}) \tag{5.128}$$

$$V_q = R_e I_q + X_{ep}I_d + V_s \cos(\delta - \theta_{vs}) \tag{5.129}$$

and finally

$$V_t = \sqrt{V_d^2 + V_q^2} \tag{5.130}$$

The quantities V_{ref} and P_C remain as inputs.

5.5   The One-Axis (Flux-Decay) Model

The reduced-order model obtained in the last section still contains the damper-winding dynamics E_d'. If T_{qo}' is sufficiently small, there is an integral manifold for this dynamic state also. A first approximation of this fast damper winding integral manifold is found by setting T_{qo}' equal to zero in (5.115)–(5.130) to obtain

$$0 = -E_d' + (X_q - X_q')I_q \qquad (5.131)$$

When used to eliminate E_d' in (5.115)–(5.130), the equations for I_d and I_q then become

$$0 = (R_s + R_e)I_d - (X_q + X_{ep})I_q + V_s \sin(\delta - \theta_{vs}) \qquad (5.132)$$

$$0 = (R_s + R_e)I_q + (X_d' + X_{ep})I_d - E_q' + V_s \cos(\delta - \theta_{vs}) \qquad (5.133)$$

As in the last sections, these two real equations can be written as one complex equation:

$$[(X_q - X_d')I_q + jE_q']e^{j(\delta - \pi/2)} =$$
$$(R_s + jX_d')(I_d + jI_q)e^{(\delta - \pi/2)}$$
$$+(R_e + jX_{ep})(I_d + jI_q)e^{(\delta - \pi/2)} + V_s e^{j\theta_{vs}} \qquad (5.134)$$

The equations for V_d and V_q remain the same as in the last sections, so that the circuit of Figure 5.3 can be constructed to reflect the algebraic constraints for this one-axis model.

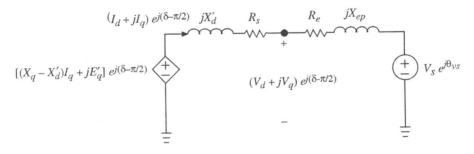

Figure 5.3: *Synchronous machine one-axis dynamic circuit*

It is easy to verify that, as in the last two sections, the "real power from" the internal source is exactly equal to the electrical torque across the air gap (T_{ELEC}) for this model.

The final form of this one-axis model, which has eliminated the stator/network and all three fast damper-winding dynamics, is obtained by substituting (5.131) into (5.115)–(5.130) to eliminate E'_d:

$$T'_{do} \frac{dE'_q}{dt} = -E'_q - (X_d - X'_d)I_d + E_{fd} \tag{5.135}$$

$$\frac{d\delta}{dt} = \omega - \omega_s \tag{5.136}$$

$$\frac{2H}{\omega_s} \frac{d\omega}{dt} = T_M - E'_q I_q - (X_q - X'_d)I_d I_q - T_{FW} \tag{5.137}$$

$$T_E \frac{dE_{fd}}{dt} = -(K_E + S_E(E_{fd}))E_{fd} + V_R \tag{5.138}$$

$$T_F \frac{dR_f}{dt} = -R_f + \frac{K_F}{T_F} E_{fd} \tag{5.139}$$

$$T_A \frac{dV_R}{dt} = -V_R + K_A R_f - \frac{K_A K_F}{T_F} E_{fd} + K_A(V_{\text{ref}} - V_t) \tag{5.140}$$

$$T_{CH} \frac{dT_M}{dt} = -T_M + P_{SV} \tag{5.141}$$

$$T_{SV} \frac{dP_{SV}}{dt} = -P_{SV} + P_C - \frac{1}{R_D}\left(\frac{\omega}{\omega_s} - 1\right) \tag{5.142}$$

with the limit constraints

$$V_R^{\min} \leq V_R \leq V_R^{\max} \tag{5.143}$$

$$0 \leq P_{SV} \leq P_{SV}^{\max} \tag{5.144}$$

and the required algebraic equations, which come from the circuit of Figure 5.3, or the solution of the following equations for I_d, I_q:

$$0 = (R_s + R_e)I_d - (X_q + X_{ep})I_q + V_s \sin(\delta - \theta_{vs}) \tag{5.145}$$

$$0 = (R_s + R_e)I_q + (X'_d + X_{ep})I_d - E'_q + V_s \cos(\delta - \theta_{vs}) \tag{5.146}$$

and then substitution into the following equations for V_d, V_q:

$$V_d = R_e I_d - X_{ep}I_q + V_s \sin(\delta - \theta_{vs}) \tag{5.147}$$

$$V_q = R_e I_q + X_{ep}I_d + V_s \cos(\delta - \theta_{vs}) \tag{5.148}$$

and finally

$$V_t = \sqrt{V_d^2 + V_q^2} \tag{5.149}$$

The quantities V_{ref} and P_C remain as inputs.

5.6   The Classical Model

The classical model is the simplest of all the synchronous machine models, but it is the hardest to justify. In an effort to derive its basis, we first state what it is. In reference to all of the dynamic circuits of this chapter, the classical model is also called the constant voltage behind the transient reactance (X'_d) model. Return to the two-axis dynamic circuit (Figure 5.2) and the dynamic model of (5.115)–(5.130). Rather than assuming $T'_{qo} = 0$, as in the last section, assume that an integral manifold exists for E'_d, E'_q, E_{fd}, R_f, and V_R, which, as a first approximation, gives E'_q equal to a constant and $(E'_d + (X'_q - X'_d)I_q)$ equal to a constant. For this constant based on the initial values E'^o_d, I^o_q, and E'^o_q, we define the constant voltage

$$E'^o \triangleq \sqrt{(E'^o_d + (X'_q - X'_d)I^o_q)^2 + (E'^o_q)^2} \tag{5.150}$$

and the constant angle

$$\delta'^o \triangleq \tan^{-1}\left(\frac{E'^o_q}{E'^o_d + (X'_q - X'_d)I^o_q}\right) - \frac{\pi}{2} \tag{5.151}$$

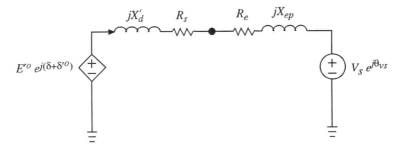

Figure 5.4: *Synchronous machine classical model dynamic circuit*

The classical model dynamic circuit is shown in Figure 5.4. Because the classical model is usually used with the assumption of constant shaft torque (initial T^o_M) and zero resistance, we assume

$$T_{CH} = \infty, \quad R_s + R_e = 0 \tag{5.152}$$

and define

$$\delta_{\text{classical}} \triangleq \delta + \delta'^o \tag{5.153}$$

The classical model is then a second-order system:

$$\frac{d\delta_{\text{classical}}}{dt} = \omega - \omega_s \tag{5.154}$$

$$\frac{2H}{\omega_s}\frac{d\omega}{dt} = T_M^o - \frac{E'^o V_s}{(X_d' + X_{ep})}\sin(\delta_{\text{classical}} - \theta_{vs}) - T_{FW} \tag{5.155}$$

This classical model can also be obtained formally from the two-axis model by setting $X_q' = X_d'$ and $T_{CH} = T_{qo}' = T_{do}' = \infty$, or from the one-axis model by setting $X_q = X_d'$, $T_{qo}' = 0$, and $T_{CH} = T_{do}' = \infty$. In the latter case, δ'^o is equal to zero, so that $\delta_{\text{classical}}$ is equal to δ and E'^o is equal to $E_q'^o$.

5.7   Damping Torques

The dynamic models proposed so far have all included a friction windage torque term T_{FW}. For opposition to rotation, such torque terms should have the form

$$T_{FW} = D_{FW}\omega \text{ or } D_{FW}'\omega^2 \tag{5.156}$$

The literature often includes damping torque terms of the form

$$T_D = D(\omega - \omega_s) \tag{5.157}$$

or

$$T_D = \frac{D'}{\omega_s}(\omega - \omega_s) \tag{5.158}$$

These damping torque terms can be positive or negative, depending on the machine speed. Values of $D' = 1$ or 2 pu have been cited as a reasonable method to account for the turbine windage damping [62]. Such terms look like induction motor slip torque terms, and have often been included to model the short circuited damper windings. Thus, if damper windings are modeled through their differential equations, then their effects need not be added in T_D. In this case, friction can be modeled through T_{FW}, if desired. If damper windings have been eliminated, as in the one-axis or classical models, all of the third damping effects have been lost. To account for their damping without including their differential equations, T_D can be added to the torque equation (added to T_{FW}), with D appropriately specified to approximate the damper-winding action. We now justify this damping torque term by returning to the two-axis model of (5.115)–(5.130), which included one damper-winding differential equation.

We would like to show that if E_d' is more accurately eliminated, a damping torque term proportional to slip speed should automatically appear in the swing equation. To show this, we propose that the integral manifold for E_d' should be found more accurately. To begin, we define the small parameter μ as

$$\mu \triangleq T_{qo}' \qquad (5.159)$$

and propose an integral manifold for E_d' of the form

$$E_d' = \phi_o + \mu\phi_1 + \mu^2\phi_2 + \ldots \qquad (5.160)$$

where each ϕ_i is a function of all remaining dynamic states. Substituting (5.160) into (5.116) gives

$$\mu \left(\frac{\partial \phi_o}{\partial E_q'} + \mu\frac{\partial \phi_1}{\partial E_q'} + \mu^2\frac{\partial \phi_2}{\partial E_q'} + \ldots \right) (-E_q' - (X_d - X_d')I_d + E_{fd})/T_{do}'$$

$$+\mu \left(\frac{\partial \phi_o}{\partial \delta} + \mu\frac{\partial \phi_1}{\partial \delta} + \mu^2\frac{\partial \phi_2}{\partial \delta} + \ldots \right) (\omega - \omega_s)$$

$$+\mu \left(\frac{\partial \phi_o}{\partial \omega} + \mu\frac{\partial \phi_1}{\partial \omega} + \mu^2\frac{\partial \phi_2}{\partial \omega} + \ldots \right) (T_M - (\phi_o + \mu\phi_1 + \mu^2\phi_2 + \ldots)I_d$$

$$-E_q'I_q - (X_q' - X_d')I_dI_q - T_{FW})/\left(\frac{2H}{\omega_s} \right)$$

$$+\mu \left(\frac{\partial \phi_o}{\partial E_{fd}} + \mu\frac{\partial \phi_1}{\partial E_{fd}} + \mu^2\frac{\partial \phi_2}{\partial E_{fd}} + \ldots \right) (-(K_E + S_E)E_{fd} + V_R)/T_E$$

$$+\mu \left(\frac{\partial \phi_o}{\partial R_f} + \mu\frac{\partial \phi_1}{\partial R_f} + \mu^2\frac{\partial \phi_2}{\partial R_f} + \ldots \right) \left(-R_f + \frac{K_F}{T_F}E_{fd} \right) /T_F$$

$$+\mu \left(\frac{\phi_o}{\partial V_R} + \mu\frac{\partial \phi_1}{\partial V_R} + \mu^2\frac{\partial \phi_2}{\partial V_R} + \frac{\partial}{\ldots} \right) \left(-V_R + K_AR_f - \frac{K_AK_F}{T_F}E_{fd} \right.$$

$$+K_A(V_{\text{ref}} - V_t)) /T_A$$

$$+\mu \left(\frac{\partial \phi_o}{\partial T_M} + \mu\frac{\partial \phi_1}{\partial T_M} + \mu^2\frac{\partial \phi_2}{\partial T_M} + \ldots \right) (-T_M + P_{SV})/T_{CH}$$

$$+\mu \left(\frac{\partial \phi_o}{\partial P_{SV}} + \mu\frac{\partial \phi_1}{\partial P_{SV}} + \mu^2\frac{\partial \phi_2}{\partial P_{SV}} + \ldots \right) \left(-P_{SV} + P_C - \frac{1}{R_D} \right.$$

$$\left. \left(\frac{\omega}{\omega_s} - 1 \right) \right) /T_{SV} = -(\phi_o + \mu\phi_1 + \mu^2\phi_2 + \ldots) + (X_q - X_q')I_q \quad (5.161)$$

where I_d and I_q are

$$
\begin{aligned}
I_d &= \frac{(R_s + R_e)(\phi_o + \mu\phi_1 + \mu^2\phi_2 + \ldots - V_s\sin(\delta - \theta_{vs}))}{\Delta_1} \\
&\quad + \frac{(X_q' + X_{ep})(E_q' - V_s\cos(\delta - \theta_{vs}))}{\Delta_1} \\
I_q &= \frac{-(X_d' + X_{ep})(\phi_o + \mu\phi_1 + \mu^2\phi_2 + \ldots - V_s\sin(\Delta - \theta_{vs}))}{\Delta_1} \\
&\quad + \frac{(R_s + R_e)(E_q' - V_s\cos(\delta - \theta_{vs}))}{\Delta_1}
\end{aligned}
$$

where

$$
\Delta_1 = (R_s + R_e)^2 + (X_d' + X_{ep})(X_q' + X_{ep}) \tag{5.162}
$$

and V_d, V_q, and V_t are as before. This partial differential equation may seem difficult to solve; on the contrary, it is actually very easy to solve for ϕ_o, ϕ_1, ϕ_2 in sequence. To see how this is done systematically, suppose that we want the simplest possible approximation of the integral manifold. This is ϕ_o, and is found by neglecting all μ terms, giving

$$
0 = -\phi_o + (X_q - X_q')I_q \mid_{\mu=0} \tag{5.163}
$$

To find ϕ_o as a function of the states, we must solve (5.162) and (5.163) neglecting all μ terms, giving

$$
\begin{aligned}
\phi_o &= \frac{(X_q - X_q')(X_d' + X_{ep})V_s\sin(\delta - \theta_{vs})}{\Delta_2} \\
&\quad + \frac{(X_q - X_q')(R_s + R_e)(E_q' - V_s\cos(\delta - \theta_{vs}))}{\Delta_2}
\end{aligned} \tag{5.164}
$$

where

$$
\Delta_2 = (R_s + R_e)^2 + (X_q + X_{ep})(X_d' + X_{ep})
$$

or

$$
\phi_o = f_{oX}(\delta) + f_{oR}(E_q', \delta) \tag{5.165}
$$

This is the first term of the series for the integral manifold for E_d', and is what would be obtained if T_{qo}' were simply set to zero. When substituted in the remaining equations, it would not reflect any of the damping due to this damper-winding (this is what was done in Sections 5.5 and 5.6). If we want to recover some of the damping due to this damper-winding without resorting to the two-axis model, we can compute ϕ_1 by returning to (5.161) and keeping μ terms, but neglecting μ^2 and higher powers of μ. Note that ϕ_o is only a function of E_q' and δ; all the partials of ϕ_o are zero except for two, giving

$$\frac{\partial \phi_o}{\partial E_q'}(-E_q' - (X_d - X_d')I_d\mid_{\mu=o} +E_{fd})/T_{do}' + \frac{\partial \phi_o}{\partial \delta}(\omega - \omega_s)$$

$$= -\phi_1 - (X_q - X_q')\left(\frac{(X_d' + X_{ep})\phi_1}{(R_s + R_e)^2 + (X_d' + X_{ep})(X_q' + X_{ep})}\right)$$

$$(5.166)$$

where I_d is evaluated at $\mu = 0$. This makes ϕ_1 some function of E_q', E_{fd}, δ, and ω, written here as

$$\phi_1 = -f_{1X}(\delta)(\omega - \omega_s) + f_{1R}(E_q', E_{fd}, \delta) \qquad (5.167)$$

Stopping with this level of accuracy, the integral manifold for E_d' is

$$E_d' = \phi_o + \mu\phi_1 + 0(\mu^2) \qquad (5.168)$$

where $0(\mu^2)$ is some error of "order" μ^2. In terms of the above functions, an approximate expression for E_d', which will capture some of the damping of the damper-winding, is

$$E_d' = f_{oX}(\delta) + f_{oR}(E_q', \delta) - T_{qo}'f_{1X}(\delta)(\omega - \omega_s)$$
$$+ T_{qo}'f_{1R}(E_q', E_{fd}, \delta) \qquad (5.169)$$

Note that f_{oR} and f_{1R} are zero if $R_s + R_e = 0$. Substitution of this approximation of the exact integral manifold for E_d' into the two-axis model gives the circuit of Figure 5.5 and the following differential equations.

$$T_{do}'\frac{dE_q'}{dt} = -E_q' - (X_d - X_d')I_d + E_{fd} \qquad (5.170)$$

$$\frac{d\delta}{dt} = \omega - \omega_s \qquad (5.171)$$

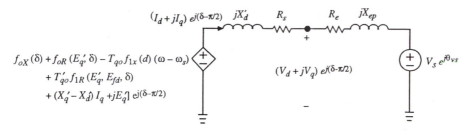

Figure 5.5: *Synchronous machine one-axis dynamic circuit with damping term*

$$\frac{2H}{\omega_s}\frac{d\omega}{dt} = T_M - I_d(f_{oX}(\delta) + f_{oR}(E_q', \delta) - T_{qo}'f_{1X}(\delta)(\omega - \omega_s)$$

$$+ T_{qo}'f_{1R}(E_q', E_{fd}, \delta)) - E_q'I_q - (X_q' - X_d')I_dI_q$$

$$- T_{FW} \tag{5.172}$$

$$T_E\frac{dE_{fd}}{dt} = -(K_E + S_E)E_{fd} + V_R \tag{5.173}$$

$$T_F\frac{dR_f}{dt} = -R_f + \frac{K_F}{T_F}E_{fd} \tag{5.174}$$

$$T_A\frac{dV_R}{dt} = -V_R + K_AR_f - \frac{K_AK_F}{T_F}E_{fd} + K_A(V_{\text{ref}} - V_t) \tag{5.175}$$

$$T_{CH}\frac{dT_M}{dt} = -T_M + P_{SV} \tag{5.176}$$

$$T_{SV}\frac{dP_{SV}}{dt} = -P_{SV} + P_C - \frac{1}{R_D}\left(\frac{\omega}{\omega_s} - 1\right) \tag{5.177}$$

with the limit constraints

$$V_R^{\min} \le V_R \le V_R^{\max} \tag{5.178}$$

$$0 \le P_{SV} \le P_{SV}^{\max} \tag{5.179}$$

and the required algebraic equations for I_d, I_q, V_t, which come from the circuit of Figure 5.5 or the solution of the following equations for I_d, I_q:

$$0 = (R_s + R_e)I_d - (X_q' + X_{ep})I_q - f_{oX}(\delta) - f_{oR}(E_q', \delta)$$

$$+ T_{qo}'f_{1X}(\delta)(\omega - \omega_s) - T_{qo}'f_{1R}(E_q', E_{fd}, \delta)$$

$$+ V_s\sin(\delta - \theta_{vs}) \tag{5.180}$$

$$0 = (R_s + R_e)I_q + (X_d' + X_{ep})I_d - E_q' + V_s\cos(\delta - \theta_{vs}) \tag{5.181}$$

and then substitution into the following for V_d, V_q:

$$V_d = R_e I_d - X_{ep} I_q + V_s \sin(\delta - \theta_{vs}) \tag{5.182}$$

$$V_q = R_e I_q + X_{ep} I_d + V_s \cos(\delta - \theta_{vs}) \tag{5.183}$$

and finally

$$V_t = \sqrt{V_d^2 + V_q^2} \tag{5.184}$$

When $R_s + R_e = 0$

$$\begin{aligned}
E_d' = {} & \frac{X_q - X_q'}{X_q + X_{ep}} V_s \sin(\delta - \theta_{vs}) - T_{qo}' \frac{(X_q' + X_{ep})(X_q - X_q')}{(X_q + X_{ep})^2} \\
& V_s \cos(\delta - \theta_{vs})(\omega - \omega_s)
\end{aligned} \tag{5.185}$$

which gives an accelerating torque of

$$\begin{aligned}
T_{\text{accel}} |_{R_s+R_e=0} = {} & T_M - \frac{E_q' V_s}{X_d' + X_{ep}} \sin(\delta - \theta_{vs}) \\
& + \frac{1}{2} \left(\frac{1}{X_d' + X_{ep}} - \frac{1}{X_q + X_{ep}} \right) V_s^2 \sin 2(\delta - \theta_{vs}) \\
& - T_{qo}' \frac{(X_q - X_q')}{(X_q + X_{ep})^2} V_s^2 \cos^2(\delta - \theta_{vs})(\omega - \omega_s) \\
& - T_{FW}
\end{aligned} \tag{5.186}$$

The T_{qo}' term reflects the damping due to the damper-winding as a slip torque term

$$T_D = T_{qo}' \frac{(X_q - X_q')}{(X_q + X_{ep})^2} V_s^2 \cos^2(\delta - \theta_{vs})(\omega - \omega_s) \tag{5.187}$$

This is often approximated as

$$T_D \approx D(\omega - \omega_s) \tag{5.188}$$

where D is a damping constant. The approximation of Equation (5.188) is normally considered valid only for small deviations about an operating point, but it has been used for large changes as well.

5.8 Synchronous Machine Saturation

The dynamic models of (5.42)–(5.65), (5.92)–(5.109), (5.115)–(5.130), (5.135)
–(5.149), and (5.154)–(5.155) began with the assumption of a linear magnetic circuit ((3.148)–(3.159)). To account for saturation, it is necessary
to repeat this analysis starting with (3.191)–(3.203), rather than (3.148)–
(3.159). Without explicit information about the saturation functions $S_d^{(2)}$,
$S_q^{(2)}$, $S_o^{(2)}$, $S_{fd}^{(2)}$, $S_{1d}^{(2)}$, $S_{1q}^{(2)}$, and $S_{2q}^{(2)}$, it is not possible to rigorously show that
these functions do not affect the fundamental assumptions about time-scale
properties. We assume that the functions do not affect the time-scale property so that the stator plus line algebraic equations for this single-machine
model become

$$0 = R_{se}I_d + \psi_{qe} + V_s \sin(\delta - \theta_{vs}) \tag{5.189}$$

$$0 = R_{se}I_q - \psi_{de} + V_s \cos(\delta - \theta_{vs}) \tag{5.190}$$

$$0 = R_{se}I_o \tag{5.191}$$

$$\psi_{de} = -X_{de}''I_d + \frac{(X_d'' - X_{ls})}{(X_d' - X_{ls})}E_q' + \frac{(X_d' - X_d'')}{(X_d' - X_{ls})}\psi_{1d}$$
$$- S_d^{(2)}(Y_2) \tag{5.192}$$

$$\psi_{qe} = -X_{qe}''I_q - \frac{(X_q'' - X_{ls})}{(X_q' - X_{ls})}E_d' + \frac{(X_q' - X_q'')}{(X_q' - X_{ls})}\psi_{2q}$$
$$- S_q^{(2)}(Y_2) \tag{5.193}$$

$$V_d = R_e I_d - X_{ep}I_q + V_s \sin(\delta - \theta_{vs}) \tag{5.194}$$

$$V_q = R_e I_q + X_{ep}I_d + V_s \cos(\delta - \theta_{vs}) \tag{5.195}$$

These equations can be written in circuit form as shown in Figure 5.6.

With this assumption, the dynamic model with saturation included but
stator/network transients eliminated is

$$T_{do}' \frac{dE_q'}{dt} = -E_q' - (X_d - X_d')[I_d - \frac{X_d' - X_d''}{(X_d' - X_{ls})^2}(\psi_{1d} + (X_d' - X_{ls})I_d$$
$$- E_q' + S_{1d}^{(2)}(Y_2))] - S_{fd}^{(2)}(Y_2) + E_{fd} \tag{5.196}$$

$$T_{do}'' \frac{d\psi_{1d}}{dt} = -\psi_{1d} + E_q' - (X_d' - X_{ls})I_d - S_{1d}^{(2)}(Y_2) \tag{5.197}$$

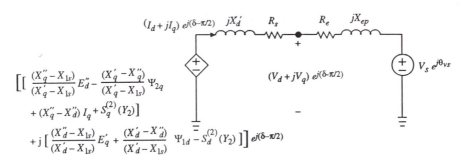

Figure 5.6: *Synchronous machine subtransient dynamic circuit including saturation*

$$T'_{qo}\frac{dE'_d}{dt} = -E'_d + (X_q - X'_q)[I_q - \frac{X'_q - X''_q}{(X'_q - X_{ls})^2}(\psi_{2q} + (X'_q - X_{ls})I_q$$

$$+E'_d + S^{(2)}_{2q}(Y_2))] + S^{(2)}_{1q}(Y_2) \qquad (5.198)$$

$$T''_{qo}\frac{d\psi_{2q}}{dt} = -\psi_{2q} - E'_d - (X'_q - X_{ls})I_q - S^{(2)}_{2q}(Y_2) \qquad (5.199)$$

$$\frac{d\delta}{dt} = \omega - \omega_s \qquad (5.200)$$

$$\frac{2H}{\omega_s}\frac{d\omega}{dt} = T_M - \frac{(X''_d - X_{ls})}{(X'_d - X_{ls})}E'_q I_q - \frac{(X'_d - X''_d)}{(X'_d - X_{ls})}\psi_{1d}I_q$$

$$-\frac{(X''_q - X_{ls})}{(X'_q - X_{ls})}E'_d I_d + \frac{(X'_q - X''_q)}{(X'_q - X_{ls})}\psi_{2q}I_d$$

$$-(X''_q - X''_d)I_d I_q + S^{(2)}_d(Y_2)I_q$$

$$-S^{(2)}_q(Y_2)I_d - T_{FW} \qquad (5.201)$$

$$T_E\frac{dE_{fd}}{dt} = -(K_E + S_E(E_{fd}))E_{fd} + V_R \qquad (5.202)$$

$$T_F\frac{dR_f}{dt} = -R_f + \frac{K_F}{T_F}E_{fd} \qquad (5.203)$$

$$T_A\frac{dV_R}{dt} = -V_R + K_A R_f - \frac{K_A K_F}{T_F}E_{fd} + K_A(V_{\text{ref}} - V_t) \qquad (5.204)$$

$$T_{CH}\frac{dT_M}{dt} = -T_M + P_{SV} \qquad (5.205)$$

$$T_{SV}\frac{dP_{SV}}{dt} = -P_{SV} + P_C - \frac{1}{R_D}\left(\frac{\omega}{\omega_s} - 1\right) \qquad (5.206)$$

with the limit constraints

$$V_R^{\min} \leq V_R \leq V_R^{\max} \qquad (5.207)$$

$$0 \leq P_{SV} \leq P_{SV}^{\max} \qquad (5.208)$$

and the required algebraic equations for V_t, I_d, and I_q, which come from the circuit of Figure 5.6:

$$0 = (R_s + R_e)I_d - (X_q'' + X_{ep})I_q - \frac{(X_q'' - X_{ls})}{(X_q' - X_{ls})}E_d'$$

$$+\frac{(X_q' - X_q'')}{(X_q' - X_{ls})}\psi_{2q} - S_q^{(2)}(Y_2) + V_s\sin(\delta - \theta_{vs}) \qquad (5.209)$$

$$0 = (R_s + R_e)I_q + (X_d'' + X_{ep})I_d - \frac{(X_d'' - X_{ls})}{(X_d' - X_{ls})}E_q'$$

$$-\frac{(X_d' - X_d'')}{(X_d' - X_{ls})}\psi_{1d} + S_d^{(2)}(Y_2) + V_s\cos(\delta - \theta_{vs}) \qquad (5.210)$$

the following equations for V_d, V_q, V_t:

$$V_d = R_eI_d - X_{ep}I_q + V_s\sin(\delta - \theta_{vs}) \qquad (5.211)$$

$$V_q = R_eI_q + X_{ep}I_d + V_s\cos(\delta - \theta_{vs}) \qquad (5.212)$$

$$V_t = \sqrt{V_d^2 + V_q^2} \qquad (5.213)$$

and finally the saturation function relations, which must be specified.

The saturation functions, in general, may be functions of I_d, I_q, and I_o. From (5.191), these functions would be evaluated at $I_o = 0$. If the saturation functions include I_d and I_q, the nonlinear algebraic equations (5.209) and (5.210) must be solved for I_d and I_q as functions of E_q', E_d', ψ_{1d}, ψ_{2q}, and δ, and then substituted into (5.211)–5.213) to obtain V_d, V_q, and V_t. If the saturation functions do not include I_d, I_q, then the currents I_d, I_q can easily be solved from (5.209) and (5.210).

To obtain a two-axis model with saturation included as above, we again assume that T_{do}'' and T_{qo}'' are sufficiently small so that an integral manifold exists for ψ_{1d} and ψ_{2q}. A first approximation of this integral manifold is found by setting T_{do}'' and T_{qo}'' equal to zero in (5.197) and (5.199) to obtain

$$0 = -\psi_{1d} + E_q' - (X_d' - X_{\ell s})I_d - S_{1d}^{(2)}(Y_2) \tag{5.214}$$

$$0 = -\psi_{2q} - E_d' - (X_q' - X_{\ell s})I_q - S_{2q}^{(2)}(Y_2) \tag{5.215}$$

We now assume that $S_{1d}^{(2)}(Y_2)$ and $S_{2q}^{(2)}(Y_2)$ are such that these two equations can be solved for ψ_{1d} and ψ_{2q} to obtain

$$\psi_{1d} = E_q' - (X_d' - X_{\ell s})I_d - S_{1d}^{(3)}(Y_3) \tag{5.216}$$

$$\psi_{2q} = -E_d' - (X_q' - X_{\ell s})I_q - S_{2q}^{(3)}(Y_3) \tag{5.217}$$

where

$$Y_3 \triangleq [I_d \; E_q' \; I_q \; E_d']^t \tag{5.218}$$

When used to eliminate ψ_{1d} and ψ_{2q} in (5.196)–(5.213), the equations for I_d and I_q become

$$0 = (R_s + R_e)I_d - (X_q' + X_{ep})I_q - E_d' - S_q^{(3)}(Y_3)$$
$$+V_s \sin(\delta - \theta_{vs}) \tag{5.219}$$

$$0 = (R_s + R_e)I_q + (X_d' + X_{ep})I_d - E_q' + S_d^{(3)}(Y_3)$$
$$V_s \cos(\delta - \theta_{vs}) \tag{5.220}$$

The saturation functions $S_d^{(3)}(Y_3)$ and $S_q^{(3)}(Y_3)$ are a combination of $S_d^{(2)}$, $S_q^{(2)}$ and $S_{1d}^{(2)}$, $S_{2q}^{(2)}$ as found by substituting (5.216) and (5.217) into (5.209) and (5.210). These two equations can be represented by the circuit shown in Figure 5.7.

The two-axis dynamic model with synchronous machine saturation is

$$T_{do}'\frac{dE_q'}{dt} = -E_q' - (X_d - X_d')I_d - S_{fd}^{(3)}(Y_3) + E_{fd} \tag{5.221}$$

$$T_{qo}'\frac{dE_d'}{dt} = -E_d' + (X_q - X_q')I_q + S_{1q}^{(3)}(Y_3) \tag{5.222}$$

$$\frac{d\delta}{dt} = \omega - \omega_s \tag{5.223}$$

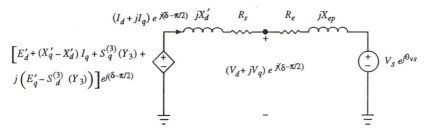

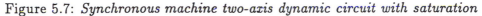

Figure 5.7: *Synchronous machine two-axis dynamic circuit with saturation*

$$\frac{2H}{\omega_s}\frac{d\omega}{dt} = T_M - E_d'I_d - E_q'I_q - (X_q' - X_d')I_dI_q$$

$$+S_d^{(3)}(Y_3)I_q - S_q^{(3)}(Y_3)I_d - T_{FW} \tag{5.224}$$

$$T_E\frac{dE_{fd}}{dt} = -(K_E + S_E(E_{fd}))E_{fd} + V_R \tag{5.225}$$

$$T_F\frac{dR_f}{dt} = -R_f + \frac{K_F}{T_F}E_{fd} \tag{5.226}$$

$$T_A\frac{dV_R}{dt} = -V_R + K_AR_f - \frac{K_AK_F}{T_F}E_{fd} + K_A(V_{\text{ref}} - V_t) \tag{5.227}$$

$$T_{CH}\frac{dT_M}{dt} = -T_M + P_{SV} \tag{5.228}$$

$$T_{SV}\frac{dP_{SV}}{dt} = -P_{SV} + P_C - \frac{1}{R_D}\left(\frac{\omega}{\omega_s} - 1\right) \tag{5.229}$$

with the limit constraints

$$V_R^{\min} \le V_R \le V_R^{\max} \tag{5.230}$$

$$0 \le P_{SV} \le P_{SV}^{\max} \tag{5.231}$$

and the required algebraic equations for I_d, I_q, which come from the circuit of Figure 5.7, or the solution of the following for I_d, I_q:

$$0 = (R_s + R_e)I_d - (X_q' + X_{ep})I_q - E_d' - S_q^{(3)}(Y_3)$$

$$+V_s\sin(\delta - \theta_{vs}) \tag{5.232}$$

$$0 = (R_s + R_e)I_q + (X_d' + X_{ep})I_d - E_q' + S_d^{(3)}(Y_3)$$

$$+V_s\cos(\delta - \theta_{vs}) \tag{5.233}$$

then substitution into the following for V_d, V_q, V_t:

$$V_d = R_e I_d - X_{ep} I_q + V_s \sin(\delta - \theta_{vs}) \tag{5.234}$$

$$V_q = R_e I_q + X_{ep} I_d + V_s \cos(\delta - \theta_{vs}) \tag{5.235}$$

$$V_t = \sqrt{V_d^2 + V_q^2} \tag{5.236}$$

If the saturation functions include a dependence on I_d and I_q, the non-linear algebraic equations (5.232) and (5.233) must be solved for I_d and I_q as functions of E_q', E_d', and δ and then substituted into (5.234)–(5.236) to obtain V_d, V_q, and V_t. If the saturation functions do not include a dependence on I_d and I_q, then I_d and I_q can easily be found from (5.232) and (5.233).

To obtain a flux–decay model with saturation included, we again assume T_{qo}' is sufficiently small so that an integral manifold exists for E_d'. A first approximation of this integral manifold is found by setting T_{qo}' equal to zero in (5.222) to obtain

$$0 = -E_d' + (X_q - X_q')I_q + S_{1q}^{(3)}(Y_3) \tag{5.237}$$

We now assume that $S_{1q}^{(3)}(Y_3)$ is such that this equation can be solved for E_d' to obtain

$$E_d' = (X_q - X_q')I_q + S_{1q}^{(4)}(Y_4) \tag{5.238}$$

where

$$Y_4 \triangleq [I_d \; E_q' \; I_q]^t \tag{5.239}$$

when used to eliminate E_d' in (5.221)–(5.236), the equations for I_d and I_q become

$$0 = (R_s + R_e)I_d - (X_q + X_{ep})I_q - S_q^{(4)}(Y_4)$$
$$V_s \sin(\delta - \theta_{vs}) \tag{5.240}$$

$$0 = (R_s + R_e)I_q + (X_d' + X_{ep})I_d - E_q' + S_d^{(4)}(Y_4)$$
$$V_s \cos(\delta - \theta_{vs}) \tag{5.241}$$

The saturation functions $S_d^{(4)}(Y_4)$ and $S_q^{(4)}(Y_4)$ are a combination of $S_d^{(3)}$, $S_q^{(3)}$, and $S_{1q}^{(4)}$ as found by substituting (5.238) into (5.219) and (5.220). These two equations can be represented by the circuit shown in Figure 5.8.

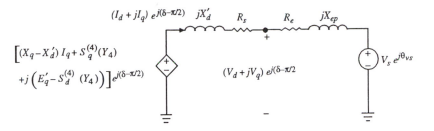

Figure 5.8: *Synchronous machine one-axis (flux-decay) dynamic circuit with saturation*

The one-axis (flux-decay) dynamic model with saturation is

$$T'_{do}\frac{dE'_q}{dt} = -E'_q - (X_d - X'_d)I_d - S^{(4)}_{fd}(Y_4) + E_{fd} \tag{5.242}$$

$$\frac{d\delta}{dt} = \omega - \omega_s \tag{5.243}$$

$$\frac{2H}{\omega_s}\frac{d\omega}{dt} = T_M - E'_q I_q - (X_q - X'_d)I_d I_q + S^{(4)}_d(Y_4)I_q$$
$$- S^{(4)}_q(Y_4)I_d - T_{FW} - T_D \tag{5.244}$$

$$T_E\frac{dE_{fd}}{dt} = -(K_E + S_E(E_{fd}))E_{fd} + V_R \tag{5.245}$$

$$T_F\frac{dR_f}{dt} = -R_f + \frac{K_F}{T_F}E_{fd} \tag{5.246}$$

$$T_A\frac{dV_R}{dt} = -V_R + K_A R_f - \frac{K_A K_F}{T_F}E_{fd} + K_A(V_{\text{ref}} - V_t) \tag{5.247}$$

$$T_{CH}\frac{dT_M}{dt} = -T_M + P_{SV} \tag{5.248}$$

$$T_{SV}\frac{dP_{SV}}{dt} = -P_{SV} + P_C - \frac{1}{R_D}\left(\frac{\omega}{\omega_s} - 1\right) \tag{5.249}$$

with the limit constraints

$$V_R^{\min} \le V_R \le V_R^{\max} \tag{5.250}$$

$$0 \le P_{SV} \le P_{SV}^{\max} \tag{5.251}$$

and the required algebraic equations, which come from the circuit of Figure

5.8:

$$0 = (R_s + R_e)I_d - (X_q + X_{ep})I_q - S_q^{(4)}(Y_4)$$
$$+V_s \sin(\delta - \theta_{vs}) \tag{5.252}$$

$$0 = (R_s + R_e)I_q + (X_d' + X_{ep})I_d - E_q' + S_d^{(4)}(Y_4)$$
$$+V_s \cos(\delta - \theta_{vs}) \tag{5.253}$$

then substitution into the following for V_d, V_q, V_t:

$$V_d = R_e I_d - X_{ep}I_q + V_s \sin(\delta - \theta_{vs}) \tag{5.254}$$

$$V_q = R_e I_q + X_{ep}I_d + V_s \cos(\delta - \theta_{vs}) \tag{5.255}$$

$$V_t = \sqrt{V_d^2 + V_q^2} \tag{5.256}$$

If the saturation functions include a dependence on I_d and I_q, the non-linear algebraic equations (5.252) and (5.253) must be solved for I_d and I_q as functions of E_q', δ and then substituted into (5.254)–(5.256) to obtain V_d, V_q, and V_t. If the saturation functions do not include a dependence on I_d and I_q, then I_d and I_q can easily be found from (5.252) and (5.253).

There is really no point in trying to incorporate saturation into the classical model, since it essentially assumes constant flux linkage.

5.9   Problems

5.1 Given the two-axis dynamic circuit of Figure 5.2, solve for I_d and I_q in terms of the circuit parameters plus δ, E_q', E_d' and the source V_s, θ_{vs}.

5.2 Given the one-axis dynamic circuit of Figure 5.3, solve for I_d and I_q in terms of the circuit parameters plus δ, E_q' and the source V_s, θ_{vs}.

5.3 Using the two-axis dynamic model of Section 5.4, derive an expression for the voltage behind X_q that gives a circuit that is equally as valid as that of Figure 5.2.

5.4 Using the one-axis dynamic model of Section 5.5, derive an expression for the voltage behind X_q that gives a circuit that is equally as valid as that of Figure 5.3.

5.5 Given a synchronous generator with a two-axis model:

$$(V_d + jV_q)e^{j(\delta - \pi/2)} = 1\angle 0 \text{ pu}$$

$$(I_d + jI_q)e^{j(\delta - \pi/2)} = 0.5\angle 30° \text{ pu}$$

$R_s = 0$ pu, $X_d = 1.2$ pu, $X_q = 1.0$ pu, $X'_d = 0.2$ pu, $X'_q = 0.2$ pu

(a) Find the steady-state values of δ, E'_q, E'_d.

(b) Find the inputs E_{fd} and T_m.

(c) Find E'^o and δ'^o for the classical model.

5.6 Given the following two-axis dynamic model of a synchronous machine connected to an infinite bus, with the parameters shown (and $R_s = 0$):

(a) If the machine is in a stable steady-state condition supplying zero real power and some reactive power Q, what can you say about the values of I_d and I_q?

(b) Find the values of T_M and E_{fd} in part (a) if $Q = 1$ pu.

(c) Describe, in as much detail as possible, the system response if E_{fd} is suddenly changed to be 2.0 pu.

$$T'_{do}\frac{dE'_q}{dt} = -E'_q - (X_d - X'_d)I_d + E_{fd}$$

$$T'_{qo}\frac{dE'_d}{dt} = -E'_d + (X_q - X'_q)I_q$$

$$\frac{d\delta}{dt} = \omega - \omega_s$$

$$\frac{2H}{\omega_s}\frac{d\omega}{dt} = T_M - E'_d I_d - E'_q I_q - (X'_q - X'_d)I_d I_q$$

$$\left[E'_d + \left(X'_q - X'_d\right)I_q + jE'_q\right]e^{j(\delta - \pi/2)} = jX'_d[I_d + jI_q]e^{j(\delta - \pi/2)} + Ve^{j\theta}$$

$$T'_{do} = 5.0 \text{ sec}, \quad X_d = 1.2 \text{ pu}, \quad X'_d = 0.3 \text{ pu}$$

$$T'_{qo} = 0.6 \text{ sec}, \quad X_q = 1.1 \text{ pu}, \quad X'_q = 0.7 \text{ pu}$$

$$\omega_s = 2\pi 60 \text{ rad/sec}, \quad H = 6.0 \text{ sec}, \quad V = 1.0 \text{ pu}, \quad \theta = 0 \text{ rad}$$

Chapter 6

MULTIMACHINE DYNAMIC MODELS

This chapter considers the dynamic model of many synchronous machines interconnected by transformers and transmission lines. For the initial analysis, loads will be considered balanced symmetrical R-L elements. For notation, we adopt the following "number" symbols:

m = number of synchronous machines (if there is an infinite bus, it is machine number 1)

n = number of system three-phase buses (excluding the datum or reference bus)

b = total number of machines plus transformers plus lines plus loads (total branches).

6.1 The Synchronously Rotating Reference Frame

It is convenient to transform all synchronous machine stator and network variables into a reference frame that converts balanced three-phase sinusoidal variations into constants. Such a transformation is

$$T_{dqos} \triangleq \frac{2}{3} \begin{bmatrix} \cos \omega_s t & \cos(\omega_s t - \frac{2\pi}{3}) & \cos(\omega_s t + \frac{2\pi}{3}) \\ -\sin \omega_s t & -\sin(\omega_s t - \frac{2\pi}{3}) & -\sin(\omega_s t + \frac{2\pi}{3}) \\ \frac{1}{2} & \frac{1}{2} & \frac{1}{2} \end{bmatrix} \qquad (6.1)$$

with

$$T_{dqos}^{-1} = \begin{bmatrix} \cos\omega_s t & -\sin\omega_s t & 1 \\ \cos(\omega_s t - \frac{2\pi}{3}) & -\sin(\omega_s t - \frac{2\pi}{3}) & 1 \\ \cos(\omega_s t + \frac{2\pi}{3}) & -\sin(\omega_s t + \frac{2\pi}{3}) & 1 \end{bmatrix} \qquad (6.2)$$

Recalling the transformation and scaling of Section 3.3, we use the subscript i for all variables and parameters to denote machine i. We now define the scaled machine stator voltages, currents, and flux linkages in the synchronously rotating reference frame as

$$\begin{bmatrix} V_{Di} \\ V_{Qi} \\ V_{Oi} \end{bmatrix} \triangleq T_{dqos} T_{dqoi}^{-1} \begin{bmatrix} V_{di} \\ V_{qi} \\ V_{oi} \end{bmatrix} = T_{qdos} \frac{V_{BABCi}}{V_{BDQi}} \begin{bmatrix} V_{ai} \\ V_{bi} \\ V_{ci} \end{bmatrix}$$

$$= \frac{1}{\sqrt{2}} T_{dqos} \begin{bmatrix} V_{ai} \\ V_{bi} \\ V_{ci} \end{bmatrix} \qquad i = 1,\ldots,m \qquad (6.3)$$

$$\begin{bmatrix} I_{Di} \\ I_{Qi} \\ I_{Oi} \end{bmatrix} \triangleq T_{dqos} T_{dqoi}^{-1} \begin{bmatrix} I_{di} \\ I_{qi} \\ I_{oi} \end{bmatrix} = T_{dqos} \frac{I_{BABCi}}{I_{BDQi}} \begin{bmatrix} I_{ai} \\ I_{bi} \\ I_{ci} \end{bmatrix}$$

$$= \frac{1}{\sqrt{2}} T_{dqos} \begin{bmatrix} I_{ai} \\ I_{bi} \\ I_{ci} \end{bmatrix} \qquad i = 1,\ldots,m \qquad (6.4)$$

$$\begin{bmatrix} \psi_{Di} \\ \psi_{Qi} \\ \psi_{Oi} \end{bmatrix} \triangleq T_{dqos} T_{dqoi}^{-1} \begin{bmatrix} \psi_{di} \\ \psi_{qi} \\ \psi_{oi} \end{bmatrix} = T_{dqos} \frac{\Lambda_{BABCi}}{\Lambda_{BDQi}} \begin{bmatrix} \psi_{ai} \\ \psi_{bi} \\ \psi_{ci} \end{bmatrix}$$

$$= \frac{1}{\sqrt{2}} T_{dqos} \begin{bmatrix} \psi_{ai} \\ \psi_{bi} \\ \psi_{ci} \end{bmatrix} \qquad i = 1,\ldots,m \qquad (6.5)$$

where T_{dqoi} is the machine i transformation of Section 3.3, and all base scaling quantities are also as previously defined. For $\delta_i = \frac{P}{2}\theta_{\text{shaft}_i} - \omega_s t$, it is easy to show that

$$T_{dqos}T_{dqoi}^{-1} = \begin{bmatrix} \sin\delta_i & \cos\delta_i & 0 \\ -\cos\delta_i & \sin\delta_i & 0 \\ 0 & 0 & 1 \end{bmatrix} \qquad i = 1,\ldots,m \qquad (6.6)$$

and

$$T_{dqoi}T_{dqos}^{-1} = \begin{bmatrix} \sin\delta_i & -\cos\delta_i & 0 \\ \cos\delta_i & \sin\delta_i & 0 \\ 0 & 0 & 1 \end{bmatrix} \qquad i = 1,\ldots,m \qquad (6.7)$$

This transformation gives

$$(V_{Di} + jV_{Qi}) = (V_{di} + jV_{qi})e^{j(\delta_i - \frac{\pi}{2})} \qquad i = 1,\ldots,m \qquad (6.8)$$

and

$$(I_{Di} + jI_{Qi}) = (I_{di} + jI_{qi})e^{j(\delta_i - \frac{\pi}{2})} \qquad i = 1,\ldots,m \qquad (6.9)$$

We now assume that all of the m machine data sets and variables have been scaled by selecting a common system-wide power base and voltage bases that are related in accordance with the interconnecting nominal transformer ratings. Applying this transformation to the general model of (3.148)–(3.159) with the same scaling of (5.37)–(5.40) with $\epsilon = 1/\omega_s$, the multimachine model (without controls) in the synchronously rotating reference frame is

$$\epsilon\frac{d\psi_{Di}}{dt} = R_{si}I_{Di} + \psi_{Qi} + V_{Di} \qquad i = 1,\ldots,m \qquad (6.10)$$

$$\epsilon\frac{d\psi_{Qi}}{dt} = R_{si}I_{Qi} - \psi_{Di} + V_{Qi} \qquad i = 1,\ldots,m \qquad (6.11)$$

$$\epsilon\frac{d\psi_{Oi}}{dt} = R_{si}I_{Oi} + V_{Oi} \qquad i = 1,\ldots,m \qquad (6.12)$$

$$T'_{doi}\frac{dE'_{qi}}{dt} = -E'_{qi} - (X_{di} - X'_{di})\left[I_{di} - \frac{(X'_{di} - X''_{di})}{(X'_{di} - X_{\ell si})^2}(\psi_{1di}\right.$$

$$\left. + (X'_{di} - X_{\ell si})I_{di} - E'_{qi})\right] + E_{fdi} \qquad i = 1,\ldots,m \quad (6.13)$$

$$T''_{doi}\frac{d\psi_{1di}}{dt} = -\psi_{1di} + E'_{qi} - (X'_{di} - X_{\ell si})I_{di} \qquad i = 1,\ldots,m \qquad (6.14)$$

$$T'_{qoi}\frac{dE'_{di}}{dt} = -E'_{di} + (X_{qi} - X'_{qi})\left[I_{qi} - \frac{(X'_{qi} - X''_{qi})}{(X'_{qi} - X_{\ell si})^2}(\psi_{2qi}\right.$$

$$\left. + (X'_{qi} - X_{\ell si})I_{qi} + E'_{di})\right] \qquad i = 1, \ldots, m \qquad (6.15)$$

$$T''_{qoi}\frac{d\psi_{2qi}}{dt} = -\psi_{2qi} - E'_{di} - (X'_{qi} - X_{\ell si})I_{qi} \qquad i = 1, \ldots, m \qquad (6.16)$$

$$T_{si}\frac{d\delta_i}{dt} = \omega_{ti} \qquad i = 1, \ldots, m \qquad (6.17)$$

$$T_{si}\frac{d\omega_{ti}}{dt} = T_{Mi} - (\psi_{di}I_{qi} - \psi_{qi}I_{di}) - T_{FWi} \qquad i = 1, \ldots, m \qquad (6.18)$$

$$\psi_{di} = -X''_{di}I_{di} + \frac{(X''_{di} - X_{\ell si})}{(X'_{di} - X_{\ell si})}E'_{qi} + \frac{(X'_{di} - X''_{di})}{(X'_{di} - X_{\ell si})}\psi_{1di}$$

$$i = 1, \ldots, m \qquad (6.19)$$

$$\psi_{qi} = -X''_{qi}I_{qi} - \frac{(X''_{qi} - X_{\ell si})}{(X'_{qi} - X_{\ell si})}E'_{di} + \frac{(X'_{qi} - X''_{qi})}{(X'_{qi} - X_{\ell si})}\psi_{2qi}$$

$$i = 1, \ldots, m \qquad (6.20)$$

$$\psi_{oi} = -X_{\ell si}I_{oi} \qquad i = 1, \ldots, m \qquad (6.21)$$

where

$$(V_{Di} + jV_{Qi}) = (V_{di} + jV_{qi})e^{j(\delta_i - \frac{\pi}{2})} \qquad i = 1, \ldots, m \qquad (6.22)$$

$$(I_{Di} + jI_{Qi}) = (I_{di} + jI_{qi})e^{j(\delta_i - \frac{\pi}{2})} \qquad i = 1, \ldots, m \qquad (6.23)$$

$$(\psi_{Di} + j\psi_{Qi}) = (\psi_{di} + j\psi_{qi})e^{j(\delta_i - \frac{\pi}{2})} \qquad i = 1, \ldots, m \qquad (6.24)$$

The terminal constraints for each machine are still unspecified. The next section gives a multimachine set of terminal constraints, which can be analyzed in a manner similar to the last chapter.

6.2 Network and R-L Load Constraints

Rather than introduce an infinite bus as a terminal constraint, we propose that all m synchronous machine terminals be interconnected by balanced symmetrical three-phase *R-L* elements. These *R-L* elements are either transmission lines where capacitive effects have been neglected, or transformers. For now, we assume that all loads are balanced symmetrical three-phase *R-L* elements, so that the multimachine dynamic model can be written in a multi-time-scale form, as in Chapter 5. We assume that all line, transformer, and load variables have been scaled by selecting the same common power base as the machines, and voltage buses that are related in accordance with the interconnecting nominal transformer ratings. The scaled voltage across the line, transformers, and loads is assumed to be related to the scaled current through them by

$$V_{ai} = -R_i I_{ai} + \frac{1}{\omega_s}\frac{d\psi_{ai}}{dt} \qquad i = m+1,\ldots,b \qquad (6.25)$$

$$V_{bi} = -R_i I_{bi} + \frac{1}{\omega_s}\frac{d\psi_{bi}}{dt} \qquad i = m+1,\ldots,b \qquad (6.26)$$

$$V_{ci} = -R_i I_{ci} + \frac{1}{\omega_s}\frac{d\psi_{ci}}{dt} \qquad i = m+1,\ldots,b \qquad (6.27)$$

with

$$\begin{bmatrix} \psi_{ai} \\ \psi_{bi} \\ \psi_{ci} \end{bmatrix} = \begin{bmatrix} X_{esi} & X_{emi} & X_{emi} \\ X_{emi} & X_{esi} & X_{emi} \\ X_{emi} & X_{emi} & X_{esi} \end{bmatrix} \begin{bmatrix} -I_{ai} \\ -I_{bi} \\ -I_{ci} \end{bmatrix}$$
$$i = m+1,\ldots,b \quad (6.28)$$

To connect the machines, these constraints must also be transformed. The scaled line, transformer, and load variables in the synchronously rotating reference frame are defined as

$$\begin{bmatrix} V_{Di} \\ V_{Qi} \\ V_{Oi} \end{bmatrix} \triangleq \frac{1}{\sqrt{2}} T_{dqos} \begin{bmatrix} V_{ai} \\ V_{bi} \\ V_{ci} \end{bmatrix} \qquad i = m+1,\ldots,b \qquad (6.29)$$

$$\begin{bmatrix} I_{Di} \\ I_{Qi} \\ I_{Oi} \end{bmatrix} \triangleq \frac{1}{\sqrt{2}} T_{dqos} \begin{bmatrix} I_{ai} \\ I_{bi} \\ I_{ci} \end{bmatrix} \qquad i = 1+m,\ldots,b \qquad (6.30)$$

$$\begin{bmatrix} \psi_{Di} \\ \psi_{Qi} \\ \psi_{Oi} \end{bmatrix} \triangleq \frac{1}{\sqrt{2}} T_{dqos} \begin{bmatrix} \psi_{ai} \\ \psi_{bi} \\ \psi_{ci} \end{bmatrix} \qquad i = m+1,\ldots,b \qquad (6.31)$$

The $\sqrt{2}$ is needed because of (6.3)–(6.5). Applying this transformation to (6.25)–(6.28) gives the network and load constraints suitable for connecting the m machines:

$$\epsilon\frac{d\psi_{Di}}{dt} = R_i I_{Di} + \psi_{Qi} + V_{Di} \quad i = m+1, \ldots, b \qquad (6.32)$$

$$\epsilon\frac{d\psi_{Qi}}{dt} = R_i I_{Qi} - \psi_{Di} + V_{Qi} \quad i = m+1, \ldots, b \qquad (6.33)$$

$$\epsilon\frac{d\psi_{Oi}}{dt} = R_i I_{Oi} + V_{Oi} \quad i = m+1, \ldots, b \qquad (6.34)$$

$$\psi_{Di} = -X_{epi} I_{Di} \quad i = m+1, \ldots, b \qquad (6.35)$$

$$\psi_{Qi} = -X_{epi} I_{Qi} \quad i = m+1, \ldots, b \qquad (6.36)$$

$$\psi_{Oi} = -X_{eoi} I_{Oi} \quad i = m+1, \ldots, b \qquad (6.37)$$

with $X_{epi} = X_{esi} - X_{emi}$ and $X_{eoi} = X_{esi} + 2X_{emi}$. The stator and network plus loads all have exactly the same form when expressed in this reference frame. The b sets of flux linkages and currents are not all independent.

6.3   Elimination of Stator/Network Transients

With this synchronous machine plus the R-L element model, it is possible to formally extend the elimination of stator/network transients of Section 5.3 to the multimachine case. To be proper, we should reduce our dynamic states to an independent set. For inductive elements, this is normally done by writing the dynamics in terms of an independent set of loop currents. To do this, we use several concepts from basic graph theory. Consider the three-node, four-branch directed graph of Figure 6.1. It is a directed graph because each branch (labeled a, b, c, d) has an arrow associated with it. This arrow is the assumed direction of the branch current, as well as the assumed polarity of the voltage.

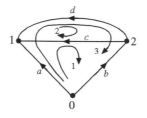

Figure 6.1: *Directed graph*

For example, the four branch voltages (all written with the same rise/drop convention) must satisfy Kirchhoff's laws (the three loops shown in Figure 6.1):

$$v_a - v_b - v_c = 0 \tag{6.38}$$

$$v_c - v_d = 0 \tag{6.39}$$

$$v_a - v_b - v_d = 0 \tag{6.40}$$

or

$$C_a^t v_{\text{branch}} = 0 \tag{6.41}$$

where C_a is the augmented branch-loop incidence matrix

$$
C_a = \begin{array}{c} \\ a \\ b \\ c \\ d \end{array}
\begin{array}{ccc} 1 & 2 & 3 \end{array}
\left(\begin{array}{ccc}
1 & 0 & 1 \\
-1 & 0 & -1 \\
-1 & 1 & 0 \\
0 & -1 & -1
\end{array}\right) \tag{6.42}
$$

In general, the branch-loop incidence matrix is $b \times \ell$, where ℓ is the total number of loops in the graph (or circuit). Clearly, C_a is not unique, since we can define loops in any direction we choose. Furthermore, electrical engineers should recall that, for a given circuit (graph), there are at most $b-n$ independent loop currents. For planar circuits, these are the mesh currents. In the graph shown, $n = 2$, since n was defined previously as the number of nodes excluding the reference node. Thus, since one of the nodes in Figure 6.1 must be the reference node, there are only two independent columns in C_a (two independent rows of C_a^t). For assumed arrows of the b branches and assumed directions of the ℓ loops, the C_a matrix can be written by inspection using the following algorithm:

$$C_a(i,j) \quad = 1 \quad \text{if branch } i \text{ is in the same direction as loop } j, \text{ and branch } i \text{ is in loop } j$$

$$C_a(i,j) \quad = -1 \quad \text{if branch } i \text{ is in the opposite direction of loop } j, \text{ and branch } i \text{ is in loop } j$$

$$C_a(i,j) \quad = 0 \quad \text{if branch } i \text{ is not in loop } j$$

Since the branch directions are the assumed branch current directions as well,

$$i_{\text{branch}} = C_a i_{\text{loop}} \tag{6.43}$$

which, for our example, gives

$$i_a = i_{loop1} + i_{loop3} \tag{6.44}$$

$$i_b = -i_{loop1} - i_{loop3} \tag{6.45}$$

$$i_c = -i_{loop1} + i_{loop2} \tag{6.46}$$

$$i_d = -i_{loop2} - i_{loop3} \tag{6.47}$$

which can be easily verified by inspection.

The $b-n$ independent loops can be determined by partitioning C_a in a special way. Begin with b branches all disconnected (as our multimachine model currently is). Connect n branches such that every node is connected, but no loops are formed. The resulting graph is called a tree. The remaining $b-n$ branches are called tree links, and are now added one at a time. Adding one tree link will create a loop. This loop should be oriented in the same direction as the tree link. Adding a second tree link will create a second loop. This loop should also be oriented in the same direction as the second tree link, and must not include the previously added tree link. When this process is completed, there will be $b-n$ loops, which correspond to the $b-n$ tree links. Thus, the incidence matrix created in this manner will have only $b-n$ columns and will be structured as

$$
C_b = \begin{matrix} 1 \\ \vdots \\ n \\ n+1 \\ \vdots \\ b \end{matrix}
\begin{array}{c} 1 \cdots \ell' \\ \left[\begin{array}{c} C_T \\ \hline I \end{array} \right] \end{array}
\tag{6.48}
$$

This $b \times \ell'$ ($\ell' = b - n$) branch loop incidence matrix is called a basic loop matrix, and does not have the subscript a. The columns of C_a that do not appear in C_b are dependent. It should be clear that we still have the relations for Kirchhoff's laws:

$$i_{branch} = C_b i_{indep.\ loop} \tag{6.49}$$

$$C_b^t v_{branch} = 0 \tag{6.50}$$

This matrix will now be used with our b sets of stator/network/load equations. We define

$$\psi_{Dbranch} \triangleq [\psi_{D1} \cdots \psi_{Db}]^t \tag{6.51}$$

$$I_{D\text{branch}} \triangleq [I_{D1} \cdots I_{Db}]^t \qquad (6.52)$$

$$V_{D\text{branch}} \triangleq [V_{D1} \cdots V_{Db}]^t \qquad (6.53)$$

and similarly for Q and O. For a system with a basic branch loop incidence matrix describing the interconnection of these b branches as C_b, we define loop flux linkages as

$$\psi_{D\text{loop}} \triangleq C_b^t \psi_{D\text{branch}} \qquad (6.54)$$

$$\psi_{Q\text{loop}} \triangleq C_b^t \psi_{Q\text{branch}} \qquad (6.55)$$

$$\psi_{O\text{loop}} \triangleq C_b^t \psi_{O\text{branch}} \qquad (6.56)$$

and write the corresponding branch currents in terms of independent loop currents as

$$I_{D\text{branch}} = C_b I_{D\text{loop}} \qquad (6.57)$$

$$I_{Q\text{branch}} = C_b I_{Q\text{loop}} \qquad (6.58)$$

$$I_{O\text{branch}} = C_b I_{O\text{loop}} \qquad (6.59)$$

By our choice of numbering, the first m branches are synchronous machines, and the last $b-m$ branches are transformers, lines, and loads. It might appear that this numbering scheme would not allow us to have two synchronous machines connected to the same bus, since the first n branches must form a tree. This clearly is not a real limitation, since the ordering of C_a can be changed arbitrarily after it is formed to suit a particular preference.

We have not shown it formally, but the topology relationships between the abc branch voltages and branch currents are the same as for the dqo branch voltages and currents. We now write the stator/network/load transients of (6.10)–(6.24) and (6.32)–(6.37) in vector/matrix form:

$$\epsilon \frac{d\psi_{D\text{branch}}}{dt} = R_{\text{branch}} I_{D\text{branch}} + \psi_{Q\text{branch}} + V_{D\text{branch}} \qquad (6.60)$$

$$\epsilon \frac{d\psi_{Q\text{branch}}}{dt} = R_{\text{branch}} I_{Q\text{branch}} - \psi_{D\text{branch}} + V_{Q\text{branch}} \qquad (6.61)$$

$$\epsilon \frac{d\psi_{O\text{branch}}}{dt} = R_{\text{branch}} I_{O\text{branch}} + V_{O\text{branch}} \qquad (6.62)$$

where additional algebraic equations relating the flux linkages and currents could be written using (6.19)–(6.24) and (6.35)–(6.37).

Multiplying these equations by C_b^t and using the properties of C_b from (6.50) and (6.57)–(6.59) gives the b-n independent sets of stator/network/load transients

$$\epsilon \frac{d\psi_{Dloop}}{dt} = C_b^t R_{branch} C_b I_{Dloop} + \psi_{Qloop} \qquad (6.63)$$

$$\epsilon \frac{d\psi_{Qloop}}{dt} = C_b^t R_{branch} C_b I_{Qloop} - \psi_{Dloop} \qquad (6.64)$$

$$\epsilon \frac{d\psi_{Oloop}}{dt} = C_b^t R_{branch} C_b I_{Oloop} \qquad (6.65)$$

As in the single machine/infinite bus case of Chapter 5, the $R_{branch} = 0$ case has a very interesting result. When resistance is zero, (6.63)–(6.65) have an exact integral manifold for ψ_{Dloop}, ψ_{Qloop}, and ψ_{Oloop}, regardless of ϵ:

$$\psi_{Dloop} \Big|_{R_{branch} = 0} = \psi_{Qloop} \Big|_{R_{branch} = 0} = \psi_{Oloop} \Big|_{R_{branch} = 0} = 0$$
$$(6.66)$$

This means that if ψ_{Dloop}, ψ_{Qloop}, and ψ_{Oloop} are initially zero and $R_{branch} = 0$, then ψ_{Dloop}, ψ_{Qloop}, and ψ_{Oloop} remain at zero for all time t regardless of the size of ϵ. If ψ_{Dloop}, ψ_{Qloop}, and ψ_{Oloop} are not initially zero and $R_{branch} = 0$, then ψ_{Dloop}, ψ_{Qloop} oscillate forever and ψ_{Oloop} remains at its initial value, according to (6.63)–(6.65). Unlike the single machine/infinite bus case, the $R_{branch} = 0$ condition is not practical, since this means that the only possible load could be one or more synchronous machines acting as motors.

For R_{branch} not equal to zero but ϵ sufficiently small, an integral manifold for ψ_{Dloop}, ψ_{Qloop}, and ψ_{Oloop} still exists, but has not been found exactly. An approximation of this integral manifold can be found by setting ϵ to zero and solving all the algebraic equations for the stator/network/load flux linkages and currents. This requires the solution of the following equations:

$$0 = C_b^t R_{branch} C_b I_{Dloop} + \psi_{Qloop} \qquad (6.67)$$

$$0 = C_b^t R_{branch} C_b I_{Qloop} - \psi_{Dloop} \qquad (6.68)$$

$$0 = C_b^t R_{branch} C_b I_{Oloop} \qquad (6.69)$$

with

$$\psi_{D\text{loop}} = C_b^t \psi_{D\text{branch}} \tag{6.70}$$

$$\psi_{Q\text{loop}} = C_b^t \psi_{Q\text{branch}} \tag{6.71}$$

$$\psi_{O\text{loop}} = C_b^t \psi_{O\text{branch}} \tag{6.72}$$

$$I_{D\text{branch}} = C_b I_{D\text{loop}} \tag{6.73}$$

$$I_{Q\text{branch}} = C_b I_{Q\text{loop}} \tag{6.74}$$

$$I_{O\text{branch}} = C_b I_{O\text{loop}} \tag{6.75}$$

and (6.19)–(6.24) plus (6.35)–(6.37). Alternatively, we solve (6.10)–(6.12), (6.19)–(6.24), and (6.32)–(6.37) evaluated at $\epsilon = 0$. The "zero" variables are decoupled and will not be carried further. Adding (6.10) plus j times (6.11) and using (6.22)–(6.24) gives

$$0 = R_{si}(I_{di} + jI_{qi})e^{j(\delta_i - \frac{\pi}{2})} - j(\psi_{di} + j\psi_{qi})e^{j(\delta - \frac{\pi}{2})}$$

$$+(V_{di} + jV_{qi})e^{j(\delta_i - \frac{\pi}{2})} \qquad i = 1, \ldots, m \tag{6.76}$$

Substituting (6.19) and (6.20) gives the following complex equation, which can be written as the circuit of Figure 6.2.

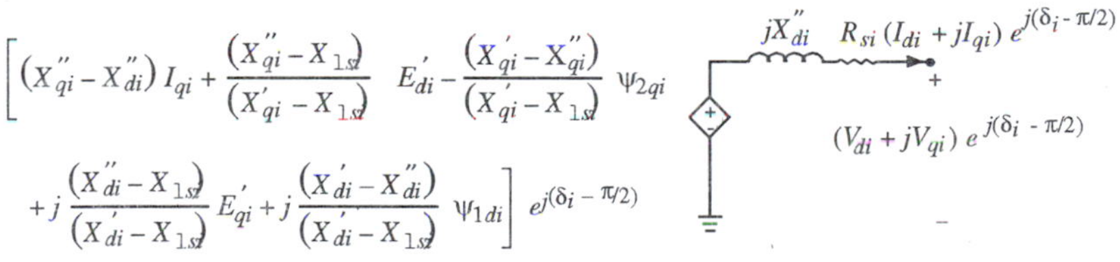

$$\left[(X_{qi}'' - X_{di}'')I_{qi} + \frac{(X_{qi}'' - X_{1si})}{(X_{qi}' - X_{1si})} E_{di}' - \frac{(X_{qi}' - X_{qi}'')}{(X_{qi}' - X_{1si})} \psi_{2qi} \right.$$

$$\left. + j\frac{(X_{di}'' - X_{1si})}{(X_{di}' - X_{1si})} E_{qi}' + j\frac{(X_{di}' - X_{di}'')}{(X_{di}' - X_{1si})} \psi_{1di} \right] e^{j(\delta_i - \pi/2)}$$

Figure 6.2: *Multimachine subtransient dynamic circuit $(i = 1, \ldots, m)$*

$$0 = (R_{si} + jX_{di}'')(I_{di} + jI_{qi})e^{j(\delta_i - \frac{\pi}{2})} + (V_{di} + jV_{qi})e^{j(\delta_i - \frac{\pi}{2})}$$

$$- \left[(X_{qi}'' - X_{di}'')I_{qi} + \frac{(X_{qi}'' - X_{\ell si})}{(X_{qi}' - X_{\ell si})} E_{di}' - \frac{(X_{qi}' - X_{qi}'')}{(X_{qi}' - X_{\ell si})} \psi_{2qi} \right.$$

$$\left. + j\frac{(X_{di}'' - X_{\ell si})}{(X_{di}' - X_{\ell si})} E_{qi}' + j\frac{(X_{di}' - X_{di}'')}{(X_{di}' - X_{\ell si})} \psi_{1di} \right] e^{j(\delta_i - \frac{\pi}{2})}$$

$$i = 1, \ldots, m \tag{6.77}$$

Adding (6.32) plus j times (6.33) and eliminating ψ_{Di}, ψ_{Qi} using (6.35) and (6.36) give the network/load equation, which can be written as the circuit of Figure 6.3.

$$0 = (R_i + jX_{epi})(I_{Di} + jI_{Qi}) + (V_{Di} + jV_{Qi}) \quad i = m+1, \ldots, b \qquad (6.78)$$

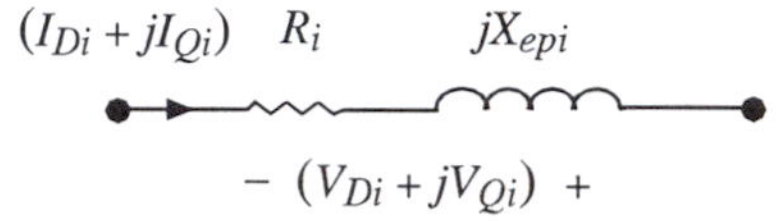

Figure 6.3: *Network plus R-L load dynamic circuit* $(i = m+1, \ldots, b)$

To convert from branch subscript notation to bus subscript notation, we make the following notational numbering of branches. Assume that loads are present at all n buses, and connected to the reference bus. These n branch voltages are then bus voltages denoted as

$$V_i e^{j\theta_i} \triangleq (V_{Di} + jV_{Qi}) \quad i = 1, \ldots, n \qquad (6.79)$$

All branches in excess of these n have branch voltages that are either equal to bus voltages or equal to the difference between two bus voltages. With this bus notation, the multimachine dynamic model is any connection of the circuits of Figures 6.2 and 6.3 into a network resulting in m machines, b branches, and n buses.

Generalization of Network and Load Dynamic Models

Before stating the final model, we now make a major assumption about the actual network and loads. We assume that an integral manifold also exists for all dynamic states associated with networks and loads that are not simple R-L elements. We assume that the integral manifold for the network dynamic states can be approximated by the same representation as above (Figure 6.3 allowing X_{ep} to be negative). We also assume that the integral manifold for the load electrical dynamic states can be approximated by sets of two algebraic equations that can be written as sets of complex equations of the following general form, using the convention of Figure 6.4.

$$(V_{Di} + jV_{Qi})(I_{LDi} - jI_{LQi}) = P_{Li}(V_i) + jQ_{Li}(V_i) \quad i = 1, \ldots, n \qquad (6.80)$$

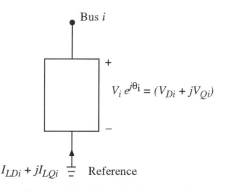

Figure 6.4: *Generalized load electrical dynamic circuit*

where $P_{Li}(V_i)$ and $Q_{Li}(V_i)$ may be nonlinear functions of the bus voltage magnitude V_i. Commonly used load models are

$$P_{Li}(V_i) = P_{Loi} + k_{P1i}V_i + k_{P2i}V_i^2 + \dots \quad i = 1, \dots, n \quad (6.81)$$

$$Q_{Li}(V_i) = Q_{Loi} + k_{Q1i}V_i + k_{Q2i}V_i^2 + \dots \quad i = 1, \dots, n \quad (6.82)$$

or any combination of terms involving any power of V_i. The constants P_{Loi} and Q_{Loi} represent a "constant power" component, k_{P1i} and k_{Q1i} represent a "constant current magnitude at constant power factor" component, and k_{P2i} and k_{Q2i} represent a "constant impedance" component. Note that P_{Li} and Q_{Li} are given as "injected" powers, meaning that P_{Li} will normally be negative for a passive load.

With this generalized model for the network and loads, we write the algebraic equations for the interconnection of m machine circuits with all transformers, lines, and loads using the standard network bus admittance matrix defined through the bus voltages and net injected currents by

$$(I_{di} + jI_{qi})e^{j(\delta_i - \frac{\pi}{2})} + (I_{LDi} + jI_{LQi})$$

$$= \sum_{k=1}^{n} Y_{ik}e^{j\alpha_{ik}}V_k e^{j\theta_k} \quad i = 1, \dots, m \quad (6.83)$$

$$(I_{LDi} + jI_{LQi}) = \sum_{k=1}^{n} Y_{ik}e^{j\alpha_{ik}}V_k e^{j\theta_k} \quad i = m+1, \dots, n \quad (6.84)$$

where all quantities are as previously defined, and $Y_{ik}e^{j\alpha_{ik}}$ is the ik^{th} entry of the network bus admittance matrix. This matrix is formed using all of the branches of the form of Figure 6.3. It has the same formulation as the admittance matrix used for load–flow analysis.

This representation gives the following dynamic model for the m machine, n bus power system after stator/network and load electrical transients have been eliminated, and using the load model of Figure 6.4 and (6.80):

$$T'_{doi}\frac{dE'_{qi}}{dt} = -E'_{qi} - (X_{di} - X'_{di})[I_{di} - \frac{(X'_{di} - X''_{di})}{(X'_{di} - X_{\ell si})^2}$$

$$(\psi_{1di} + (X'_{di} - X_{\ell si})I_{di} - E'_{qi})] + E_{fdi}$$

$$i = 1, \ldots, m \qquad (6.85)$$

$$T''_{doi}\frac{d\psi_{1di}}{dt} = -\psi_{1di} + E'_{qi} - (X'_{di} - X_{\ell si})I_{di} \quad i = 1, \ldots, m \qquad (6.86)$$

$$T'_{qoi}\frac{dE'_{di}}{dt} = -E'_{di} + (X_{qi} - X'_{qi})\left[I_{qi} - \frac{(X'_{qi} - X''_{qi})}{(X'_{qi} - X_{\ell si})^2}\right.$$

$$\left.(\psi_{2qi} + (X'_{qi} - X_{\ell si})I_{qi} + E'_{di})\right] \quad i = 1, \ldots, m \quad (6.87)$$

$$T''_{qoi}\frac{d\psi_{2qi}}{dt} = -\psi_{2qi} - E'_{di} - (X'_{qi} - X_{\ell si})I_{qi} \quad i = 1, \ldots, m \qquad (6.88)$$

$$\frac{d\delta_i}{dt} = \omega_i - \omega_s \quad i = 1, \ldots, m \qquad (6.89)$$

$$\frac{2H_i}{\omega_s}\frac{d\omega_i}{dt} = T_{Mi} - \frac{(X''_{di} - X_{\ell si})}{(X'_{di} - X_{\ell si})}E'_{qi}I_{qi} - \frac{(X'_{di} - X''_{di})}{(X'_{di} - X_{\ell si})}\psi_{1di}I_{qi}$$

$$-\frac{(X''_{qi} - X_{\ell si})}{(X'_{qi} - X_{\ell si})}E'_{di}I_{di} + \frac{(X'_{qi} - X''_{qi})}{(X'_{qi} - X_{\ell si})}\psi_{2qi}I_{di}$$

$$-(X''_{qi} - X''_{di})I_{di}I_{qi} - T_{FWi} \quad i = 1, \ldots, m \qquad (6.90)$$

$$T_{Ei}\frac{dE_{fdi}}{dt} = -(K_{Ei} + S_{Ei}(E_{fdi}))E_{fdi} + V_{Ri} \quad i = 1, \ldots, m \quad (6.91)$$

$$T_{Fi}\frac{dR_{fi}}{dt} = -R_{fi} + \frac{K_{Fi}}{T_{Fi}}E_{fdi} \quad i = 1, \ldots, m \qquad (6.92)$$

$$T_{Ai}\frac{dV_{Ri}}{dt} = -V_{Ri} + K_{Ai}R_{fi} - \frac{K_{Ai}K_{Fi}}{T_{Fi}}E_{fdi}$$

$$+K_{Ai}(V_{\mathrm{ref}i} - V_i) \quad i = 1, \ldots, m \qquad (6.93)$$

$$T_{CHi}\frac{dT_{Mi}}{dt} = -T_{Mi} + P_{SVi} \quad i = 1, \ldots, m \qquad (6.94)$$

$$T_{SVi}\frac{dP_{SVi}}{dt} = -P_{SVi} + P_{Ci} - \frac{1}{R_{Di}}\left(\frac{\omega_i}{\omega_s} - 1\right) \quad i = 1, \ldots, m \quad (6.95)$$

with the limit constraints

$$V_{Ri}^{\min} \leq V_{Ri} \leq V_{Ri}^{\max} \quad i = 1, \ldots, m \quad (6.96)$$

$$0 \leq P_{SVi} \leq P_{SVi}^{\max} \quad i = 1, \ldots, m \quad (6.97)$$

and the algebraic constraints

$$0 = V_i e^{j\theta_i} + (R_{si} + jX_{di}'')(I_{di} + jI_{qi})e^{j(\delta_i - \frac{\pi}{2})}$$

$$- \left[(X_{qi}'' - X_{di}'')I_{qi} + \frac{(X_{qi}'' - X_{\ell si})}{(X_{qi}' - X_{\ell si})}E_{di}' - \frac{(X_{qi}' - X_{qi}'')}{(X_{qi}' - X_{\ell si})}\psi_{2qi}\right.$$

$$\left. + j\frac{(X_{di}'' - X_{\ell si})}{(X_{di}' - X_{\ell si})}E_{qi}' + j\frac{(X_{di}' - X_{di}'')}{(X_{di}' - X_{\ell si})}\psi_{1di}\right]e^{j(\delta_i - \frac{\pi}{2})}$$

$$i = 1, \ldots, m \quad (6.98)$$

$$V_i e^{j\theta_i}(I_{di} - jI_{qi})e^{-j(\delta_i - \frac{\pi}{2})} + P_{Li}(V_i) + jQ_{Li}(V_i)$$

$$= \sum_{k=1}^{n} V_i V_k Y_{ik} e^{j(\theta_i - \theta_k - \alpha_{ik})} \quad i = 1, \ldots, m \quad (6.99)$$

$$P_{Li}(V_i) + jQ_{Li}(V_i) = \sum_{k=1}^{n} V_i V_k Y_{ik} e^{j(\theta_i - \theta_k - \alpha_{ik})}$$

$$i = m + 1, \ldots, n \quad (6.100)$$

The voltage regulator input voltage is V_i, which is automatically defined through the network algebraic constraints. Also, for given functions $P_{Li}(V_i)$ and $Q_{Li}(V_i)$, the $n + m$ complex algebraic equations must be solved for V_i, θ_i ($i = 1, \ldots, n$), I_{di}, I_{qi} ($i = 1, \ldots, m$) in terms of the states δ_i, E_{di}', ψ_{2qi}, E_{qi}', ψ_{1di} ($i = 1, \ldots, m$). The currents clearly can be explicitly eliminated by solving either (6.98) or (6.99) and substituting into the differential equations and remaining algebraic equations. This would leave only n complex algebraic equations to be solved for the n complex voltages $V_i e^{j\theta_i}$.

The Special Case of "Impedance Loads"

In some cases, the above dynamic model can be put in explicit closed form without algebraic equations. We make the special assumptions about the load representations:

$$P_{Li}(V_i) = k_{P2i}V_i^2 \quad i = 1, \ldots, n \qquad (6.101)$$

$$Q_{Li}(V_i) = k_{Q2i}V_i^2 \quad i = 1, \ldots, n \qquad (6.102)$$

In this case, the loads and $R_{si} + jX_{di}''$ can be added into the bus admittance matrix diagonal entries to obtain the following algebraic equations:

$$(I_{di} + jI_{qi})e^{j(\delta_i - \frac{\pi}{2})} = \frac{1}{(R_{si} + jX_{di}'')}\left[(X_{qi}'' - X_{di}'')I_{qi} + \frac{(X_{qi}'' - X_{\ell si})}{(X_{qi}' - X_{\ell si})}E_{di}'\right.$$

$$- \frac{(X_{qi}' - X_{qi}'')}{(X_{qi}' - X_{\ell si})}\psi_{2qi} + j\frac{(X_{di}'' - X_{\ell si})}{(X_{di}' - X_{\ell si})}E_{qi}'$$

$$\left. + j\frac{(X_{di}' - X_{di}'')}{(X_{di}' - X_{\ell si})}\psi_{1di}\right]e^{j(\delta_i - \frac{\pi}{2})} - \left(\frac{1}{R_{si} + jX_{di}''}\right)V_i e^{j\theta_i}$$

$$i = 1, \ldots, m \qquad (6.103)$$

$$0 = -\left(\frac{1}{R_{si} + jX_{di}''}\right)\left[(X_{qi}'' - X_{di}'')I_{qi} + \frac{(X_{qi}'' - X_{\ell si})}{(X_{qi}' - X_{\ell si})}E_{di}'\right.$$

$$- \frac{(X_{qi}' - X_{qi}'')}{(X_{qi}' - X_{\ell si})}\psi_{2qi} + j\frac{(X_{di}'' - X_{\ell si})}{(X_{di}' - X_{\ell si})}E_{qi}'$$

$$\left. + j\frac{(X_{di}' - X_{di}'')}{(X_{di}' - X_{\ell si})}\psi_{1di}\right]e^{j(\delta_i - \frac{\pi}{2})}$$

$$+ \sum_{k=1}^{n}(G_{ik}'' + jB_{ik}'')V_k e^{j\theta_k} \quad i = 1, \ldots, m \qquad (6.104)$$

$$0 = \sum_{k=1}^{n}(G_{ik}'' + jB_{ik}'')V_k e^{j\theta_k} \quad i = m+1, \ldots, n \qquad (6.105)$$

where $G_{ik}'' + jB_{ik}''$ is the ik^{th} entry of the bus admittance matrix, including all "constant impedance" loads *and* $1/(R_{si} + jX_{di}'')$ on the ith diagonal.

Clearly, all of the $V_k e^{j\theta_k}$ ($k = 1, \ldots, n$) can be eliminated by solving (6.104) and (6.105) and substituting into (6.103) to obtain m complex equations that are linear in I_{di}, I_{qi} ($i = 1, \ldots, m$). These can, in principle, be solved through the inverse of a matrix that would be a function of δ_i, E'_{di}, ψ_{2qi}, E'_{qi}, ψ_{1di} ($i = 1, \ldots, m$). This inverse can be avoided if the following simplification is made:

$$X''_{qi} = X''_{di} \qquad i = 1, \ldots, m \qquad (6.106)$$

With this assumption, (6.104) and (6.105) can be solved for $V_k e^{j\theta_k}$ and substituted into (6.103) to obtain

$$
\begin{aligned}
I_{di} + jI_{qi} = \sum_{k=1}^{m} (G''_{\mathrm{red}\,ik} + jB''_{\mathrm{red}\,ik}) \Bigg[& \frac{(X''_{qk} - X_{\ell sk})}{(X'_{qk} - X_{\ell sk})} E'_{dk} \\
& - \frac{(X'_{qk} - X''_{qk})}{(X'_{qk} - X_{\ell sk})} \psi_{2qk} + j\frac{(X''_{dk} - X_{\ell sk})}{(X'_{dk} - X_{\ell sk})} E'_{qk} \\
& + j\frac{(X'_{dk} - X''_{dk})}{(X'_{dk} - X_{\ell sk})} \psi_{1dk} \Bigg] e^{j(\delta_k - \delta_i)} \qquad i = 1, \ldots, m \quad (6.107)
\end{aligned}
$$

where $G''_{\mathrm{red}\,ik} + jB''_{\mathrm{red}\,ik}$ is the ik^{th} entry of an $m \times m$ admittance matrix (often called the matrix reduced to "internal nodes"). This can easily be solved for I_{di}, I_{qi} ($i = 1, \ldots, m$) and substituted into the differential equations (6.85)–(6.95). Substitution of I_{di} I_{qi} into (6.98) also gives V_i as a function of the states (needed for (6.93)), so the resulting dynamic model is in explicit form without algebraic equations.

6.4 Multimachine Two-Axis Model

The reduced-order model of the last section still contains the damper-winding dynamics ψ_{1di} and ψ_{2qi}. If T''_{doi} and T''_{qoi} are sufficiently small, there is an integral manifold for these dynamic states. A first approximation of the fast damper-winding integral manifold is found by setting T''_{doi} and T''_{qoi} equal to zero in (6.86)–(6.88) to obtain

$$0 = -\psi_{1di} + E'_{qi} - (X'_{di} - X_{\ell si})I_{di} \qquad i = 1, \ldots, m \qquad (6.108)$$

$$0 = -\psi_{2qi} - E'_{di} - (X'_{qi} - X_{\ell si})I_{qi} \qquad i = 1, \ldots, m \qquad (6.109)$$

When used to eliminate ψ_{1di} and ψ_{2qi}, the synchronous machine dynamic circuit is changed from Figure 6.2 to Figure 6.5, giving the following multi-machine two-axis model:

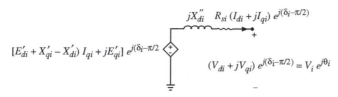

Figure 6.5: *Synchronous machine two-axis model dynamic circuit ($i = 1, \ldots, m$)*

$$T'_{doi}\frac{dE'_{qi}}{dt} = -E'_{qi} - (X_{di} - X'_{di})I_{di} + E_{fdi} \quad i = 1, \ldots, m \quad (6.110)$$

$$T'_{qoi}\frac{dE'_{di}}{dt} = -E'_{di} + (X_{qi} - X'_{qi})I_{qi} \quad i = 1, \ldots, m \quad (6.111)$$

$$\frac{d\delta_i}{dt} = \omega_i - \omega_s \quad i = 1, \ldots, m \quad (6.112)$$

$$\frac{2H_i}{\omega_s}\frac{d\omega_i}{dt} = T_{Mi} - E'_{di}I_{di} - E'_{qi}I_{qi}$$

$$-(X'_{qi} - X'_{di})I_{di}I_{qi} - T_{FWi} \quad i = 1, \ldots, m \quad (6.113)$$

$$T_{Ei}\frac{dE_{fdi}}{dt} = -(K_{Ei} + S_{Ei}(E_{fdi}))E_{fdi} + V_{Ri} \quad i = 1, \ldots, m \quad (6.114)$$

$$T_{Fi}\frac{dR_{fi}}{dt} = -R_{fi} + \frac{K_{Fi}}{T_{Fi}}E_{fdi} \quad i = 1, \ldots, m \quad (6.115)$$

$$T_{Ai}\frac{dV_{Ri}}{dt} = -V_{Ri} + K_{Ai}R_{fi} - \frac{K_{Ai}K_{Fi}}{T_{Fi}}E_{fdi}$$

$$+K_{Ai}(V_{\text{ref}_i} - V_i) \quad i = 1, \ldots, m \quad (6.116)$$

$$T_{CHi}\frac{dT_{Mi}}{dt} = -T_{Mi} + P_{SVi} \quad i = 1, \ldots, m \quad (6.117)$$

$$T_{SVi}\frac{dP_{SVi}}{dt} = -P_{SVi} + P_{Ci} - \frac{1}{R_{Di}}\left(\frac{\omega_i}{\omega_s} - 1\right) \quad i = 1, \ldots, m \quad (6.118)$$

with the limit constraints

$$V_{Ri}^{\min} \leq V_{Ri} \leq V_{Ri}^{\max} \quad i = 1, \ldots, m \quad (6.119)$$

$$0 \leq P_{SVi} \leq P_{SVi}^{\max} \quad i = 1, \ldots, m \quad (6.120)$$

and the algebraic constraints

$$0 = V_i e^{j\theta_i} + (R_{si} + jX'_{di})(I_{di} + jI_{qi})e^{j(\delta_i - \frac{\pi}{2})}$$

$$-[E'_{di} + (X'_{qi} - X'_{di})I_{qi} + jE'_{qi}]e^{j(\delta_i - \frac{\pi}{2})}$$

$$i = 1, \ldots, m \tag{6.121}$$

$$V_i e^{j\theta_i} \; (I_{di} - jI_{qi})e^{-j(\delta_i - \frac{\pi}{2})} + P_{Li}(V_i) + jQ_{Li}(V_i)$$

$$= \sum_{k=1}^{n} V_i V_k Y_{ik} e^{j(\theta_i - \theta_k - \alpha_{ik})} \qquad i = 1, \ldots, m \tag{6.122}$$

$$P_{Li}(V_i) + jQ_{Li}(V_i) = \sum_{k=1}^{n} V_i V_k Y_{ik} e^{j(\theta_i - \theta_k - \alpha_{ik})}$$

$$i = m+1, \ldots, n \tag{6.123}$$

As before, for given functions $P_{Li}(V_i)$ and $Q_{Li}(V_i)$, the $n + m$ complex algebraic equations must be solved for V_i, θ_i $(i = 1, \ldots, m)$ and I_{di}, I_{qi} $(i = 1, \ldots, m)$ in terms of the states δ_i, E'_{di}, E'_{qi} $(i = 1, \ldots, m)$. The currents can clearly be explicitly eliminated by solving either (6.121) or (6.122) and substituting into the differential equations and remaining algebraic equations. This would leave only n complex algebraic equations to be solved for the n complex voltages $V_i e^{j\theta_i}$.

The Special Case of "Impedance Loads"

In some cases, this two-axis dynamic model can be put in explicit closed form without algebraic equations. We make the special assumptions about the load representations

$$P_{Li}(V_i) = k_{P2i}V_i^2 \qquad i = 1, \ldots, n \tag{6.124}$$

$$Q_{Li}(V_i) = k_{Q2i}V_i^2 \qquad i = 1, \ldots, n \tag{6.125}$$

In this case, the loads and $R_{si} + jX'_{di}$ can be added into the bus admittance matrix diagonal entries to obtain the following simplified algebraic equations for the two-axis model with "constant impedance" loads

$$(I_{di} + jI_{qi})e^{j(\delta_i - \frac{\pi}{2})} = \left(\frac{1}{R_{si} + jX'_{di}}\right)[E'_{di} + (X'_{qi} - X'_{di})I_{qi}$$

$$+ jE'_{qi}]e^{j(\delta_i - \frac{\pi}{2})} - \left(\frac{1}{R_{si} + jX'_{di}}\right)V_i e^{j\theta_i} \qquad i = 1, \ldots, m \tag{6.126}$$

$$0 = -\left(\frac{1}{R_{si} + jX'_{di}}\right)[E'_{di} + (X'_{qi} - X'_{di})I_{qi} + jE'_{qi}]e^{j(\delta_i - \frac{\pi}{2})}$$

$$+ \sum_{k=1}^{n}(G'_{ik} + jB'_{ik})V_k e^{j\theta_k} \qquad i = 1,\ldots,m \qquad (6.127)$$

$$0 = \sum_{k=1}^{n}(G'_{ik} + jB'_{ik})V_k e^{j\theta_k} \qquad i = m+1,\ldots,n \qquad (6.128)$$

where $G'_{ik} + jB'_{ik}$ is the ik^{th} entry of the admittance matrix, including all "constant impedance" loads *and* $1/(R_{si} + jX'_{di})$ on the ith diagonal.

Clearly, all of the $V_k e^{j\theta_k}$ $(k = 1,\ldots,n)$ can be eliminated by solving (6.127) and (6.128) and substituting into (6.126) to obtain m complex equations that are linear in I_{di}, I_{qi} $(i = 1,\ldots,m)$. These can, in principle, be solved through the inverse of a matrix that would be a function of the states δ_i, E'_{qi}, E'_{di} $(i = 1,\ldots,m)$. This inverse can be avoided if the following simplification is made:

$$X'_{qi} = X'_{di} \qquad i = 1,\ldots,m \qquad (6.129)$$

With this assumption, (6.127) and (6.128) can be solved for $V_k e^{j\theta_k}$ and substituted into (6.126) to obtain

$$I_{di} + jI_{qi} = \sum_{k=1}^{m}\left(G'_{\substack{red \\ ik}} + jB'_{\substack{red \\ ik}}\right)[E'_{dk} + jE'_{qk}]e^{j(\delta_k - \delta_i)}$$

$$i = 1,\ldots,m \qquad (6.130)$$

where $G'_{\substack{red \\ ik}} + jB'_{\substack{red \\ ik}}$ is the ik^{th} entry of an $m \times m$ admittance matrix (often called the matrix reduced to "internal nodes"). This can easily be solved for I_{di}, I_{qi} $(i = 1,\ldots,m)$ and substituted into the differential equations (6.110)–(6.18). Substitution of I_{di}, I_{qi} into (6.121) also gives V_i as a function of the states (needed for (6.116)), so the resulting dynamic model is in explicit form without algebraic equations.

6.5   Multimachine Flux–Decay Model

The reduced-order model of the last section still contains the damper-winding dynamics E'_{di}. If T'_{qoi} for all $i = 1,\ldots,m$ are sufficiently small, there is an integral manifold for these dynamic states. A first approximation of the remaining fast damper-winding integral manifold is found by setting T'_{qoi} equal to zero in (6.111) to obtain

$$0 = -E'_{di} + (X_{qi} - X'_{qi})I_{qi} \quad i = 1, \ldots, m \qquad (6.131)$$

When used to eliminate E'_{di}, the synchronous machine dynamic circuit is changed from Figure 6.5 to Figure 6.6, giving the following multimachine one-axis or flux-decay model:

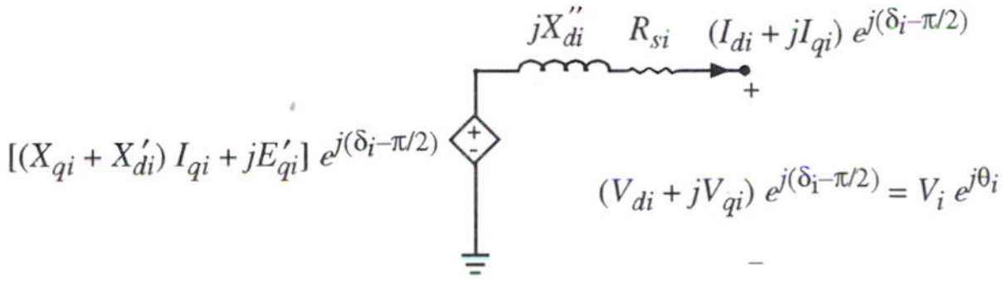

Figure 6.6: *Synchronous machine flux-decay model dynamic circuit (i=1, ...,m)*

$$T'_{doi}\frac{dE'_{qi}}{dt} = -E'_{qi} - (X_{di} - X'_{di})I_{di} + E_{fdi} \quad i = 1, \ldots, m \quad (6.132)$$

$$\frac{d\delta_i}{dt} = \omega_i - \omega_s \quad i = 1, \ldots, m \qquad (6.133)$$

$$\frac{2H_i}{\omega_s}\frac{d\omega_i}{dt} = T_{Mi} - E'_{qi}I_{qi} - (X_{qi} - X'_{di})I_{di}I_{qi} - T_{FWi}$$
$$i = 1, \ldots, m \qquad (6.134)$$

$$T_{Ei}\frac{dE_{fdi}}{dt} = -(K_{Ei} + S_{Ei}(E_{fdi}))E_{fdi} + V_{Ri} \quad i = 1, \ldots, m \quad (6.135)$$

$$T_{Fi}\frac{dR_{fi}}{dt} = -R_{fi} + \frac{K_{Fi}}{T_{Fi}}E_{fdi} \quad i = 1, \ldots, m \qquad (6.136)$$

$$T_{Ai}\frac{dV_{Ri}}{dt} = -V_{Ri} + K_{Ai}R_{fi} - \frac{K_{Ai}K_{fi}}{T_{Fi}}E_{fdi}$$
$$K_{Ai}(V_{\text{ref}i} - V_i) \quad i = 1, \ldots, m \qquad (6.137)$$

$$T_{CHi}\frac{dT_{Mi}}{dt} = -T_{Mi} + P_{SVi} \quad i = 1, \ldots, m \qquad (6.138)$$

$$T_{SVi}\frac{dP_{SVi}}{dt} = -P_{SVi} + P_{Ci} - \frac{1}{R_{Di}}\left(\frac{\omega_i}{\omega_s} - 1\right) \quad i = 1, \ldots, m \quad (6.139)$$

with the limit constraints

$$V_{Ri}^{\min} \leq V_{Ri} \leq V_{Ri}^{\max} \quad i = 1, \ldots, m \qquad (6.140)$$

$$0 \leq P_{SVi} \leq P_{SVi}^{\max} \quad i = 1, \ldots, m \qquad (6.141)$$

and the algebraic constraints

$$0 = V_i e^{j\theta_i} + (R_{si} + jX_{di}')(I_{di} + jI_{qi})e^{j(\delta_i - \frac{\pi}{2})}$$

$$- [(X_{qi} - X_{di}')I_{qi} + jE_{qi}']e^{j(\delta_i - \frac{\pi}{2})}$$

$$i = 1, \ldots, m \qquad (6.142)$$

$$V_i e^{j\theta_i} \ (I_{di} - jI_{qi})e^{-j(\delta_i - \frac{\pi}{2})} + P_{Li}(V_i) + jQ_{Li}(V_i)$$

$$= \sum_{k=1}^{n} V_i V_k Y_{ik} e^{j(\theta_i - \theta_k - \alpha_{ik})} \quad i = 1, \ldots, m \qquad (6.143)$$

$$P_{Li}(V_i) + jQ_{Li}(V_i) = \sum_{k=1}^{n} V_i V_k Y_{ik} e^{j(\theta_i - \theta_k - \alpha_{ik})}$$

$$i = m + 1, \ldots, n \qquad (6.144)$$

As before, for given functions $P_{Li}(V_i)$ and $Q_{Li}(V_i)$, the $n + m$ complex algebraic equations must be solved for V_i, θ_i $(i = 1, \ldots, n)$, I_{di}, I_{qi} $(i = 1, \ldots, m)$ in terms of the states δ_i, E_{qi}' $(i = 1, \ldots, m)$. The currents can clearly be explicitly eliminated by solving either (6.142) or (6.143) and substituting into the differential equations and remaining algebraic equations. This would leave only n complex algebraic equations to be solved for the n complex voltages $V_i e^{j\theta_i}$.

The Special Case of "Impedance Loads"

In some cases, this flux-decay dynamic model can be put in explicit closed form without algebraic equations. We make the special assumptions about the load representations

$$P_{Li}(V_i) = k_{P2i}V_i^2 \quad i = 1, \ldots, n \qquad (6.145)$$

$$Q_{Li}(V_i) = k_{Q2i}V_i^2 \quad i = 1, \ldots, n \qquad (6.146)$$

In this case, the loads and $R_{si} + jX_{di}'$ can be added into the bus admittance matrix diagonal entries to obtain the following simplified algebraic equations for the flux-decay model with "constant impedance" loads

$$(I_{di} + jI_{qi})e^{j(\delta_i - \frac{\pi}{2})} = \frac{1}{(R_{si} + jX'_{di})}[(X_{qi} - X'_{di})I_{qi} + jE'_{qi}]e^{j(\delta_i - \frac{\pi}{2})}$$

$$-\frac{1}{(R_{si} + jX'_{di})}V_i e^{j\theta_i} \quad i = 1, \ldots, m \qquad (6.147)$$

$$0 = -\frac{1}{(R_{si} + jX'_{di})}[(X_{qi} - X'_{di})I_{qi} + jE'_{qi}]e^{j(\delta_i - \frac{\pi}{2})}$$

$$+ \sum_{k=1}^{n}(G'_{ik} + jB'_{ik})V_k e^{j\theta_k} \quad i = 1, \ldots, m \quad (6.148)$$

$$0 = \sum_{k=1}^{n}(G'_{ik} + jB'_{ik})V_k e^{j\theta_k} \quad i = m+1, \ldots, n \quad (6.149)$$

where $G'_{ik} + jB'_{ik}$ is the ik^{th} entry of the admittance matrix, including all "constant impedance" loads *and* $1/(R_{si} + jX'_{di})$ on the ith diagonal.

Clearly, all of the $V_k e^{j\theta_k}$ $(k = 1, \ldots, n)$ can be eliminated by solving (6.148) and (6.149) and substituting into (6.147) to obtain m complex equations that are linear in I_{di}, I_{qi} $(i = 1, \ldots, m)$. These can, in principle, be solved through the inverse of a matrix that would be a function of the states δ_i, E'_{qi} $(i = 1, \ldots, m)$. This inverse can be avoided if the following simplification (usually not considered valid) is made:

$$X_{qi} = X'_{di} \quad i = 1, \ldots, m \qquad (6.150)$$

With this simplification, (6.148) and (6.149) can be solved for $V_k e^{j\theta_k}$ and substituted into (6.147) to obtain

$$I_{di} + jI_{qi} = \sum_{k=1}^{m}(G'_{\text{red}_{ik}} + jB'_{\text{red}_{ik}})E'_{qk}e^{j(\delta_k - \delta_i)} \quad i = 1, \ldots, m \qquad (6.151)$$

where $G'_{\text{red}_{ik}} + jB'_{\text{red}_{ik}}$ is an $m \times m$ admittance matrix (often called the matrix reduced to "internal nodes"). This can easily be solved for I_{di}, I_{qi} $(i = 1, \ldots, m)$ and substituted into the differential equations (6.132)–(6.139). Substitution of I_{di}, I_{qi} into (6.142) also gives V_i as a function of the states (needed for (6.137)) so that the resulting dynamic model is in explicit form without algebraic equations. We emphasize that the simplification of (6.150) is usually not considered valid for most machines.

6.6   Multimachine Classical Model

As in the single machine/infinite bus case of Chapter 5, the derivation of the classical model requires assumptions that cannot be rigorously supported. Returning to the multimachine two-axis model, rather than assuming $T'_{qoi} = 0$ $(i = 1, \ldots, m)$, we assume that an integral manifold exists for E'_{di}, E'_{qi}, E_{fdi}, R_{fi}, V_{Ri} $(i = 1, \ldots, m)$ that as a first approximation, gives each E'_{qi} equal to a constant and each $(E'_{di} + (X'_{qi} - X'_{di})I_{qi})$ equal to a constant. For this constant based on initial values E'^o_{di}, I^o_{qi}, E'^o_{qi}, we define the constant voltage

$$E'^o_i \triangleq \sqrt{(E'^o_{di} + (X'_{qi} - X'_{di})I^o_{qi})^2 + (E'^o_{qi})^2} \qquad (6.152)$$

and the constant angle

$$\delta'^o_i \triangleq \tan^{-1}\left(\frac{E'^o_{qi}}{E'^o_{di} + (X'_{qi} - X'_{di})I^o_{qi}}\right) - \frac{\pi}{2} \qquad (6.153)$$

The classical model dynamic circuit is then shown in Figure 6.7.

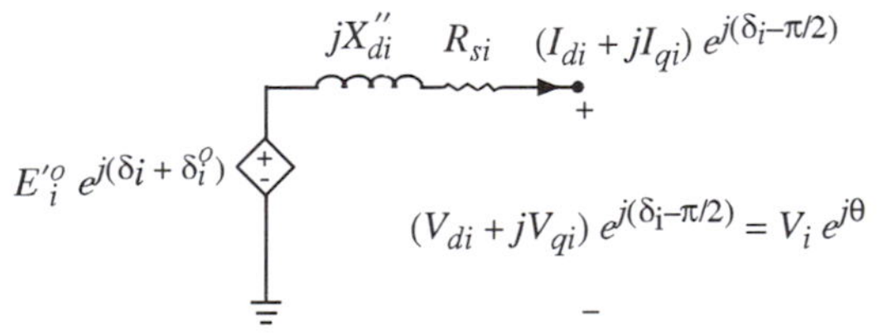

Figure 6.7: *Synchronous machine classical model dynamic circuit (i = 1, …, m)*

Because the classical model is usually used with the assumption of constant shaft torque, we assume

$$T_{CHi} = \infty \qquad (6.154)$$

The classical model is then a 2m-order system (obtained from (6.110)–(6.123))

$$\frac{d\delta_i}{dt} = \omega_i - \omega_s \quad i = 1, \ldots, m \qquad (6.155)$$

$$\frac{2H_i}{\omega_s}\frac{d\omega_i}{dt} = T^o_{Mi} - \mathrm{Real}[(E'^o_i e^{j(\delta_i + \delta'^o_i)})(I_{di} - jI_{qi})e^{-j(\delta_i - \frac{\pi}{2})}]$$

$$-T_{FWi} \quad i = 1, \ldots, m \qquad (6.156)$$

and the algebraic constraints

$$0 = V_i e^{j\theta_i} + (R_{si} + jX'_{di})(I_{di} + jI_{qi})e^{j(\delta_i - \frac{\pi}{2})}$$

$$- E'^o_i e^{j(\delta_i + \delta'^o_i)} \quad i = 1, \ldots, m \tag{6.157}$$

$$V_i e^{j\theta_i} \ (I_{di} - jI_{qi})e^{-j(\delta_i - \frac{\pi}{2})} + P_{Li}(V_i) + jQ_{Li}(V_i) =$$

$$\sum_{k=1}^{n} V_i V_k Y_{ik} e^{j(\theta_i - \theta_k - \alpha_{ik})} \tag{6.158}$$

$$P_{Li}(V_i) + jQ_{Li}(V_i) = \sum_{k=1}^{n} V_i V_k Y_{ik} e^{j(\theta_i - \theta_k - \alpha_{ik})}$$

$$i = m + 1, \ldots, n \tag{6.159}$$

As before, for given functions $P_{Li}(V_i)$ and $Q_{Li}(V_i)$, the $n + m$ complex algebraic equations must be solved for V_i, θ_i ($i = 1, \ldots, n$) and I_{di}, I_{qi} ($i = 1, \ldots, m$) in terms of the states δ_i. The currents can easily be explicitly eliminated by solving either (6.157) or (6.158) and substituting into the differential equations and remaining algebraic equations. This would leave only n complex equations to be solved for the n complex voltages $V_i e^{j\theta_i}$.

The Special Case of "Impedance Loads"

In some cases, this classical model can be put in explicit closed form without algebraic equations. We make the special assumptions about the load representations

$$P_{Li}(V_i) = k_{P2i}V_i^2 \quad i = 1, \ldots, n \tag{6.160}$$

$$Q_{Li}(V_i) = k_{Q2i}V_i^2 \quad i = 1, \ldots, n \tag{6.161}$$

In this case, the loads and $R_{si} + jX'_{di}$ can be added into the bus admittance matrix diagonal entries to obtain the following simplified algebraic equations for the classical model with "constant impedance" loads

$$(I_{di} + jI_{qi})e^{j(\delta_i - \frac{\pi}{2})} = \frac{1}{(R_{si} + jX'_{di})} E'^o_i e^{j(\delta_i + \delta'^o_i)}$$

$$- \frac{1}{(R_{si} + jX'_{di})} V_i e^{j\theta_i} \quad i = 1, \ldots, m \tag{6.162}$$

$$0 = -\frac{1}{(R_{si} + jX'_{di})}E_i'^{lo}e^{j(\delta_i + \delta_i'^o)} + \sum_{k=1}^{n}(G'_{ik} + jB'_{ik})V_k e^{j\theta_k}$$

$$i = 1, \ldots, m \tag{6.163}$$

$$0 = \sum_{k=1}^{n}(G'_{ik} + jB'_{ik})V_k e^{j\theta_k} \qquad i = m+1, \ldots, n \tag{6.164}$$

where $G'_{ik} + jB'_{ik}$ is the ik^{th} entry of the admittance matrix, including all "constant impedance" loads *and* $1/(R_{si} + jX'_{di})$ on the ith diagonal.

Clearly, all of the $V_k e^{j\theta_k}(k = 1, \ldots, n)$ can be eliminated by solving (6.163) and (6.164) and substituting into (6.162) to obtain m complex equations of the form

$$(I_{di} + jI_{qi})e^{j(\delta_i - \frac{\pi}{2})} = \sum_{k=1}^{m}(G'_{\text{red}_{ik}} + jB'_{\text{red}_{ik}})E_k'^{lo}e^{j(\delta_k + \delta_k'^o)} \tag{6.165}$$

where $G'_{\text{red}_{ik}} + jB'_{\text{red}_{ik}}$ is the ik^{th} entry of an $m \times m$ admittance matrix (often called the matrix reduced to "internal nodes"). Defining

$$\delta_{\text{classical}} \overset{\Delta}{=} \delta_i + \delta_i'^o \quad i = 1, \ldots, m \tag{6.166}$$

the multimachine classical model with constant impedance loads is found by substituting (6.165) into (6.156):

$$\frac{d\delta_{\text{classical}}}{dt} = \omega_i - \omega_s \quad i = 1, \ldots, m \tag{6.167}$$

$$\frac{2H_i}{\omega_s}\frac{d\omega_i}{dt} = T_{Mi} - \sum_{k=1}^{m}E_i'^{lo}E_k'^{lo}G'_{\text{red}_{ik}}\cos(\delta_{\text{classical}_i} - \delta_{\text{classical}_k})$$

$$- \sum_{k=1}^{m}E_i'^{lo}E_k'^{lo}B'_{\text{red}_{ik}}\sin(\delta_{\text{classical}_i} - \delta_{\text{classical}_k})$$

$$- T_{FWi} \quad i = 1, \ldots, m \tag{6.168}$$

The classical model can also be obtained formally from the two-axis model by setting $X'_{qi} = X'_{di}$, and $T_{CHi} = T'_{qoi} = T'_{doi} = \infty$ $(i = 1, \ldots, m)$, or from the one-axis model by setting $X_{qi} = X'_{di}$, and $T'_{qoi} = 0$, $T_{CHi} = T'_{doi} = \infty$ $(i = 1, \ldots, m)$. In this latter case, $\delta_i'^o$ is equal to zero, so that $\delta_{\text{classical}}$ is equal to δ_i and $E_i'^{lo}$ is equal to $E_{qi}'^{lo}$ $(i = 1, \ldots, m)$.

6.7   Multimachine Damping Torques

As in the single machine/infinite bus case of Chapter 5, the multimachine flux-decay and classical models do not have explicit speed-damping torques. The friction and windage torque term could be specified as

$$T_{FW} = D_{FW}\omega \text{ or } D'_{FW}\omega^2 \tag{6.169}$$

While this would provide some damping, the effect of all damper windings has been lost in the reduction process. As in the single machine/infinite bus case, it is possible to justify the addition of speed-damping torques. The mathematical derivation of these damping torque terms proceeds in a direct extension to the multimachine case. For example, the integral manifold for each machine E'_{di} must be approximated more accurately, as done in Section 5.7. The resulting improved integral manifolds have the following general form:

$$E'_{di} = \sum_{k=1}^{m} f_{oik}(\delta_i - \delta_k, E'_{qi}, E'_{qk})$$

$$-T'_{qoi} \sum_{k=1}^{m} f_{1ik}(\delta_i - \delta_k, E'_{qi}, E'_{qk}, E_{fdi}, E_{fdk}, \omega_i, \omega_k) \tag{6.170}$$

where f_{oik} is the same as (6.131) written after elimination of I_{qi}. The functions f_{1ik} contain all the speeds of all machines. Rather than use the complicated form of (6.170), it is customary to simply recognize that this additional term involving speeds will contribute speed-damping torques, which can be approximated by

$$T_{Di} = \sum_{k=1}^{m} D_{ik}(\omega_k - \omega_s) \tag{6.171}$$

with D_{ik} treated as a constant. The swing equation for each machine would then be

$$\frac{2H_i}{\omega_s}\frac{d\omega_i}{dt} = T_{Mi} - T_{ELEC_i} - T_{Di} - T_{FWi} \tag{6.172}$$

with T_{ELEC_i} and all other dynamics evaluated using the simple integral manifold approximation for E'_{di} in (6.131). In order of accuracy, it is emphasized that the most accurate model would keep all damper-winding dynamics. The second most accurate model would use a good integral manifold approximation for the damper windings ((6.170)). The least accurate model would use the crude integral manifold for the damper windings ((6.131)), together with a constant D_{ik} cross-damping torque term ((6.171)).

6.8   Multimachine Models with Saturation

The dynamic models of (6.10)–(6.24), (6.85)–(6.100), (6.110)–(6.123), (6.132) –(6.144), and (6.155)–(6.159) began with the assumption of a linear magnetic circuit. In order to account for saturation, it is necessary to repeat this analysis with the saturation functions included. For this case, we assume the saturation functions of [37], given in (3.204)–(3.209). With these functions and the assumption that saturation does not significantly affect time scales, the stator/network transients can still be formally eliminated to obtain the same dynamic circuits of Figures 6.2 to 6.4 and the following multimachine dynamic model:

$$T'_{doi}\frac{dE'_{qi}}{dt} = -E'_{qi} - (X_{di} - X'_{di})\left[I_{di} - \frac{(X'_{di} - X''_{di})}{(X'_{di} - X_{\ell si})^2}(\psi_{1di}\right.$$

$$\left. + (X'_{di} - X_{\ell si})I_{di} - E'_{qi})\right] - S^{(2)}_{fdi} + E_{fdi}$$

$$i = 1,\ldots,m \tag{6.173}$$

$$T''_{doi}\frac{d\psi_{1di}}{dt} = -\psi_{1di} + E'_{qi} - (X'_{di} - X_{\ell si})I_{di} \quad i = 1,\ldots,m \tag{6.174}$$

$$T'_{qoi}\frac{dE'_{di}}{dt} = -E'_{di} + (X_{qi} - X'_{qi})\left[I_{qi} - \frac{(X'_{qi} - X''_{qi})}{(X'_{qi} - X_{\ell si})^2}\right.$$

$$\left. (\psi_{2qi} + (X'_{qi} - X_{\ell si})I_{qi} + E'_{di})\right] + S^{(2)}_{1qi}$$

$$i = 1,\ldots,m \tag{6.175}$$

$$T''_{qoi}\frac{d\psi_{2qi}}{dt} = -\psi_{2qi} - E'_{di} - (X'_{qi} - X_{\ell si})I_{qi} \quad i = 1,\ldots,m \tag{6.176}$$

$$\frac{d\delta_i}{dt} = \omega_i - \omega_s \quad i = 1,\ldots,m \tag{6.177}$$

$$\frac{2H_i}{\omega_s}\frac{d\omega_i}{dt} = T_{Mi} - \frac{(X''_{di} - X_{\ell si})}{(X'_{di} - X_{\ell si})}E'_{qi}I_{qi} - \frac{(X'_{di} - X''_{di})}{(X'_{di} - X_{\ell si})}\psi_{1di}I_{qi}$$

$$-\frac{(X''_{qi} - X_{\ell si})}{(X'_{qi} - X_{\ell si})}E'_{di}I_{di} + \frac{(X'_{qi} - X''_{qi})}{(X'_{qi} - X_{\ell si})}\psi_{2qi}I_{di}$$

$$-(X''_{qi} - X''_{di})I_{di}I_{qi} - T_{FWi} \quad i = 1,\ldots,m \tag{6.178}$$

$$T_{Ei}\frac{dE_{fdi}}{dt} = -(K_{Ei} + S_{Ei}(E_{fdi}))E_{fdi} + V_{Ri} \quad i = 1,\ldots,m \tag{6.179}$$

$$T_{Fi}\frac{dR_{fi}}{dt} = -R_{Fi} + \frac{K_{Fi}}{T_{Fi}}E_{fdi} \quad i = 1, \ldots, m \tag{6.180}$$

$$T_{Ai}\frac{dV_{Ri}}{dt} = -V_{Ri} + K_{Ai}R_{fi} - \frac{K_{Ai}K_{Fi}}{T_{Fi}}E_{fdi}$$
$$+ K_{Ai}(V_{\mathrm{ref}i} - V_i) \quad i = 1, \ldots, m \tag{6.181}$$

$$T_{CHi}\frac{dT_{Mi}}{dt} = -T_{Mi} + P_{SVi} \quad i = 1, \ldots, m \tag{6.182}$$

$$T_{SVi}\frac{dP_{SVi}}{dt} = -P_{SVi} + P_{Ci} - \frac{1}{R_{Di}}\left(\frac{\omega_i}{\omega_s} - 1\right) \quad i = 1, \ldots, m \tag{6.183}$$

with the limit constraints

$$V_{Ri}^{\min} \le V_{Ri} \le V_{Ri}^{\max} \quad i = 1, \ldots, m \tag{6.184}$$

$$0 \le P_{SVi} \le P_{SVi}^{\max} \quad i = 1, \ldots, m \tag{6.185}$$

and the algebraic constraints

$$0 = V_i e^{j\theta_i} + (R_{si} + jX_{di}'')(I_{di} + jI_{qi})e^{j(\delta_i - \frac{\pi}{2})}$$
$$- \left[(X_{qi}'' - X_{di}'')I_{qi} + \frac{(X_{qi}'' - X_{\ell si})}{(X_{qi}' - X_{\ell si})}E_{di}' - \frac{(X_{qi}' - X_{qi}'')}{(X_{qi}' - X_{\ell si})}\psi_{2qi} \right.$$
$$\left. + j\frac{(X_{di}'' - X_{\ell si})}{(X_{di}' - X_{\ell si})}E_{qi}' + j\frac{(X_{di}' - X_{di}'')}{(X_{di}' - X_{\ell si})}\psi_{1di} \right] e^{j(\delta_i - \frac{\pi}{2})}$$
$$i = 1, \ldots, m \tag{6.186}$$

$$V_i e^{j\theta_i} \quad (I_{di} - jI_{qi})e^{-j(\delta_i - \frac{\pi}{2})} + P_{Li}(V_i) + jQ_{Li}(V_i)$$
$$= \sum_{k=1}^{n} V_i V_k Y_{ik} e^{j(\theta_i - \theta_k - \alpha_{ik})} \quad i = 1, \ldots, m \tag{6.187}$$

$$P_{Li}(V_i) + jQ_{Li}(V_i) = \sum_{k=1}^{n} V_i V_k Y_{ik} e^{j(\theta_i - \theta_k - \alpha_{ik})}$$
$$i = m+1, \ldots, n \tag{6.188}$$

and finally the saturation function relations

$$S_{fdi}^{(2)} \triangleq \frac{\psi_{di}''}{|\psi_i''|} S_{smi}(|\psi_i''|) \quad i = 1, \ldots, m \tag{6.189}$$

$$S_{1qi}^{(2)} \triangleq \frac{\psi_{qi}''(X_{qi} - X_{\ell si})}{|\psi_i''|(X_{di} - X_{\ell si})} S_{smi}(|\psi_i''|) \quad i = 1, \ldots, m \tag{6.190}$$

where

$$|\psi_i''| \triangleq (\psi_{di}''^2 + \psi_{qi}''^2)^{\frac{1}{2}} \quad i = 1, \ldots, m \tag{6.191}$$

$$\psi_{di}'' \triangleq \frac{(X_{di}'' - X_{\ell si})}{(X_{di}' - X_{\ell si})} E_{qi}' + \frac{(X_{di}' - X_{di}'')}{(X_{di}' - X_{\ell si})} \psi_{1di} \quad i = 1, \ldots, m \tag{6.192}$$

$$\psi_{qi}'' \triangleq -\frac{(X_{qi}'' - X_{\ell si})}{(X_{qi}' - X_{\ell si})} E_{di}' + \frac{(X_{qi}' - X_{qi}'')}{(X_{qi}' - X_{\ell si})} \psi_{2qi} \quad i = 1, \ldots, m \tag{6.193}$$

and S_{smi} are given nonlinear functions that are zero at $|\psi_i''| = 0$ and increase exponentially with $|\psi_i''|$.

With our choice of saturation functions, the special case of impedance loads still leads to an explicit set of differential equations without algebraic equations. This should be clear, since the only added terms are $S_{fdi}^{(2)}$ and $S_{1qi}^{(2)}$, which are assumed functions of dynamic states. This will not impair the elimination of currents and voltages needed to obtain the reduced admittance matrix formulation. If saturation of network transformers were considered, this elimination would involve nonlinearities in algebraic states that would make an explicit formulation difficult, if not impossible.

The Multimachine Two-Axis Model with Synchronous Machine Saturation

To obtain a two-axis model with saturation included as above, we again assume that T_{doi}'' and T_{qoi}'' are sufficiently small so that an integral manifold exists for each ψ_{1di} and ψ_{2qi}. A first approximation of the fast damper-winding integral manifold is found by setting T_{doi}'' and T_{qoi}'' equal to zero in (6.174) and (6.176) to obtain

$$0 = -\psi_{1di} + E_{qi}' - (X_{di}' - X_{\ell si})I_{di} \quad i = 1, \ldots, m \tag{6.194}$$

$$0 = -\psi_{2qi} - E_{di}' - (X_{qi}' - X_{\ell si})I_{qi} \quad i = 1, \ldots, m \tag{6.195}$$

When used to eliminate ψ_{1di} and ψ_{2qi}, the synchronous machine dynamic circuit is changed from Figure 6.2 to Figure 6.5, giving the following multimachine two-axis dynamic model with synchronous machine saturation:

$$T_{doi}'\frac{dE_{qi}'}{dt} = -E_{qi}' - (X_{di} - X_{di}')I_{di} - S_{fdi}^{(3)} + E_{fdi}$$

$$i = 1, \ldots, m \tag{6.196}$$

$$T'_{qoi}\frac{dE'_{di}}{dt} = -E'_{di} + (X_q - X'_{qi})I_{qi} + S_{1qi}^{(3)} \quad i = 1, \ldots, m \quad (6.197)$$

$$\frac{d\delta_i}{dt} = \omega_i - \omega_s \quad i = 1, \ldots, m \quad (6.198)$$

$$\frac{2H_i}{\omega_s}\frac{d\omega_i}{dt} = T_{Mi} - E'_{di}I_{di} - E'_{qi}I_{qi}$$
$$-(X'_{qi} - X'_{di})I_{di}I_{qi} - T_{FWi} \quad i = 1, \ldots, m \quad (6.199)$$

$$T_{Ei}\frac{dE_{fdi}}{dt} = -(K_{Ei} + S_{Ei}(E_{fdi}))E_{fdi} + V_{Ri} \quad i = 1, \ldots, m \quad (6.200)$$

$$T_{Fi}\frac{dR_{fi}}{dt} = -R_{fi} + \frac{K_{Fi}}{T_{Fi}}E_{fdi} \quad i = 1, \ldots, m \quad (6.201)$$

$$T_{Ai}\frac{dV_{Ri}}{dt} = -V_{Ri} + K_{Ai}R_{fi} - \frac{K_{Ai}K_{Fi}}{T_{Fi}}E_{fdi}$$
$$+ K_{Ai}(V_{\text{ref}i} - V_i) \quad i = 1, \ldots, m \quad (6.202)$$

$$T_{CHi}\frac{dT_{Mi}}{dt} = -T_M + P_{SVi} \quad i = 1, \ldots, m \quad (6.203)$$

$$T_{SVi}\frac{dP_{SVi}}{dt} = -P_{SVi} + P_{Ci} - \frac{1}{R_{Di}}\left(\frac{\omega_i}{\omega_s} - 1\right) \quad i = 1, \ldots, m \quad (6.204)$$

with the limit constraints

$$V_{Ri}^{\min} \le V_{Ri} \le V_{Ri}^{\max} \quad i = 1, \ldots, m \quad (6.205)$$

$$0 \le P_{SVi} \le P_{SVi}^{\max} \quad i = 1, \ldots, m \quad (6.206)$$

and the algebraic constraints

$$0 = V_i e^{j\theta_i} + (R_{si} + jX'_{di})(I_{di} + jI_{qi})e^{j(\delta_i - \frac{\pi}{2})}$$
$$-[E'_{di} + (X'_{qi} - X'_{di})I_{qi} + jE'_{qi}]e^{j(\delta_i - \frac{\pi}{2})}$$
$$i = 1, \ldots, m \quad (6.207)$$

$$V_i e^{j\theta_i}\,(I_{di} - jI_{qi})e^{-j(\delta_i - \frac{\pi}{2})} + P_{Li}(V_i) + jQ_{Li}(V_i) =$$
$$\sum_{k=1}^{n} V_i V_k Y_{ik}e^{j(\theta_i - \theta_k - \propto_{ik})} \quad i = 1, \ldots, m \quad (6.208)$$

$$P_{Li}(V_i) + jQ_{Li}(V_i) = \sum_{k=1}^{n} V_i V_k Y_{ik}e^{j(\theta_i - \theta_k - \propto_{ik})}$$
$$i = m + 1, \ldots, n \quad (6.209)$$

and finally the saturation function relations

$$S_{fdi}^{(3)} \triangleq \frac{E_{qi}'}{|\psi_i'|} S_{smi}(|\psi_i'|) \quad i = 1, \ldots, m \tag{6.210}$$

$$S_{1qi}^{(3)} \triangleq \frac{-E_{di}'(X_{qi} - X_{\ell si})}{|\psi_i'|(X_{di} - X_{\ell si})} S_{smi}(|\psi_i'|) \quad i = 1, \ldots, m \tag{6.211}$$

where

$$|\psi_i'| \triangleq (E_{qi}'^2 + E_{di}'^2)^{\frac{1}{2}} \quad i = 1, \ldots, m \tag{6.212}$$

We have made the assumption that when ψ_{1di} and ψ_{2qi} are eliminated by the approximate integral manifold of (6.194) and (6.195), the saturation function can be approximated by using $\psi_{di}'' \approx E_{qi}'$, $\psi_{qi}'' \approx -E_{di}'$. This prevents I_{di} or I_{qi} from entering the saturation function explicitly.

As before, the special case of impedance loads still leads to a dynamic model in explicit form without algebraic equations, because the saturation functions are assumed to be functions of only the dynamic states E_{di}' and E_{qi}'.

The Multimachine Flux-Decay Model with Synchronous Machine Saturation

To obtain a flux-decay model with saturation included, we again assume that T_{qoi}' is sufficiently small so that an integral manifold exists for each E_{di}'. A first approximation of this integral manifold is found by setting T_{qoi}' equal to zero in (6.197) to obtain:

$$0 = -E_{di}' + (X_{qi} - X_{qi}')I_{qi} + S_{1qi}^{(3)} \quad i = 1, \ldots, m \tag{6.213}$$

As in the single machine/infinite bus case, there is now a problem, since (6.213) must be solved for each E_{di}' (recall that each $S_{1qi}^{(3)}$ is a function of E_{di}' and E_{qi}'). While we could simply carry (6.213) along as another algebraic equation, we can again recognize that since we have already approximated the integral manifold (and, of course, neglected all off-manifold dynamics) we may as well simplify further and set $S_{1qi}^{(3)} \approx 0$. This gives the following flux-decay model with saturation only in the field axis.

$$T_{doi}' \frac{dE_{qi}'}{dt} = -E_{qi}' - (X_{di} - X_{di}')I_{di} - S_{fdi}^{(4)} + E_{fdi}$$
$$i = 1, \ldots, m \tag{6.214}$$

$$\frac{d\delta_i}{dt} = \omega_i - \omega_s \quad i = 1, \ldots, m \tag{6.215}$$

$$\frac{2H_i}{\omega_s}\frac{d\omega_i}{dt} = T_{Mi} - E'_{qi}I_{qi} - (X_{qi} - X'_{di})I_{di}I_{qi} - T_{FWi}$$
$$i = 1, \ldots, m \tag{6.216}$$

$$T_{Ei}\frac{dE_{fdi}}{dt} = -(K_{Ei} + S_{Ei}(E_{fdi}))E_{fdi} + V_{Ri} \quad i = 1, \ldots, m \tag{6.217}$$

$$T_{Fi}\frac{dR_{fi}}{dt} = -R_{fi} + \frac{K_{Fi}}{T_{Fi}}E_{fdi} \quad i = 1, \ldots, m \tag{6.218}$$

$$T_{Ai}\frac{dV_{Ri}}{dt} = -V_{Ri} + K_{Ai}R_{fi} - \frac{K_{Ai}K_{Fi}}{T_{Fi}}E_{fdi}$$
$$+ K_{Ai}(V_{\text{ref}_i} - V_i) \quad i = 1, \ldots, m \tag{6.219}$$

$$T_{CHi}\frac{dT_{Mi}}{dt} = -T_{Mi} + P_{SVi} \quad i = 1, \ldots, m \tag{6.220}$$

$$T_{SVi}\frac{dP_{SVi}}{dt} = -P_{SVi} + P_{Ci} - \frac{1}{R_{Di}}\left(\frac{\omega_i}{\omega_s} - 1\right) \quad i = 1, \ldots, m \tag{6.221}$$

with the limit constraints

$$V_{Ri}^{\min} \leq V_{Ri} \leq V_{Ri}^{\max} \quad i = 1, \ldots, m \tag{6.222}$$

$$0 \leq P_{SVi} \leq P_{SVi}^{\max} \quad i = 1, \ldots, m \tag{6.223}$$

and the algebraic constraints

$$0 = V_i e^{j\theta_i} + (R_{si} + jX'_{di})(I_{di} + jI_{qi})e^{j(\delta_i - \frac{\pi}{2})}$$
$$- [(X_{qi} - X'_{di})I_{qi} + jE'_{qi}]e^{j(\delta_i - \frac{\pi}{2})} \quad i = 1, \ldots, m \tag{6.224}$$

$$V_i e^{j\theta_i}(I_{di} - jI_{qi})e^{-j(\delta_i - \frac{\pi}{2})} + P_{Li}(V_i) + jQ_{Li}(V_i) =$$
$$\sum_{k=1}^{n} V_i V_k Y_{ik} e^{j(\theta_i - \theta_k - \alpha_{ik})} \quad i = 1, \ldots, m \tag{6.225}$$

$$P_{Li}(V_i) + jQ_{Li}(V_i) = \sum_{k=1}^{n} V_i V_k Y_{ik} e^{j(\theta_i - \theta_k - \alpha_{ik})} \quad i = 1, \ldots, m \tag{6.226}$$

and finally the saturation function relation

$$S_{fdi}^{(4)} = S_{smi}(E'_{qi}) \quad i = 1, \ldots, m \tag{6.227}$$

As before, the special case of impedance loads still leads to a dynamic model without algebraic equations, since the additional saturation terms are functions only of the dynamic states E'_{qi}.

There is little point in trying to incorporate saturation into the multimachine classical model, since it essentially assumes constant flux linkage.

6.9 Frequency During Transients

The steady-state relationship between voltages and currents in loads clearly depends on frequency. For inductive loads, the reactance increases with frequency. For induction motors, the nominal speed increases with frequency. During transients, frequency does not have any meaning, since voltage and current waveforms are not pure sinusoids. It is, however, possible to define a quantity during transients that reflects the concept of frequency and is equal to frequency in the sinusoidal steady state. This can be done by considering the algebraic variables V_{Di}, V_{Qi} introduced earlier in the synchronously rotating reference frame together with their polar forms:

$$V_{Di} + jV_{Qi} = V_i e^{j\theta_i} \quad i = 1, \ldots, n \tag{6.228}$$

From the inverse transformation, the scaled abc voltages are (with $V_{Oi} = 0$):

$$V_{ai} = \sqrt{2}V_i \cos(\omega_s t + \theta_i) \quad i = 1, \ldots, n \tag{6.229}$$

$$V_{bi} = \sqrt{2}V_i \cos\left(\omega_s t + \theta_i - \frac{2\pi}{3}\right) \quad i = 1, \ldots, n \tag{6.230}$$

$$V_{ci} = \sqrt{2}V_i \cos\left(\omega_s t + \theta_i + \frac{2\pi}{3}\right) \quad i = 1, \ldots, n \tag{6.231}$$

We emphasize that these forms are valid for both transient and steady-state analyses. In general, V_i and θ_i will both change during a transient. A logical definition of a "dynamic frequency" is

$$\omega_{di} \overset{\Delta}{=} \omega_s + \frac{d\theta_i}{dt} \quad i = 1, \ldots, n \tag{6.232}$$

If the multimachine system is in synchronism with all machines turning at a constant speed, the system frequency is equal to this dynamic frequency (possibly above or below ω_s). During transients, each bus will have a dynamic frequency determined by $d\theta_i/dt$. We emphasize that this definition is only one of many possible such quantities that have the property of being equal to true frequency in steady state.

Adding frequency dependence into the load model by using $P_{Li}(V_i, \omega_{di})$, $Q_{Li}(V_i, \omega_{di})$ instead of $P_{Li}(V_i)$, and $Q_{Li}(V_i)$ requires that $\theta_i (i = 1, \ldots, n)$ become dynamic state variables. This is done by simply adding the following n additional differential equations:

$$\frac{d\theta_i}{dt} = \omega_{di} - \omega_s \quad i = 1, \ldots, n \qquad (6.233)$$

The n dynamic frequencies ω_{di} are algebraic variables that must be eliminated together with all other algebraic variables (I_{di}, I_{qi}, V_i). Since this normally cannot be done easily, it is customary to avoid making the θ_i dynamic states by approximating the frequency dependence. This is done by keeping each ω_{di} constant over one time step of numerical integration (Δt) as

$$\omega_{di}(t) \approx \frac{\theta_i(t) - \theta_i(t - \Delta t)}{\Delta t} + \omega_s \qquad (6.234)$$

where t is the current time. In this method, the θ_i remain as algebraic variables and are simply monitored at each time step to update the frequency-dependent terms of $P_{Li}(V_i, \omega_{di})$ and $Q_{Li}(V_i, \omega_{di})$ for the next time step.

6.10 Angle References and an Infinite Bus

The multimachine dynamic models of the last sections have at least one more differential equation than is needed to solve an m machine, n bus problem, because every rotational system must have a reference for angles. To illustrate this, we define the angles relative to machine 1 as

$$\delta_i' \overset{\Delta}{=} \delta_i - \delta_1 \quad i = 1, \ldots, m \qquad (6.235)$$

$$\theta_i' \overset{\Delta}{=} \theta_i - \delta_1 \quad i = 1, \ldots, n \qquad (6.236)$$

with derivatives for the new dynamic states

$$\frac{d\delta_1'}{dt} = 0 \qquad (6.237)$$

$$\frac{d\delta_i'}{dt} = \omega_i - \omega_1 \quad i = 2, \ldots, m \qquad (6.238)$$

Inspection of the algebraic equations that accompany all of the previous models will reveal that all angles can be written in terms of δ_i' and θ_i'. This means that the full system models have the same form as before, with the following modifications:

- Replace δ_i with δ_i' (and $\delta_1' = 0$).

- Replace θ_i with θ_i'.

- Replace ω_s in the time derivative of δ_i' with ω_1.

With this formulation, the order of the system is formally reduced by 1, since

$$\delta_1' = 0 \tag{6.239}$$

While δ_1' remains at zero for all time, ω_1 changes during a transient. The original δ_1 can be obtained by integrating $\omega_1 - \omega_s$ over time. The original δ_i $(i = 2, \ldots, m)$ and θ_i $(i = 1, \ldots, n)$ can be recovered easily from δ_i', θ_i and δ_1' through (6.235)–(6.236), if desired.

The dynamic system order can be further reduced either if H_1 is set to infinity (machine 1 has constant speed) or if speeds do not explicitly appear on the right-hand side of any dynamic equations (no speed-damping torques and no governor action). This could be formally done by defining

$$\omega_i' \overset{\Delta}{=} \omega_i - \omega_1 \quad i = 1, \ldots, m \tag{6.240}$$

so that

$$\frac{d\delta_i'}{dt} = \omega_i' \quad i = 1, \ldots, m \tag{6.241}$$

and ω_i' replaces ω_i as a dynamic state. In this case, there is no need to include either the angle or the speed equation for machine 1, since all other dynamics would depend only on δ_i' $(i = 2, \ldots, m)$ and ω_i' $(i = 2, \ldots, m)$. This situation also arises when the only speed terms on the right-hand side appear in the swing equations with uniform damping $(H_i/D_i = H_k/D_k \;\; i, k = 1, \ldots, m)$.

A common transformation used in transient stability analysis is the center-of-inertia (COI) reference. Rather than reference each angle to a specific machine (i.e., δ_1), the COI reference transformation defines the COI angle and speeds as

$$\delta_{COI} \overset{\Delta}{=} \frac{1}{M_T} \sum_{i=1}^{m} M_i \delta_i \tag{6.242}$$

$$\omega_{COI} \overset{\Delta}{=} \frac{1}{M_T} \sum_{i=1}^{m} M_i \omega_i \tag{6.243}$$

where

$$M_T \overset{\Delta}{=} \sum_{i=1}^{m} M_i \tag{6.244}$$

and

$$M_i \triangleq \frac{2H_i}{\omega_s} \tag{6.245}$$

The COI-referenced angles and speeds are

$$\hat{\delta}_i \triangleq \delta_i - \delta_{COI} \quad i = 1, \ldots, m \tag{6.246}$$

$$\hat{\omega}_i \triangleq \omega_i - \omega_{COI} \quad i = 1, \ldots, m \tag{6.247}$$

With the introduction of δ_{COI} and ω_{COI}, it is possible to use these as dynamic states together with any other $m - 1$ pairs of COI-referenced mechanical pairs. Choosing $2, \ldots, m$, the new mechanical state-space would consist of δ_{COI}, ω_{COI}, $\hat{\delta}_i$, $\hat{\omega}_i$ $(i = 2, \ldots, m)$. This would require the elimination of $\hat{\delta}_1$ and $\hat{\omega}_1$ in terms of δ_{COI}, ω_{COI} and $\hat{\delta}_i$, $\hat{\omega}_i$ $(i = 2, \ldots, m)$. The resulting system would have M_T multiplying the time derivative of ω_{COI}. Since M_T represents the total system inertia, it is usually quite large relative to any single M_i. For this reason, it is often taken to be infinity, in which case the COI mechanical pair is eliminated, reducing the dynamic order by 2. It is important to emphasize that simply using COI-referenced variables does not, in itself, reduce the dynamic order. The reduction requires the use of COI-referenced variables together with the approximation that M_T is infinity. The resulting swing equations are complicated by the COI reference, since an inertia-weighted form of "system" acceleration is subtracted from each machine's true acceleration.

Chapter 7

MULTIMACHINE SIMULATION

In this chapter, we consider simulation techniques for a multimachine power system using a two-axis machine model with no saturation and neglecting both the stator and the network transients. The resulting differential-algebraic model is systematically derived. Both the partitioned-explicit (PE) and the simultaneous-implicit (SI) methods for integration are discussed. The SI method is preferred in both research grade programs and industry programs, since it can handle "stiff" equations very well. After explaining the SI method consistent with our analytical development so far, we then explain the equivalent but notationally different method, the well-known EPRI-ETMSP (Extended Transient Midterm Stability Program) [70]. A numerical example to illustrate the systematic computation of initial conditions is presented.

7.1 Differential-Algebraic Model

We first rewrite the two-axis model of Section 6.4 in a form suitable for simulation after neglecting the subtransient reactances and saturation. We also neglect the turbine governor dynamics resulting in T_{Mi} being a constant. The limit constraints on V_{Ri} are also deleted, since we wish to concentrate on modeling and simulation. We assume a linear damping term $T_{FWi} = D_i(\omega_i - \omega_s)$. The resulting differential-algebraic equations follow from (6.196)–(6.209) for the m machine, n bus system with the IEEE-Type I exciter as

1. Differential Equations

$$T'_{doi}\frac{dE'_{qi}}{dt} = -E'_{qi} - (X_{di} - X'_{di})I_{di} + E_{fdi} \qquad i = 1, \ldots, m \qquad (7.1)$$

$$T'_{qoi}\frac{dE'_{di}}{dt} = -E'_{di} + (X_{qi} - X'_{qi})I_{qi} \qquad i = 1, \ldots, m \qquad (7.2)$$

$$\frac{d\delta_i}{dt} = \omega_i - \omega_s \qquad i = 1, \ldots, m \qquad (7.3)$$

$$\frac{2H_i}{\omega_s}\frac{d\omega_i}{dt} = T_{Mi} - E'_{di}I_{di} - E'_{qi}I_{qi} - (X'_{qi} - X'_{di})I_{di}I_{qi}$$

$$-D_i(\omega_i - \omega_s) \qquad i = 1, \ldots, m \qquad (7.4)$$

$$T_{Ei}\frac{dE_{fdi}}{dt} = -(K_{Ei} + S_{Ei}(E_{fdi}))E_{fdi} + V_{Ri} \qquad i = 1, \ldots, m \qquad (7.5)$$

$$T_{Fi}\frac{dR_{fi}}{dt} = -R_{fi} + \frac{K_{Fi}}{T_{Fi}}E_{fdi} \qquad i = 1, \ldots, m \qquad (7.6)$$

$$T_{Ai}\frac{dV_{Ri}}{dt} = -V_{Ri} + K_{Ai}R_{fi} - \frac{K_{Ai}K_{Fi}}{T_{Fi}}E_{fdi} + K_{Ai}(V_{\text{ref}i} - V_i)$$

$$i = 1, \ldots, m \qquad (7.7)$$

Equation (7.4) has dimensions of torque in per-unit. When the stator transients were neglected, the electrical torque became equal to the per-unit power associated with the internal voltage source.

2. Algebraic Equations

The algebraic equations consist of the stator algebraic equations and the network equations. The stator algebraic equations directly follow from the dynamic equivalent circuit of Figure 6.5, which is reproduced in Figure 7.1. Application of Kirchhoff's Voltage Law (KVL) to Figure 7.1 yields the stator algebraic equations:

(a) Stator algebraic equations

$$0 = V_i e^{j\theta_i} + (R_{si} + jX'_{di})(I_{di} + jI_{qi})e^{j(\delta_i - \frac{\pi}{2})}$$

$$-[E'_{di} + (X'_{qi} - X'_{di})I_{qi} + jE'_{qi}]e^{j(\delta_i - \frac{\pi}{2})}$$

$$i = 1, \ldots, m \qquad (7.8)$$

(b) Network equations

The dynamic circuit, together with the static network and the loads, is shown in Figure 7.2. The network equations written at the n buses are in complex form. From (6.208) and (6.209), these network equations are

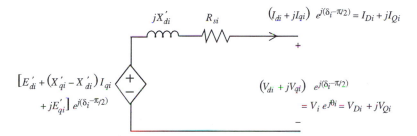

Figure 7.1: *Synchronous machine two-axis model dynamic circuit (i =* $1, \ldots, m$*)*

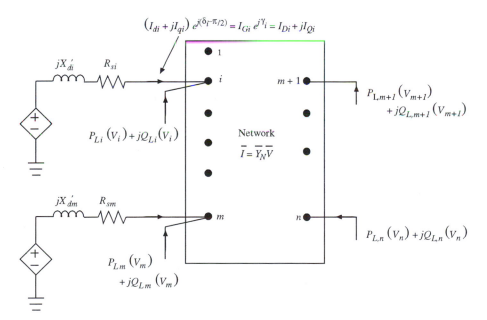

Figure 7.2: *Interconnection of synchronous machine dynamic circuit and the rest of the network*

Generator Buses

$$V_i e^{j\theta_i}(I_{di} - jI_{qi})e^{-j(\delta_i - \frac{\pi}{2})} + P_{Li}(V_i) + jQ_{Li}(V_i) = \sum_{k=1}^{n} V_i V_k Y_{ik} e^{j(\theta_i - \theta_k - \alpha_{ik})}$$

$$i = 1, \ldots, m \qquad\qquad\qquad\qquad\qquad (7.9)$$

Load Buses

$$P_{Li}(V_i) + jQ_{Li}(V_i) = \sum_{k=1}^{n} V_i V_k Y_{ik} e^{j(\theta_i - \theta_k - \alpha_{ik})} \quad i = m+1, \ldots, n \quad (7.10)$$

In (7.9), $V_i e^{j\theta_i}(I_{di} - jI_{qi})e^{-j(\delta_i - \pi/2)} \triangleq P_{Gi} + jQ_{Gi}$ is the complex power "injected" into bus i due to the generator. Thus, (7.9) and (7.10) are only the real and reactive power balance equation at all the n buses. Equation (7.9), which constitutes the power balance equations at the generator buses, shows the interaction of the algebraic variables and the state variables δ_i, E'_{qi}, and E'_{di}. We thus have

1. Seven differential equations (d.e.'s) for each machine, i.e., $7m$ d.e.'s ((7.1)–(7.7)).

2. One complex stator algebraic equation (7.8) (two real equations) for each machine, i.e., $2m$ real equations.

3. One complex network equation (7.9) and (7.10) (two real equations) at each network bus, i.e., $2n$ real equations.

We have $7m + 2m + 2n$ equations with $x = [x_1^t \ldots x_m^t]^t$ as the *state* vector where $x_i = [E'_{qi}\ E'_{di}\ \delta_i\ \omega_i\ E_{fdi}\ R_{fi}\ V_{Ri}]^t$ as the state vector for each machine. $y = [I_{d-q}^t\ V^t\ \theta^t]^t$ is the set of algebraic variables where

$$I_{d-q} = [I_{d1}\ I_{q1} \ldots I_{dm} I_{qm}]^t$$
$$V = [V_1 \ldots V_n]^t,\ \theta = [\theta_1 \ldots \theta_n]^t,\ \overline{V} = [\overline{V}_1 \ldots \overline{V}_n]^t$$

Functionally, therefore, the differential equations (7.1)–(7.7), together with the stator algebraic equations (7.8) and the network equations (7.9)–(7.10), form a set of differential-algebraic equations of the form

$$\dot{x} = f(x, y, u) \qquad\qquad\qquad (7.11)$$

$$0 = g(x, y) \qquad\qquad\qquad (7.12)$$

$u = [u_1^t \ldots u_m^t]^t$ with $u_i = [\omega_s\ T_{mi}\ V_{\text{ref}_i}]^t$ as the input vector for each machine. We now formally put (7.1)–(7.10) in the form (7.11) and (7.12).

7.2 Stator Algebraic Equations

There are several different ways of writing the stator algebraic equations (7.8) as two real equations for computational purposes. The idea is to express I_{di}, I_{qi} in terms of the state and network variables. Both the polar form and the rectangular form will be explained.

7.2.1 Polar form

In this form, the network voltages appear in polar form. If we multiply (7.8) by $e^{-j(\delta_i - \frac{\pi}{2})}$ and equate the real and imaginary parts, we obtain

$$E'_{di} - V_i \sin(\delta_i - \theta_i) - R_{si}I_{di} + X'_{qi}I_{qi} = 0 \quad i = 1, \ldots, m \quad (7.13)$$

$$E'_{qi} - V_i \cos(\delta_i - \theta_i) - R_{si}I_{qi} - X'_{di}I_{di} = 0 \quad i = 1, \ldots, m \quad (7.14)$$

We define

$$\begin{bmatrix} R_{si} & -X'_{qi} \\ X'_{di} & R_{si} \end{bmatrix} \triangleq Z_{d-q,i}$$

Then, from (7.13) and (7.14):

$$\begin{bmatrix} I_{di} \\ I_{qi} \end{bmatrix} = [Z_{d-q,i}]^{-1} \begin{bmatrix} E'_{di} - V_i \sin(\delta_i - \theta_i) \\ E'_{qi} - V_i \cos(\delta_i - \theta_i) \end{bmatrix} \quad i = 1, \ldots, m \quad (7.15)$$

Equations (7.13) and (7.14) are implicit in I_{di}, I_{qi}, whereas in (7.15) they are expressed explicitly in terms of the state variables x_i and the algebraic variables V_i, θ_i. Thus

$$\begin{bmatrix} I_{di} \\ I_{qi} \end{bmatrix} = h_{pi}(x_i, V_i, \theta_i) \quad i = 1, \ldots, m \quad (7.16)$$

7.2.2 Rectangular form

This can be easily derived by recognizing the fact that

$$\overline{V}_i = V_{Di} + jV_{Qi} = V_i e^{j\theta_i} = V_i \cos\theta_i + jV_i \sin\theta_i \quad (7.17)$$

By expanding (7.13) and (7.14) and noting from (7.17) that $V_{Di} = V_i \cos\theta_i$ and $V_{Qi} = V_i \sin\theta_i$, we obtain the implicit form in rectangular coordinates as

$$E'_{di} - V_{Di} \sin\delta_i + V_{Qi} \cos\delta_i - R_{si}I_{di} + X'_{qi}I_{qi} = 0 \quad (7.18)$$

$$E'_{qi} - V_{Di} \cos\delta_i - V_{Qi} \sin\delta_i - R_{si}I_{qi} - X'_{di}I_{di} = 0 \quad (7.19)$$

To obtain the explicit form, I_{di}, I_{qi} in (7.18) and (7.19) can be expressed in terms of E'_{di}, E'_{qi}, δ_i, V_{Di}, and V_{Qi}. Alternatively, the right-hand side of (7.15) is expanded as

$$\begin{bmatrix} I_{di} \\ I_{qi} \end{bmatrix} = [Z_{d-q,i}]^{-1} \begin{bmatrix} E'_{di} \\ E'_{qi} \end{bmatrix} - [Z_{d-q,i}]^{-1} \begin{bmatrix} V_i(\sin \delta_i \cos \theta_i - \cos \delta_i \sin \theta_i) \\ V_i(\cos \delta_i \cos \theta_i + \sin \delta_i \sin \theta_i) \end{bmatrix}$$

(7.20)

Using the fact from (7.17) that $V_{Di} = V_i \cos \theta_i$ and $V_{Qi} = V_i \sin \theta_i$, (7.20) becomes

$$\begin{bmatrix} I_{di} \\ I_{qi} \end{bmatrix} = [Z_{d-q,i}]^{-1} \begin{bmatrix} E'_{di} \\ E'_{qi} \end{bmatrix} - [Z_{d-q,i}]^{-1} \begin{bmatrix} \sin \delta_i & -\cos \delta_i \\ \cos \delta_i & \sin \delta_i \end{bmatrix} \begin{bmatrix} V_{Di} \\ V_{Qi} \end{bmatrix} \quad (7.21)$$

$$= h_{ri}(x_i, V_{Di}, V_{Qi}) \quad i = 1, \ldots, m \quad (7.22)$$

Note that (7.21) can be obtained directly from (7.18) and (7.19). Symbolically, (7.16) or (7.22) can be expressed for all machines as

$$I_{d-q} = h_p(x, V, \theta) \text{ or } h_r(x, V_D, V_Q)$$
$$\triangleq h(x, \overline{V}) \quad (7.23)$$

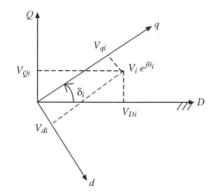

Figure 7.3: *Graphical representation*

7.2.3   Alternate form of stator algebraic equations

In much of the literature, a block diagram representation of stator equations is done through an "interface" block that reflects the machine–network transformation. The machine–network transformation is given by

$$\begin{bmatrix} F_{di} \\ F_{qi} \end{bmatrix} = \begin{bmatrix} \sin \delta_i & -\cos \delta_i \\ \cos \delta_i & \sin \delta_i \end{bmatrix} \begin{bmatrix} F_{Di} \\ F_{Qi} \end{bmatrix} \quad i = 1, \ldots, m \qquad (7.24)$$

and

$$\begin{bmatrix} F_{Di} \\ F_{Qi} \end{bmatrix} = \begin{bmatrix} \sin \delta_i & \cos \delta_i \\ -\cos \delta_i & \sin \delta_i \end{bmatrix} \begin{bmatrix} F_{di} \\ F_{qi} \end{bmatrix} \quad i = 1, \ldots, m \qquad (7.25)$$

where F may be either I or V. Figure 7.3 is a graphical representation of (7.24) and (7.25) illustrated for the voltage $\overline{V}_i = V_i e^{j\theta_i}$. Using (7.24) in (7.21), we obtain

$$\begin{bmatrix} I_{di} \\ I_{qi} \end{bmatrix} = [Z_{d-q,i}]^{-1} \begin{bmatrix} E'_{di} - V_{di} \\ E'_{qi} - V_{qi} \end{bmatrix} \quad i = 1, \ldots, m \qquad (7.26)$$

Thus

$$\begin{bmatrix} E'_{di} - V_{di} \\ E'_{qi} - V_{qi} \end{bmatrix} = [Z_{d-q,i}] \begin{bmatrix} I_{di} \\ I_{qi} \end{bmatrix} \quad i = 1, \ldots, m \qquad (7.27)$$

The interface block in the block diagram in Figure 7.5 is now consistent with (7.24), (7.25), and (7.26). Note that, in this formulation, algebraic equation (7.26) or (7.27) is in machine reference only, whereas (7.24) and (7.25) act as an "interface" between the machine and the network.

7.3   Network Equations

The network equations can be expressed either in power-balance or current-balance form. The latter form is more popular with the industry software packages. We discuss both of them now.

7.3.1   Power-balance form

The network equations for the generator buses ((7.9)) are separated into real and imaginary parts for $i = 1, \ldots, m$

$$I_{di}V_i \sin(\delta_i - \theta_i) + I_{qi}V_i \cos(\delta_i - \theta_i) + P_{Li}(V_i)$$

$$-\sum_{k=1}^{n} V_i V_k Y_{ik} \cos(\theta_i - \theta_k - \alpha_{ik}) = 0 \qquad (7.28)$$

$$I_{di}V_i \cos(\delta_i - \theta_i) - I_{qi}V_i \sin(\delta_i - \theta_i) + Q_{Li}(V_i)$$

$$-\sum_{k=1}^{n} V_i V_k Y_{ik} \sin(\theta_i - \theta_k - \alpha_{ik}) = 0 \qquad (7.29)$$

For the load buses, a similiar procedure using (7.10) gives for $i = m+1, \ldots, n$

$$P_{Li}(V_i) - \sum_{k=1}^{n} V_i V_k Y_{ik} \cos(\theta_i - \theta_k - \alpha_{ik}) = 0 \qquad (7.30)$$

$$Q_{Li}(V_i) - \sum_{k=1}^{n} V_i V_k Y_{ik} \sin(\theta_i - \theta_k - \alpha_{ik}) = 0 \qquad (7.31)$$

Note that load can be present at the generator as well as at the load buses. The network equations (7.28)–(7.31) can be rearranged so that the real power equations appear first and the reactive power equations appear next, as follows.

Real Power Equations

$$I_{di}V_i \sin(\delta_i - \theta_i) + I_{qi}V_i \cos(\delta_i - \theta_i) + P_{Li}(V_i)$$

$$-\sum_{k=1}^{n} V_i V_k Y_{ik} \cos(\theta_i - \theta_k - \alpha_{ik}) = 0 \qquad i = 1, \ldots, m \,(7.32)$$

$$P_{Li}(V_i) - \sum_{k=1}^{n} V_i V_k Y_{ik} \cos(\theta_i - \theta_k - \alpha_{ik}) = 0 \qquad i = m+1, \ldots, n(7.33)$$

Reactive Power Equations

$$I_{di}V_i \cos(\delta_i - \theta_i) - I_{qi}V_i \sin(\delta_i - \theta_i) + Q_{Li}(V_i)$$

$$-\sum_{k=1}^{n} V_i V_k Y_{ik} \sin(\theta_i - \theta_k - \alpha_{ik}) = 0 \qquad (7.34)$$

$$Q_{Li}(V_i) - \sum_{k=1}^{n} V_i V_k Y_{ik} \sin(\theta_i - \theta_k - \alpha_{ik}) = 0 \quad i = m+1, \ldots, n \quad (7.35)$$

Thus, the differential-algebraic equation (DAE) model is:

1. The differential equations (7.1)–(7.7)

2. The stator algebraic equations of the form (7.23) in the polar form

3. The network equations (7.32)–(7.35) in the power-balance form

The differential-algebraic equations are now written symbolically as

$$\dot{x} = f_o(x, I_{d-q}, \overline{V}, u) \tag{7.36}$$

$$I_{d-q} = h(x, \overline{V}) \tag{7.37}$$

$$0 = g_o(x, I_{d-q}, \overline{V}) \tag{7.38}$$

Substitution of (7.37) into (7.36) and (7.38) gives

$$\dot{x} = f_1(x, \overline{V}, u) \tag{7.39}$$

$$0 = g_1(x, \overline{V}) \tag{7.40}$$

Note that (7.40) is in the power-balance form. This is the differential-algebraic equation (DAE) analytical model with the network algebraic variables in the polar form. We prefer this form, since in load-flow equations the voltages are generally in polar form. Simplified forms of this model result from the reduced-order model of the synchronous machine as well as the exciter, which will be discussed later.

7.3.2 Current-balance form

Instead of the power-balance form of (7.32)–(7.35), one can have the current-balance form, which is essentially the nodal set of equations

$$\overline{I} = \overline{Y}_N \overline{V} \tag{7.41}$$

where $\overline{Y}_N$ is the $n \times n$ bus admittance matrix of the network with elements $\overline{Y}_{ik} = Y_{ik} e^{j\alpha_{ik}} = G_{ik} + jB_{ik}$, $\overline{I}$ is the net injected current vector and $\overline{V}$ is the bus voltage vector. Depending on how $\overline{I}$ is expressed, it can take different forms, as discussed below. Equation (7.41) can also be derived from (7.9)–(7.10) by dividing both sides of the equation by $V_i e^{j\theta_i}$ and then taking the complex conjugate as follows.

$$(I_{di} + jI_{qi})e^{j(\delta_i - \pi/2)} + \frac{P_{Li}(V_i) - jQ_{Li}(V_i)}{V_i e^{-j\theta_i}} = \sum_{k=1}^{n} Y_{ik} e^{j\alpha_{ik}} V_k e^{j\theta_k}$$

$$i = 1, \ldots, m \qquad (7.42)$$

$$\frac{P_{Li}(V_i) - jQ_{Li}(V_i)}{V_i e^{-j\theta_i}} = \sum_{k=1}^{n} Y_{ik} e^{j\alpha_{ik}} V_k e^{j\theta_k} \quad i = m+1, \ldots, n \qquad (7.43)$$

These equations are the same as (6.83) and (6.84). Equations (7.42) and (7.43) can be symbolically denoted in matrix form as

$$\overline{I}_o(I_{d-q}, x, \overline{V}) = \overline{Y}_N \overline{V} \qquad (7.44)$$

The other algebraic equation is

$$I_{d-q} = h(x, \overline{V}) \qquad (7.45)$$

Substitution of (7.45) in (7.36) and (7.44) leads to the DAE model

$$\begin{aligned} \dot{x} &= f_1(x, \overline{V}, u) \\ \overline{I}_1(x, \overline{V}) &= \overline{Y}_N \overline{V} \end{aligned} \qquad (7.46)$$

Example 7.1

We illustrate the DAE models discussed in the previous section with a numerical example. We consider the popular Western System Coordinating Council (WSCC) 3-machine, 9-bus system [73] shown in Figure 7.4. This is also the system appearing in [74] and widely used in the literature. The base MVA is 100, and system frequency is 60 Hz. The converged load-flow data obtained using the EPRI-IPFLOW program [75] is given in Table 7.1.

The $\overline{Y}_{bus}$ for the network (also denoted as $\overline{Y}_N$) can be written by inspection from Figure 7.4 and is shown in Table 7.2. The machine data and the exciter data are given in Table 7.3. The exciter is assumed to be identical for all the machines and is of the IEEE-Type I. Define $\frac{2H_i}{\omega_s} \triangleq M_i$. Assume that $\frac{D_1}{M_1} = 0.1$, $\frac{D_2}{M_2} = 0.2$, and $\frac{D_3}{M_3} = 0.3$ (all in pu).

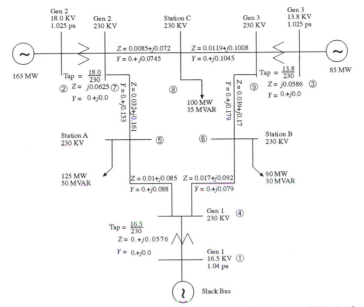

Figure 7.4: *WSCC 3-machine, 9-bus system; the value of Y is half the line charging (Copyright 1977. Electric Power Research Institute. EPRI EL-0484. Power System Dynamic Analysis, Phase I. Reprinted with Permission.).*

Table 7.1: Load-Flow Results of the WSCC 3-Machine, 9-Bus System

	Bus #	Voltage (pu)	P_G (pu)	Q_G (pu)	$-P_L$ (pu)	$-Q_L$ (pu)
1	(swing)	1.04	0.716	0.27	–	–
2	(P-V)	$1.025\angle 9.3°$	1.63	0.067	–	–
3	(P-V)	$1.025\angle 4.7°$	0.85	-0.109	–	–
4	(P-Q)	$1.026\angle -2.2°$	–	–	–	–
5	(")	$0.996\angle -4.0°$	–	–	1.25	0.5
6	(")	$1.013\angle -3.7°$	–	–	0.9	0.3
7	(")	$1.026\angle 3.7°$	–	–	–	–
8	(")	$1.016\angle 0.7°$	–	–	1.00	0.35
9	(")	$1.032\angle 2.0°$	–	–	–	–

Table 7.2: $\overline{Y}_N$ for the Network in Figure 7.4

	1	2	3	4	5	6	7	8	9
1	$-j17.361$	0	0	$j17.361$	0	0	0	0	0
2	0	$-j16$	0	0	0	0	$j16$	0	0
3	0	0	$-j17.065$	0	0	0	0	0	$j17.065$
4	$j17.361$	0	0	3.307 $-j39.309$	-1.365 $+j11.604$	-1.942 $+j10.511$	0	0	0
5	0	0	0	-1.365 $+j11.604$	2.553 $-j17.338$	0	-1.188 $+j5.975$	0	0
6	0	0	0	-1.942 $+j10.511$	0	3.224 $-j15.841$	0	0	-1.282 $+j5.588$
7	0	$j16$	0	0	-1.188 $+j5.975$	0	2.805 $-j35.4460$	-1.617 $+j13.698$	0
8	0	0	0	0	0	0	-1.617 $+j13.698$	2.772 $-j23.303$	-1.155 $+j9.784$
9	0	0	$j17.065$	0	0	-1.282 $+j5.588$	0	-1.155 $+j9.784$	2.437 $-j32.1540$

Table 7.3: Machine and Exciter Data

Machine Data

Parameters	M/C 1	M/C 2	M/C 3
$H(\text{secs})$	23.64	6.4	3.01
$X_d(\text{pu})$	0.146	0.8958	1.3125
$X_d'(\text{pu})$	0.0608	0.1198	0.1813
$X_q(\text{pu})$	0.0969	0.8645	1.2578
$X_q'(\text{pu})$	0.0969	0.1969	0.25
$T_{do}'(\text{sec})$	8.96	6.0	5.89
$T_{qo}'(\text{sec})$	0.31	0.535	0.6

Exciter Data

Parameters	Exciter 1	Exciter 2	Exciter 3
K_A	20	20	20
$T_A(\text{sec})$	0.2	0.2	0.2
K_E	1.0	1.0	1.0
$T_E(\text{sec})$	0.314	0.314	0.314
K_F	0.063	0.063	0.063
$T_F(\text{sec})$	0.35	0.35	0.35

$$S_{Ei}(E_{fdi}) = 0.0039e^{1.555E_{fdi}} \quad i = 1, 2, 3$$

The differential equations corresponding to (7.1)–(7.7) are

$$
\begin{bmatrix} \dot{E}'_{qi} \\ \dot{E}'_{di} \\ \dot{\delta}_i \\ \dot{\omega}_i \\ \dot{E}_{fdi} \\ \dot{R}_{fi} \\ \dot{V}_{Ri} \end{bmatrix} = [A_i] \begin{bmatrix} E'_{qi} \\ E'_{di} \\ \delta_i \\ \omega_i \\ E_{fdi} \\ R_{fi} \\ V_{Ri} \end{bmatrix} + R_i(E'_{qi}, E'_{di}, E_{fdi}, I_{di}, I_{qi}, V_i) + C_i u_i
$$

$$i = 1, 2, 3$$

$$(7.47)$$

where

$$
A_i = \begin{bmatrix}
\frac{-1}{T'_{doi}} & 0 & 0 & 0 & \frac{1}{T'_{doi}} & 0 & 0 \\
0 & \frac{-1}{T'_{qoi}} & 0 & 0 & 0 & 0 & 0 \\
0 & 0 & 0 & 1 & 0 & 0 & 0 \\
0 & 0 & 0 & \frac{-D_i}{M_i} & 0 & 0 & 0 \\
0 & 0 & 0 & 0 & -\frac{K_{Ei}}{T_{Ei}} & 0 & \frac{1}{T_{Ei}} \\
0 & 0 & 0 & 0 & \frac{K_{Fi}}{T^2_{Fi}} & \frac{-1}{T_{Fi}} & 0 \\
0 & 0 & 0 & 0 & \frac{-K_{Ai}K_{Fi}}{T_{Ai}T_{Fi}} & \frac{K_{Ai}}{T_{Ai}} & \frac{-1}{T_{Ai}}
\end{bmatrix}
$$

$$i = 1, 2, 3$$

$$(7.48)$$

$$
R_i = \begin{bmatrix}
\frac{-(X_{di}-X'_{di})I_{di}}{T'_{doi}} \\
\frac{(X_{qi}-X'_{qi})I_{qi}}{T'_{qoi}} \\
0 \\
\frac{-\omega_s}{2H_i}[(E'_{di}I_{di} + E'_{qi}I_{qi}) + (X'_{qi} - X'_{di})I_{di}I_{qi}] \\
-\frac{S_{Ei}(E_{fdi})}{T_{Ei}} \\
0 \\
\frac{-K_{Ai}}{T_{Ai}}V_i
\end{bmatrix}
$$

$$
C_i \;=\; \begin{bmatrix} 0 & 0 & 0 \\ 0 & 0 & 0 \\ -1 & 0 & 0 \\ \frac{D_i}{M_i} & \frac{1}{M_i} & 0 \\ 0 & 0 & 0 \\ 0 & 0 & 0 \\ 0 & 0 & \frac{K_{Ai}}{T_{Ai}} \end{bmatrix}
\qquad
u_i \;=\; \begin{bmatrix} \omega_s \\ T_{Mi} \\ V_{\mathrm{ref}i} \end{bmatrix}
\qquad i = 1,2,3 \qquad (7.49)
$$

Substituting the numerical values, we obtain

$$
A_1 = \begin{bmatrix}
-0.112 & 0 & 0 & 0 & 0.112 & 0 & 0 \\
0 & -3.226 & 0 & 0 & 0 & 0 & 0 \\
0 & 0 & 0 & 1 & 0 & 0 & 0 \\
0 & 0 & 0 & -0.1 & 0 & 0 & 0 \\
0 & 0 & 0 & 0 & -3.185 & 0 & 3.185 \\
0 & 0 & 0 & 0 & 0.514 & -2.86 & 0 \\
0 & 0 & 0 & 0 & -18 & 100 & -5
\end{bmatrix}
$$

$$
A_2 = \begin{bmatrix}
-0.167 & 0 & 0 & 0 & 0.167 & 0 & 0 \\
0 & -1.87 & 0 & 0 & 0 & 0 & 0 \\
0 & 0 & 0 & 1 & 0 & 0 & 0 \\
0 & 0 & 0 & -0.2 & 0 & 0 & 0 \\
0 & 0 & 0 & 0 & -3.185 & 0 & 3.185 \\
0 & 0 & 0 & 0 & 0.514 & -2.86 & 0 \\
0 & 0 & 0 & 0 & -18 & 100 & -5
\end{bmatrix}
$$

$$
A_3 = \begin{bmatrix}
-0.17 & 0 & 0 & 0 & 0.17 & 0 & 0 \\
0 & -1.67 & 0 & 0 & 0 & 0 & 0 \\
0 & 0 & 0 & 1 & 0 & 0 & 0 \\
0 & 0 & 0 & -0.3 & 0 & 0 & 0 \\
0 & 0 & 0 & 0 & -3.185 & 0 & 3.185 \\
0 & 0 & 0 & 0 & 0.514 & -2.86 & 0 \\
0 & 0 & 0 & 0 & -18 & 100 & -5
\end{bmatrix}
\qquad (7.50)
$$

$$R_1 = \begin{bmatrix} -0.0095I_{d1} \\ 0 \\ 0 \\ -8(E'_{d1}I_{d1} + E'_{q1}I_{q1}) \\ -0.29I_{d1}I_{q1} \\ 0.0124e^{1.555E_{fd1}} \\ 0 \\ -100V_1 \end{bmatrix}, \quad R_2 = \begin{bmatrix} -0.13I_{d2} \\ 1.25I_{q2} \\ 0 \\ -29.5(E'_{d2}I_{d2} + E'_{q2}I_{q2}) \\ -2.27I_{d2}I_{q2} \\ 0.0124e^{1.555E_{fd2}} \\ 0 \\ -100V_2 \end{bmatrix},$$

$$R_3 = \begin{bmatrix} -0.19I_{d3} \\ 1.7I_{q3} \\ 0 \\ -62.6(E'_{d3}I_{d3} + E'_{q3}I_{q3}) \\ -4.3I_{d3}I_{q3} \\ 0.0124e^{1.555E_{fd3}} \\ 0 \\ -100V_3 \end{bmatrix} \tag{7.51}$$

$$C_1 = \begin{bmatrix} 0 & 0 & 0 \\ 0 & 0 & 0 \\ -1 & 0 & 0 \\ 0.1 & 8 & 0 \\ 0 & 0 & 0 \\ 0 & 0 & 0 \\ 0 & 0 & 100 \end{bmatrix}, \quad C_2 = \begin{bmatrix} 0 & 0 & 0 \\ 0 & 0 & 0 \\ -1 & 0 & 0 \\ 0.2 & 29.5 & 0 \\ 0 & 0 & 0 \\ 0 & 0 & 0 \\ 0 & 0 & 100 \end{bmatrix},$$

$$C_3 = \begin{bmatrix} 0 & 0 & 0 \\ 0 & 0 & 0 \\ -1 & 0 & 0 \\ 0.3 & 62.6 & 0 \\ 0 & 0 & 0 \\ 0 & 0 & 0 \\ 0 & 0 & 100 \end{bmatrix} \tag{7.52}$$

The stator algebraic equations corresponding to (7.13) and (7.14) (assuming $R_{si} \equiv 0$) are

$$E'_{d1} - V_1 \sin(\delta_1 - \theta_1) + 0.0969I_{q1} = 0$$

$$E'_{q1} - V_1 \cos(\delta_1 - \theta_1) - 0.0608I_{d1} = 0$$

$$E'_{d2} - V_2 \sin(\delta_2 - \theta_2) + 0.1969I_{q2} = 0$$

$$E'_{q2} - V_2 \cos(\delta_2 - \theta_2) - 0.1198I_{d2} = 0$$

$$E'_{d3} - V_3 \sin(\delta_3 - \theta_3) + 0.2500I_{q3} = 0$$

$$E'_{q3} - V_3 \cos(\delta_3 - \theta_3) - 0.1813I_{d3} = 0 \qquad (7.53)$$

The network equations are (with the notation $\theta_{ij} = \theta_i - \theta_j$) as follows. The constant power loads are treated as injected into the buses.

Real Power Equations

$$I_{d1}V_1 \sin(\delta_1 - \theta_1) + I_{q1}V_1 \cos(\delta_1 - \theta_1)$$
$$- 17.36V_1V_4 \sin\theta_{14} = 0$$
$$I_{d2}V_2 \sin(\delta_2 - \theta_2) + I_{q2}V_2 \cos(\delta_2 - \theta_2)$$
$$- 16.00\ V_2V_7 \sin\theta_{27} = 0$$
$$I_{d3}V_3 \sin(\delta_3 - \theta_3) + I_{q3}V_3 \cos(\delta_3 - \theta_3)$$
$$- 17.06V_3V_9 \sin\theta_{39} = 0$$
$$-17.36V_4V_1 \sin\theta_{41} - 3.31V_4^2 + 1.36V_4V_5 \cos\theta_{45} - 11.6V_4V_5 \sin\theta_{45}$$
$$+ 1.942V_4V_6 \cos\theta_{46} - 10.51V_4V_6 \sin\theta_{46} = 0$$
$$-1.25 + 1.36V_5V_4 \cos\theta_{54} - 11.6V_5V_4 \sin\theta_{54} + 1.19V_5V_7 \cos\theta_{57}$$
$$- 5.97V_5V_7 \sin\theta_{57} - 2.55V_5^2 = 0$$
$$-0.9 + 1.94V_6V_4 \cos\theta_{64} - 10.51V_6V_4 \sin\theta_{64} - 3.22V_6^2$$
$$+ 1.28V_6V_9 \cos\theta_{69} - 5.59V_6V_9 \sin\theta_{69} = 0$$
$$-16V_7V_2 \sin\theta_{72} + 1.19V_7V_5 \cos\theta_{75} - 5.98V_7V_5 \sin\theta_{75}$$
$$- 2.8V_7^2 + 1.62V_7V_8 \cos\theta_{78}$$
$$- 13.7V_7V_8 \sin\theta_{78} = 0$$
$$-1 + 1.62V_8V_7 \cos\theta_{87} - 13.7V_8V_7 \sin\theta_{87} - 2.77V_8^2$$
$$+ 1.16V_8V_9 \cos\theta_{89} - 9.8V_8V_9 \sin\theta_{89} = 0$$
$$-17.065V_9V_3 \sin\theta_{93} + 1.28V_9V_6 \cos\theta_{96} - 5.59V_9V_6 \sin\theta_{96}$$

$$+ 1.15 V_9 V_8 \cos \theta_{98} - 9.78 V_9 V_8 \sin \theta_{98}$$
$$- 2.4 V_9^2 = 0 \tag{7.54}$$

Reactive Power Equations

$$I_{d1} V_1 \cos(\delta_1 - \theta_1) - I_{q1} V_1 \sin(\delta_1 - \theta_1)$$
$$+ 17.36 V_1 V_4 \cos \theta_{14} - 17.36 V_1^2 = 0$$

$$I_{d2} V_2 \cos(\delta_2 - \theta_2) - I_{q2} V_2 \sin(\delta_2 - \theta_2)$$
$$+ 16 V_2 V_7 \cos \theta_{27} - 16 V_2^2 = 0$$

$$I_{d3} V_3 \cos(\delta_3 - \theta_3) - I_{q3} V_3 \sin(\delta_3 - \theta_3)$$
$$+ 17.07 V_3 V_9 \cos \theta_{39} - 17.07 V_3^2 = 0$$

$$17.36 V_4 V_1 \cos \theta_{41} - 39.3 V_4^2 + 1.36 V_4 V_5 \sin \theta_{45}$$
$$+ 11.6 V_4 V_5 \cos \theta_{45} + 1.94 V_4 V_6 \sin \theta_{46}$$
$$+ 10.52 V_4 V_6 \cos \theta_{46} = 0$$

$$-0.5 + 1.37 V_5 V_4 \sin \theta_{54} + 11.6 V_5 V_4 \cos \theta_{54} - 17.34 V_5^2$$
$$+ 1.19 V_5 V_7 \sin \theta_{57} + 5.98 V_5 V_7 \cos \theta_{57} = 0$$

$$-0.3 + 1.94 V_6 V_4 \sin \theta_{64} + 10.51 V_6 V_4 \cos \theta_{64} - 15.84 V_6^2$$
$$+ 1.28 V_6 V_9 \sin \theta_{69} + 5.59 V_6 V_9 \cos \theta_{69} = 0$$

$$16 V_7 V_2 \cos \theta_{72} + 1.19 V_7 V_5 \sin \theta_{75} + 5.98 V_7 V_5 \cos \theta_{75}$$
$$- 35.45 V_7^2 + 1.62 V_7 V_8 \sin \theta_{78} + 13.67 V_7 V_8 \cos \theta_{78} = 0$$

$$-0.35 + 1.62 V_8 V_7 \sin \theta_{87} + 13.67 V_8 V_7 \cos \theta_{87} - 23.3 V_8^2 + 1.15 V_8 V_9 \sin \theta_{89}$$
$$+ 9.78 V_8 V_9 \cos \theta_{89} = 0$$

$$17.065 V_9 V_3 \cos \theta_{93} + 1.28 V_9 V_6 \sin \theta_{96} + 5.59 V_9 V_6 \cos \theta_{96}$$
$$+ 1.16 V_9 V_8 \sin \theta_{98} + 9.78 V_9 V_8 \cos \theta_{98} - 32.15 V_9^2 = 0 \tag{7.55}$$

It is easy to solve (7.53) for I_{di}, I_{qi} ($i = 1,2,3$), substitute them in (7.47) and (7.54)–(7.55), and obtain the equations $\dot{x} = f_1(x, \overline{V}, u)$ and $0 = g_1(x, \overline{V})$. This is left as an exercise for the reader.

□

Example 7.2

In this example, we put the DAE model with the network equations in the current-balance form.

The differential equations (7.47) and the stator algebraic equations (7.53) are unchanged. The network equations are

$$\overline{I}_o(I_{d-q}, x, \overline{V}) = \overline{Y}_N \overline{V} \tag{7.56}$$

where

$$\overline{I}_o(I_{d-q}, x, \overline{V}) = \begin{bmatrix} (I_{d1} + jI_{q1})e^{j(\delta_1 - \pi/2)} \\ (I_{d2} + jI_{q2})e^{j(\delta_2 - \pi/2)} \\ (I_{d3} + jI_{q3})e^{j(\delta_3 - \pi/2)} \\ 0 + j0 \\ (-1.25 + j0.5)/\overline{V}_5^* \\ (-0.9 + j0.3)/\overline{V}_6^* \\ 0 + j0 \\ (-1 + j0.35)/\overline{V}_8^* \\ 0 + j0 \end{bmatrix} \tag{7.57}$$

It is an easy exercise to substitute I_{d-q} from (7.53) in (7.57) to obtain the network equations for the form $\overline{I}_1(x, \overline{V}) = \overline{Y}_N \overline{V}$.

□

7.4 Industry Model

We now present an equivalent but alternative formulation that is used widely in commercial power system simulation packages [72]. The principal difference is in terms of suitably rearranging the equations from a programming point of view. This will be referred to as the *industry* model. From (7.15) we have

$$\begin{bmatrix} I_{di} \\ I_{qi} \end{bmatrix} = [Z_{d-q,i}]^{-1} \begin{bmatrix} E'_{di} - V_i \sin(\delta_i - \theta_i) \\ E'_{qi} - V_i \cos(\delta_i - \theta_i) \end{bmatrix} \quad i = 1, \ldots, m \tag{7.58}$$

Also from (7.21):

$$
\begin{bmatrix} I_{di} \\ I_{qi} \end{bmatrix} = [Z_{d-q,i}]^{-1} \begin{bmatrix} E'_{di} \\ E'_{qi} \end{bmatrix} - [Z_{d-q,i}]^{-1} \begin{bmatrix} \sin\,\delta_i & -\cos\,\delta_i \\ \cos\,\delta_i & \sin\,\delta_i \end{bmatrix} \begin{bmatrix} V_{Di} \\ V_{Qi} \end{bmatrix}
$$
$$
i = 1, \ldots, m \tag{7.59}
$$

Hence, I_{di}, I_{qi} are functions of either $(E'_{di}, E'_{qi}, \delta_i, V_i, \theta_i)$ or $(E'_{di},\ E'_{qi},\ \delta_i, V_{Di}, V_{Qi})$. For ease in programming, the electric power output of machine i in (7.4) is defined as

$$
\begin{aligned}
P_{ei} &\triangleq E'_{di}I_{di} + E'_{qi}I_{qi} + (X'_{qi} - X'_{di})I_{di}I_{qi} \\
&= P_{epi}(E'_{di}, E'_{qi}, \delta_i, V_i, \theta_i) \text{ or } P_{eri}(E'_{di}, E'_{qi}, \delta_i, V_{Di}, V_{Qi})
\end{aligned} \tag{7.60}
$$

if we substitute for I_{di} and I_{qi} from (7.58) or (7.59). The terminal voltage V_i is

$$
V_i = \sqrt{V_{Di}^2 + V_{Qi}^2} = \sqrt{V_{di}^2 + V_{qi}^2} \tag{7.61}
$$

We define two vectors

$$
E \triangleq [E'_{d1}\ E'_{q1} \ldots E'_{dm}\ E'_{qm}\ \delta_1 \ldots \delta_m]^t \tag{7.62}
$$

and

$$
W \triangleq [I_{d1}\ I_{q1} \ldots I_{dm}\ I_{qm}\ P_{e1} \ldots P_{em}\ V_1 \ldots V_m]^t \tag{7.63}
$$

E is a subset of the state vector x and W is a vector of algebraic variables. With these definitions, we can express the differential equations (7.1)–(7.7) as

$$
\dot{x} = F(x, W, u) \tag{7.64}
$$

The stator algebraic equations (7.58) or (7.59), together with (7.60) and (7.61), are expressed as

$$
W = G(E, \overline{V}) \tag{7.65}
$$

The network equations from (7.46) are

$$\overline{I}_2(E, \overline{V}) = \overline{Y}_N \overline{V} \tag{7.66}$$

since E is a subset of the vector x. Thus, assembling (7.64), (7.65), and (7.66) results in

$$\dot{x} = F(x, W, u) \tag{7.67}$$
$$W = G(E, \overline{V}) \tag{7.68}$$
$$\overline{I}_2(E, \overline{V}) = \overline{Y}_N \overline{V} \tag{7.69}$$

Equation (7.67) has the structure

$$\dot{x} = A(x)x + BW + Cu \tag{7.70}$$

The only dependency of $A(x)$ on x comes through the saturation function in the exciter. $A(x)$, B, and C are matrices having a block structure with each block belonging to a machine. G contains the stator algebraic equations and intermediate equations for P_{ei} and V_i in terms of $(E, \overline{V})$. This formulation is easy to program. Alternatively, we can substitute (7.68) in (7.67) to obtain

$$\dot{x} = f_1(x, \overline{V}, u)$$
$$\overline{I}_1(x, \overline{V}) = \overline{Y}_N \overline{V} \tag{7.71}$$

which is precisely equal to (7.46).

The DAE model with the network equations in the current-balance form is the preferred industry model, since the network-admittance matrix has to be refactored only if there is a network change. Otherwise, the initial factorization will remain. The DAE model with network equations in power–balance form (discussed in Section 7.3.1) has the advantage that the Jacobian of the network equations contains the power flow Jacobian, a fact useful in small-signal analysis and voltage collapse studies, as discussed in Chapter 8.

Equations (7.67)–(7.69) can be interpreted as a block diagram, as shown in Figure 7.5.

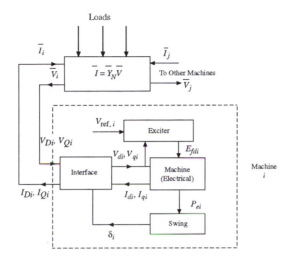

Figure 7.5: *Block diagram conceptualization of (7.67)–(7.69)*

Example 7.3

We will express the 3-machine system in Example 7.1 in the form of (7.67)–(7.69). Thus (7.67)–(7.68) become

$$
\begin{bmatrix} \dot{x}_1 \\ \dot{x}_2 \\ \dot{x}_3 \end{bmatrix} = \begin{bmatrix} A_1 & 0 & 0 \\ 0 & A_2 & 0 \\ 0 & 0 & A_3 \end{bmatrix} \begin{bmatrix} x_1 \\ x_2 \\ x_3 \end{bmatrix} + \begin{bmatrix} B_1 & 0 & 0 \\ 0 & B_2 & 0 \\ 0 & 0 & B_3 \end{bmatrix} \begin{bmatrix} W_1 \\ W_2 \\ W_3 \end{bmatrix} + Cu
$$

$$(7.72)$$

where

$$
A_i = \begin{bmatrix}
\frac{-1}{T'_{doi}} & 0 & 0 & 0 & \frac{1}{T'_{doi}} & 0 & 0 \\
0 & \frac{-1}{T'_{qoi}} & 0 & 0 & 0 & 0 & 0 \\
0 & 0 & 0 & 1 & 0 & 0 & 0 \\
0 & 0 & 0 & \frac{-D_i}{M_i} & 0 & 0 & 0 \\
0 & 0 & 0 & 0 & \frac{-(K_{Ei}+S_{Ei}(E_{fdi}))}{T_{Ei}} & 0 & \frac{1}{T_{Ei}} \\
0 & 0 & 0 & 0 & \frac{K_{Fi}}{T^2_{Fi}} & \frac{-1}{T_{Fi}} & 0 \\
0 & 0 & 0 & 0 & \frac{-K_{Ai}K_{Fi}}{T_{Fi}T_{Ai}} & \frac{K_{Ai}}{T_{Ai}} & \frac{-1}{T_{Ai}}
\end{bmatrix} \quad i = 1, 2, 3
$$

$$(7.73)$$

$$
B_i = \begin{bmatrix} \frac{-(X_{di}-X'_{di})}{T'_{doi}} & 0 & 0 & 0 \\ 0 & \frac{(X_{qi}-X'_{qi})}{T'_{qoi}} & 0 & 0 \\ 0 & 0 & 0 & 0 \\ 0 & 0 & \frac{-1}{M_i} & 0 \\ 0 & 0 & 0 & 0 \\ 0 & 0 & 0 & 0 \\ 0 & 0 & 0 & -\frac{K_{Ai}}{T_{Ai}} \end{bmatrix} \quad i = 1, 2, 3
$$

$$
W_i = \begin{bmatrix} I_{di} \\ I_{qi} \\ P_{ei} \\ V_i \end{bmatrix} \quad i = 1, 2, 3 \tag{7.74}
$$

C is $\mathrm{Diag}(C_i)$ and $u = \mathrm{Diag}(u_i)$, where C_i and u_i are given by (7.49). The stator algebraic equations and the intermediate equations corresponding to (7.68) are

$$
W_i = \begin{bmatrix} [Z_{d-q,i}]^{-1} \begin{bmatrix} E'_{di} \\ E'_{qi} \end{bmatrix} - [Z_{d-q,i}]^{-1} \begin{bmatrix} \sin \delta_i & -\cos \delta_i \\ \cos \delta_i & \sin \delta_i \end{bmatrix} \begin{bmatrix} V_{Di} \\ V_{Qi} \end{bmatrix} \\ E'_{di}I_{di} + E'_{qi}I_{qi} + (X'_{qi} - X'_{di})I_{di}I_{qi} \\ \sqrt{V_{Di}^2 + V_{Qi}^2} \end{bmatrix}
$$

$$
\qquad\qquad\qquad\qquad\qquad\qquad\qquad\qquad\qquad i = 1, 2, 3 \tag{7.75}
$$

If I_{di} and I_{qi} in P_{ei} are expressed from the first two equations, then $W_i = G_i(E_i, \overline{V}_i)$. A similar substitution for I_{di}, I_{qi} in (7.56) and (7.57) leads to

$$
\overline{I}_1(x, \overline{V}) = \overline{Y}_N \overline{V} \tag{7.76}
$$

Substitution of numerical values in the above equations is trivial, and is left as an exercise for the reader.

$\square$

Before we discuss the important issue of initial condition computation, we consider two simplifications that yield models far simpler than the two-axis model. They are also amenable to a network theoretic approach. Using these simplifications, we develop later (1) the structure-preserving flux-decay model, (2) the structure-preserving classical model, and (3) the internal node model using the classical machine model and constant impedance loads. Both (1) and (2) can have nonlinear load representations.

7.5   Simplification of the Two-Axis Model

Two simplifications can be made in the two-axis model—one regarding X'_d and X'_q, and the other regarding the nature of loads. These can be done independently or together, resulting in simplified models.

<u>Simplification #1</u> (neglecting transient saliency in the synchronous machine)

Transient saliency corresponds to different values of X'_{di} and X'_{qi}. If $X'_{qi} = X'_{di}$, then the stator algebraic equation (7.8) is simplified as

$$0 = V_i e^{j\theta_i} + (R_{si} + jX'_{di})(I_{di} + jI_{qi})e^{j(\delta_i - \pi/2)}$$
$$-(E'_{di} + jE'_{qi})e^{j(\delta_i - \pi/2)} \tag{7.77}$$

The equivalent circuit for machine i is shown in (Figure 7.6).

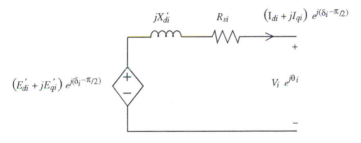

Figure 7.6: *Equivalent circuit with $X'_{di} = X'_{qi}$*

The machine–network transformation gives

$$(E'_{di} + jE'_{qi})e^{j(\delta_i - \pi/2)} = E'_{Di} + jE'_{Qi} \tag{7.78}$$
$$(I_{di} + jI_{qi})e^{j(\delta_i - \pi/2)} = I_{Di} + jI_{Qi} \tag{7.79}$$

Thus, with all the quantities in the synchronous reference frame, we have the equivalent circuit (Figure 7.7). Writing the KVL equation for Figure 7.7, we get $(E'_{Di} + jE'_{Qi}) = (R_{si} + jX'_{di})(I_{Di} + jI_{Qi}) + V_i e^{j\theta_i}$, which is equivalent to (7.77).

In the differential equations, the expression for electric power P_{ei} given by (7.60) is simplified as

$$P_{ei} = E'_{di}I_{di} + E'_{qi}I_{qi} \tag{7.80}$$

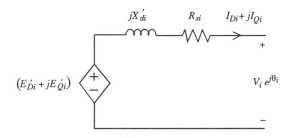

Figure 7.7: *Equivalent circuit (all quantities in network reference frame)*

It can be verified that the right-hand side of (7.80) is also equal to $E'_{Di}I_{Di} + E'_{Qi}I_{Qi}$ by taking the complex conjugate of (7.79), multiplying it by (7.78), and equating the real part.

This assumption of $X'_{di} = X'_{qi}$ is often called "neglecting transient salien-cy." The advantage is that, in the equivalent circuit, all variables are in the network reference frame, and it is particularly useful in the current-balance form of Section 7.3.2. In (7.46), the algebraic equations that are equivalent to (7.42) and (7.43) can be directly expressed in terms of I_{Di}, I_{Qi}, and $\overline{V}_i$ using $(I_{di} + jI_{qi})e^{j(\delta_i - \pi/2)} = I_{Di} + jI_{Qi}$.

Simplification #2 (constant impedance load in the transmission system)

Here, the loads are assumed to be of the constant impedance type, i.e.,

$$P_{Li}(V_i) = k_{P2i}V_i^2 \tag{7.81}$$

$$Q_{Li}(V_i) = k_{Q2i}V_i^2 \tag{7.82}$$

Then

$$P_{Li}(V_i) + jQ_{Li}(V_i) = (k_{P2i} + jk_{Q2i})V_i^2 \tag{7.83}$$

Since $(P_{Li} + jQ_{Li}) = \overline{V}_i\overline{I}_{Li}^*$, where $\overline{I}_{Li}$ is the injected current,

$$\overline{V}_i\overline{I}_{Li}^* = V_i^2(k_{P2i} + jk_{Q2i}) \tag{7.84}$$

But $V_i^2 = \overline{V}_i\overline{V}_i^*$ and conjugating (7.84), we obtain

$$(k_{P2i} - jk_{Q2i}) = \frac{\overline{I}_{Li}}{\overline{V}_i} = -\overline{y}_{ii} \tag{7.85}$$

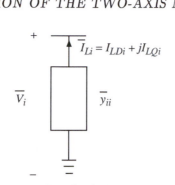

Figure 7.8: *Load admittance representation*

Because of the orientation of $\overline{V}_i$ and $\overline{I}_{Li}$, there is a negative sign in front of the load admittance $\overline{y}_{ii}$ in (7.85). From Figure 7.8, we can verify that

$$\frac{P_{Li}(V_i) - jQ_{Li}(V_i)}{\overline{V}_i^*} = \overline{I}_{Li} = -\overline{y}_{ii}\overline{V}_i \tag{7.86}$$

Hence

$$\overline{y}_{ii} = -\frac{(P_{Li}(V_i) - jQ_{Li}(V_i))}{V_i^2}$$

This is consistent with the fact that P_{Li} and Q_{Li} are injected loads. $\overline{y}_{ii}$ is the complex admittance due to the loads. Since $(I_{di} + jI_{qi})e^{j(\delta_i - \pi/2)} = I_{Di} + jI_{Qi}$, for $i = 1, \ldots, m$ and using (7.86), the network equations (7.42) and (7.43) become

$$(I_{Di} + jI_{Qi}) = \sum_{k=1}^{n} V_k e^{j\theta_k} Y_{ik} e^{j\alpha_{ik}} + \overline{y}_{ii} V_i e^{j\theta_i} \quad i = 1, \ldots, m \tag{7.87}$$

$$0 = \sum_{k=1}^{n} V_k e^{j\theta_k} Y_{ik} e^{j\alpha_{ik}} + \overline{y}_{ii} V_i e^{j\theta_i} \quad i = m+1, \ldots, n$$

i.e.,

$$I_{Di} + jI_{Qi} = \sum_{k=1}^{n} V_k e^{j\theta_k} Y'_{ik} e^{j\alpha'_{ik}} \quad i = 1, \ldots, m$$

$$0 = \sum_{k=1}^{n} V_k e^{j\theta_k} Y'_{ik} e^{j\alpha'_{ik}} \quad i = m+1, \ldots, n \tag{7.88}$$

where

$$Y'_{ik} e^{j\alpha'_{ik}} = Y_{ik} e^{j\alpha_{ik}} \qquad i \neq k$$

$$Y'_{ii} e^{j\alpha'_{ii}} = Y_{ii} e^{j\alpha_{ii}} + \bar{y}_{ii}$$

Equation (7.88) can be written as

$$
\begin{bmatrix} \bar{I}_1 \\ \vdots \\ \bar{I}_m \\ 0 \\ \vdots \\ 0 \end{bmatrix} = [\bar{Y}'] \begin{bmatrix} \bar{V}_1 \\ \vdots \\ \vdots \\ \bar{V}_n \end{bmatrix}
\tag{7.89}
$$

where $\bar{I}_1, \ldots, \bar{I}_m$ are the complex injected generator currents at the generator buses. Let the modified $\bar{Y}_{\text{bus}}$ denoted as $\bar{Y}'$ be partitioned as

$$
[\bar{Y}'] = \begin{array}{c} \\ m \\ \\ n-m \end{array} \overset{\displaystyle m \quad | \quad n-m}{\begin{pmatrix} \bar{Y}_1 & | & \bar{Y}_2 \\ - & - & - \\ \bar{Y}_3 & | & \bar{Y}_4 \end{pmatrix}}
\tag{7.90}
$$

Since there are no injections at buses $m+1, \ldots, n$, we can eliminate them to get

$$
\begin{bmatrix} \bar{I}_1 \\ \vdots \\ \bar{I}_m \end{bmatrix} = [\bar{Y}_{\text{red}}] \begin{bmatrix} \bar{V}_1 \\ \vdots \\ \bar{V}_m \end{bmatrix}
\tag{7.91}
$$

where $\bar{Y}_{\text{red}} = (\bar{Y}_1 - \bar{Y}_2 \bar{Y}_4^{-1} \bar{Y}_3)$.

Note that we can make either or both of the above two simplifications. If we make only the constant impedance approximation, then we can obtain a passive reduced network, as in Figure 7.9, but the source will be a dependent one, as in Figure 7.1. On the other hand, if we make the assumption $X'_{di} = X'_{qi}$ only, then we obtain the source representation, as in Figure 7.9, but the network equations will remain in (7.42) and (7.43). If we make both assumptions, then we obtain a passive reduced network, as in Figure 7.9, and the voltage $\bar{E}'_i = E'_{Di} + j E'_{Qi}$.

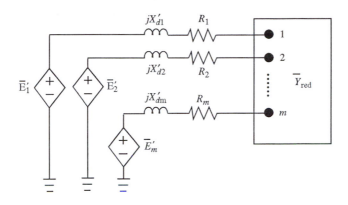

Figure 7.9: *Network reduced at generator nodes and* $X'_{di} = X'_{qi}$

Example 7.4

Based on the load-flow results in Table 7.1, convert the loads as constant admittances, and then obtain a $\overline{Y}_{\text{red}}$ representation (all in pu).

Solution

Note that the load specified in Table 7.1 is the negative of the injected load. Hence, at buses 5, 6, and 8, $\overline{y}_{ii}$ is given by

$$\overline{y}_{55} = \frac{1.25 - j0.5}{(0.9956)^2} = 1.26 - j0.504 \ , \quad \overline{y}_{66} = \frac{0.9 - j0.3}{(1.013)^2} = 0.877 - j0.292$$

$$\overline{y}_{88} = \frac{1 - j0.35}{(1.016)^2} = 0.969 - j0.339$$

These elements are now added to the $\overline{Y}_{55}$, $\overline{Y}_{66}$, and $\overline{Y}_{88}$ elements of the $\overline{Y}_{\text{bus}}$ matrix in Table 7.2, resulting in $\overline{Y}'$ as

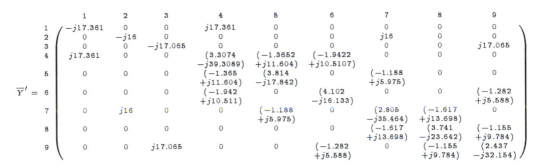

$\overline{Y}_{\text{red}}$ is obtained by eliminating the buses 4 to 9 from $\overline{Y}'$. The resulting $\overline{Y}_{\text{red}}$ is given by

$$\overline{Y}_{\text{red}} = \begin{bmatrix} 1.105 - j4.695 & 0.096 + j2.253 & 0.004 + j2.275 \\ 0.096 + j2.257 & 0.735 - j5.114 & 0.123 + j2.826 \\ 0.004 + j2.275 & 0.123 + j2.826 & 0.721 - j5.023 \end{bmatrix}$$

With this simplification, we have $\overline{I} = \overline{Y}_{\text{red}}\overline{V}$.

$$\square$$

7.6   Initial Conditions (Full Model)

The initial conditions of the state variables for the model in Section 7.1 are computed by systematically solving the load-flow equations of the network first, and then computing the other algebraic and state variables. The load-flow equations are part of the network equations, as shown below.

Load-Flow Formulation

We now revert to the formulation of the network equations that were written as real and reactive power-balance equations at the nodes, i.e., (7.32)–(7.35). We reproduce them below by writing real power equations first, followed by the reactive power equations.

$$I_{di}V_i \sin(\delta_i - \theta_i) + I_{qi}V_i \cos(\delta_i - \theta_i) + P_{Li}(V_i)$$
$$- \sum_{k=1}^{n} V_iV_kY_{ik} \cos(\theta_i - \theta_k - \alpha_{ik}) = 0 \quad i = 1, \ldots, m \quad (7.92)$$

$$P_{Li}(V_i) - \sum_{k=1}^{n} V_iV_kY_{ik} \cos(\theta_i - \theta_k - \alpha_{ik}) = 0 \quad i = m+1, \ldots, n \quad (7.93)$$

$$I_{di}V_i \cos(\delta_i - \theta_i) - I_{qi}V_i \sin(\delta_i - \theta_i) + Q_{Li}(V_i)$$
$$- \sum_{k=1}^{n} V_iV_kY_{ik} \sin(\theta_i - \theta_k - \alpha_{ik}) = 0 \quad i = 1, \ldots, m$$
$$(7.94)$$

$$Q_{Li}(V_i) - \sum_{k=1}^{n} V_i V_k Y_{ik} \sin(\theta_i - \theta_k - \alpha_{ik}) = 0 \quad i = m+1, \ldots, n \quad (7.95)$$

It follows from the dynamic circuit (Figure 7.10) that

$$P_{Gi} + jQ_{Gi} = \overline{V}_i \overline{I}_{Gi}^* = V_i e^{j\theta_i}(I_{di} - jI_{qi})e^{-j(\delta_i - \pi/2)}$$
$$= V_i(\cos\theta_i + j\sin\theta_i)(I_{di} - jI_{qi})(\sin\delta_i + j\cos\delta_i) \quad (7.96)$$

Now we equate the real and imaginary parts along with the use of trigonometric identities. It can be shown from (7.96) that

$$P_{Gi} = I_{di}V_i \sin(\delta_i - \theta_i) + I_{qi}V_i \cos(\delta_i - \theta_i) \quad (7.97)$$

and

$$Q_{Gi} = I_{di}V_i \cos(\delta_i - \theta_i) - I_{qi}V_i \sin(\delta_i - \theta_i) \quad (7.98)$$

$$\overline{I}_{Gi} \triangleq I_{Gi} e^{j\gamma_i} = (I_{di} + jI_{qi})e^{j(\delta_i - \pi/2)} \quad (7.99)$$

is the injected generator current at the generator bus in the synchronous reference frame. Figure 7.10 explains the equations at the generator bus.

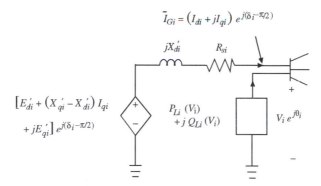

Figure 7.10: *Synchronous machine dynamic circuit*

We further define net injected power at a bus as

$$P_i(\delta_i, I_{di}, I_{qi}, V_i, \theta_i) + jQ_i(\delta_i, I_{di}, I_{qi}, V_i, \theta_i) =$$
$$(P_{Gi} + P_{Li}(V_i)) + j(Q_{Gi} + Q_{Li}(V_i)) \quad i = 1, \ldots, m \quad (7.100)$$

Thus, the power-balance equations at the buses $1, \ldots, n$ are

$$P_i(\delta_i, I_{di}, I_{qi}, V_i, \theta_i) = \sum_{k=1}^{n} V_i V_k Y_{ik} \cos\left(\theta_i - \theta_k - \alpha_{ik}\right)$$

$$i = 1, \ldots, m \tag{7.101}$$

$$P_{Li}(V_i) = \sum_{k=1}^{n} V_i V_k Y_{ik} \cos\left(\theta_i - \theta_k - \alpha_{ik}\right)$$

$$i = m+1, \ldots, n \tag{7.102}$$

$$Q_i(\delta_i, I_{di}, I_{qi}, V_i, \theta_i) = \sum_{k=1}^{n} V_i V_k Y_{ik} \sin\left(\theta_i - \theta_k - \alpha_{ik}\right)$$

$$i = 1, \ldots, m \tag{7.103}$$

$$Q_{Li}(V_i) = \sum_{k=1}^{n} V_i V_k Y_{ik} \sin\left(\theta_i - \theta_k - \alpha_{ik}\right)$$

$$i = m+1, \ldots, n \tag{7.104}$$

Standard Load Flow

Load flow has been the traditional mechanism for computing a proposed steady-state operating point. We now define standard load flow as follows, using (7.101)–(7.104). Loads are of the constant power type.

1. Specify bus voltage magnitudes numbered 1 to m.

2. Specify bus voltage angle at bus number 1 (slack bus).

3. Specify net injected real power P_i at buses numbered 2 to m.

4. Specify load powers P_{Li} and Q_{Li} at buses numbered $m+1$ to n.

The following equations result from (7.101), (7.102), and (7.104) chosen according to criteria (3) and (4). These are known as the *load-flow equations*. Thus, we have

$$0 = -P_i + \sum_{k=1}^{n} V_i V_k Y_{ik} \cos\left(\theta_i - \theta_k - \alpha_{ik}\right) \quad \overset{i=2,\ldots,m}{(\text{PV buses})} \tag{7.105}$$

$$0 = -P_{Li} + \sum_{k=1}^{n} V_i V_k Y_{ik} \cos(\theta_i - \theta_k - \alpha_{ik}) \quad \begin{matrix} i=m+1,\ldots,n \\ \text{(PQ buses)} \end{matrix} \quad (7.106)$$

$$0 = -Q_{Li} + \sum_{k=1}^{n} V_i V_k Y_{ik} \sin(\theta_i - \theta_k - \alpha_{ik}) \quad \begin{matrix} i=m+1,\ldots,n \\ \text{(PQ buses)} \end{matrix} \quad (7.107)$$

where $P_i(i = 2, \ldots, m)$, $V_i(i = 1, \ldots, m)$, $P_{Li}(i = m+1, \ldots, n)$, $Q_{Li}(i = m+1, \ldots, n)$, and θ_1 are specified numbers. The standard load flow solves (7.105)–(7.107) for $\theta_2, \ldots, \theta_n, V_{m+1}, \ldots, V_n$. After the load-flow solution, we compute the net injected powers at the slack bus and the net injected reactive power at the generator buses as

$$P_1 + jQ_1 = \sum_{k=1}^{n} V_1 V_k Y_{1k} e^{j(\theta_1 - \theta_k - \alpha_{1k})} \quad (7.108)$$

$$Q_i = \sum_{k=1}^{n} V_i V_k Y_{ik} \sin(\theta_i - \theta_k - \alpha_{ik}) \quad i = 2, \ldots, m \quad (7.109)$$

The generator powers are given by $P_{G1} = P_1 - P_{L1}$ and $Q_{Gi} = Q_i - Q_{Li}$ $(i = 1, \ldots, m)$. This standard load flow has many variations, including the addition of other devices such as tap-changing-under-load (TCUL) transformers, switching VAR sources, HVDC converters, and nonlinear load representation. It can also include inequality constraints on quantities such as Q_{Gi} at the generators, and can also have more than one slack bus. For details, refer to [15, 16, 18].

One important point about load flow should be emphasized. Load flow is normally used to evaluate operation at a specific load level (specified by a given set of powers). For a specified load and generation schedule, the solution is independent of the actual load model. That is, it is certainly possible to evaluate the voltage at a constant impedance load for a specific case where that impedance load consumes a specific amount of power. Thus, the use of "constant power" in load-flow analysis does not require or even imply that the load is truly a constant power device. It merely gives the voltage at the buses when the loads (any type) consume a specific amount of power. The load characteristic is important when the analyst wants to study the system in response to a change, such as contingency analysis or dynamic analysis. For these purposes, standard load flow is computed on the basis of constant PQ loads and usually provides the "initial conditions" for the dynamic system.

Initial Conditions for Dynamic Analysis

We use the model in the power-balance form from (7.36)–(7.38).

$$\dot{x} = f_o(x, I_{d-q}, \overline{V}, u) \tag{7.110}$$

$$I_{d-q} = h(x, \overline{V}) \tag{7.111}$$

$$0 = g_o(x, I_{d-q}, \overline{V}) \tag{7.112}$$

It is necessary to compute the initial values of all the dynamic states and the fixed inputs T_{Mi} and V_{ref_i} ($i = 1, \ldots, m$). In power system dynamic analysis, the fixed inputs and initial conditions are normally found from a base case load-flow solution. That is, the values of V_{ref_i} are computed such that the m generator voltages are as specified in the load flow. The values of T_{Mi} are computed such that the m generator real power outputs P_{Gi} are as specified and computed in the load flow for rated speed ω_s. To see how this is done, we assume that a load-flow solution (as defined in previous section) has been found, i.e., solution of (7.105)–(7.107). The first step in computing the initial conditions is normally the calculation of generator currents from (7.96), as $\overline{I}_{Gi} = I_{Gi}e^{j\gamma_i} = (P_{Gi} - jQ_{Gi})/\overline{V}_i^*$.

Step 1

Since $P_{Gi} = P_i - P_{Li}$ and $Q_{Gi} = Q_i - Q_{Li}$:

$$I_{Gi}e^{j\gamma_i} = ((P_i - P_{Li}) - j(Q_i - Q_{Li}))/(V_i e^{-j\theta_i}), \quad i = 1, \ldots, m \tag{7.113}$$

This current is in the network reference frame and is equal to $(I_{di} + jI_{qi})e^{j(\delta_i - \pi/2)}$.

In steady state, all the derivatives are zero in the differential equations (7.1)–(7.7). The first step is to calculate the rotor angles δ_i at all the machines. We use the complex stator algebraic equation (7.8) and the algebraic equation obtained from (7.2) by setting $\dot{E}_{di} = 0$. From the latter, we obtain

$$E'_{di} = (X_{qi} - X'_{qi})I_{qi} \quad i = 1, \ldots, m \tag{7.114}$$

Substitution of (7.114) in (7.8) results in

$$
\begin{aligned}
V_i e^{j\theta_i} &+ (R_{si} + jX'_{di})(I_{di} + jI_{qi})e^{j(\delta_i - \pi/2)} \\
&- [(X_{qi} - X'_{qi})I_{qi} + (X'_{qi} - X'_{di})I_{qi} + jE'_{qi}]e^{j(\delta_i - \pi/2)} = 0 \\
&i = 1, \ldots, m
\end{aligned}
\tag{7.115}
$$

i.e.,

$$
\begin{aligned}
V_i e^{j\theta_i} &+ R_{si}(I_{di} + jI_{qi})e^{j(\delta_i - \pi/2)} \\
&+ jX'_{di}(I_{di} + jI_{qi})e^{j(\delta_i - \pi/2)} \\
&- [(X_{qi} - X'_{di})I_{qi} + jE'_{qi}]e^{j(\delta_i - \pi/2)} = 0 \quad i = 1, \ldots, m
\end{aligned}
\tag{7.116}
$$

i.e.,

$$
\begin{aligned}
V_i e^{j\theta_i} &+ R_{si}(I_{di} + jI_{qi})e^{j(\delta_i - \pi/2)} \\
&+ jX'_{di}I_{di}e^{j(\delta_i - \pi/2)} - X_{qi}I_{qi}e^{j(\delta_i - \pi/2)} \\
&- jE'_{qi}e^{j(\delta_i - \pi/2)} = 0, \quad i = 1, \ldots, m
\end{aligned}
\tag{7.117}
$$

Adding and subtracting $jX_{qi}I_{di}e^{j(\delta_i - \pi/2)}$ from the left-hand side of (7.117) results in

$$
\begin{aligned}
V_i e^{j\theta_i} &+ (R_{si} + jX_{qi})(I_{di} + jI_{qi})e^{j(\delta_i - \pi/2)} \\
&- j[(X_{qi} - X'_{di})I_{di} + E'_{qi}]e^{j(\delta_i - \pi/2)} = 0 \quad i = 1, \ldots, m
\end{aligned}
\tag{7.118}
$$

Replace $(I_{di} + jI_{qi})e^{j(\delta_i - \pi/2)}$ by $I_{Gi}e^{j\gamma_i}$, which is already calculated in (7.113):

$$
\begin{aligned}
V_i e^{j\theta_i} &+ (R_{si} + jX_{qi})I_{Gi}e^{j\gamma_i} \\
&= ((X_{qi} - X'_{di})I_{di} + E'_{qi})e^{j\delta_i} \quad i = 1, \ldots, m
\end{aligned}
\tag{7.119}
$$

The right-hand side of (7.119) is a voltage behind the impedance $(R_{si} + jX_{qi})$ and has an angle δ_i. The voltage has a magnitude $(E'_{qi} + (X_{qi} - X'_{di})I_{di})$ and an angle $\delta_i = $ angle of $(V_i e^{j\theta_i} + (R_{si} + jX_{qi})I_{Gi}e^{j\gamma_i})$. The complex number representation for computing δ_i from (7.119) is shown in Figure 7.11. This representation is generally known as "phasor diagram" in the literature and "locating the q axis" of the machine.

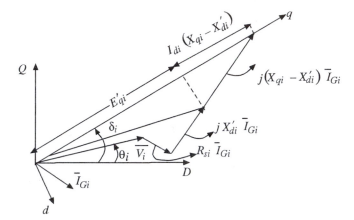

Figure 7.11: *Representation of stator algebraic equations in steady state*

Step 2

δ_i is computed as $\delta_i = \text{angle on } (V_i e^{j\theta_i} + (R_{si} + jX_{qi})I_{Gi}e^{j\gamma_i})$.

Step 3

Compute $I_{di}, I_{qi}, V_{di}, V_{qi}$ for the machines as

$$I_{di} + jI_{qi} = I_{Gi}e^{j(\gamma_i - \delta_i + \pi/2)} \qquad i = 1, \ldots, m \qquad (7.120)$$

$$V_{di} + jV_{qi} = V_i e^{j(\theta_i - \delta_i + \pi/2)} \qquad i = 1, \ldots, m \qquad (7.121)$$

Step 4

Compute E'_{di} from (7.27):

$$E'_{di} = V_{di} + R_{si}I_{di} - X'_{qi}I_{qi} \quad i = 1, \ldots, m \qquad (7.122)$$

which from (7.114) is also equal to $(X_{qi} - X'_{qi})I_{qi}$. This serves as a check on the calculations.

Step 5

Compute E'_{qi} from (7.27)

$$E'_{qi} = V_{qi} + R_{si}I_{qi} + X'_{di}I_{di} \quad i = 1, \ldots, m \qquad (7.123)$$

Step 6

Compute E_{fdi} from (7.1) (after setting the derivative equal to zero):

$$E_{fdi} = E'_{qi} + (X_{di} - X'_{di})I_{di} \qquad i = 1, \ldots, m \qquad (7.124)$$

Step 7

With the field voltage E_{fdi} known, the other variables R_{fi}, V_{Ri} and $V_{\text{ref}i}$ can be found from (7.5)–(7.7) (after setting the derivatives equal to zero):

$$V_{Ri} = (K_{Ei} + S_{Ei}(E_{fdi}))E_{fdi} \qquad i = 1, \ldots, m \qquad (7.125)$$

$$R_{fi} = \frac{K_{Fi}}{T_{Fi}} E_{fdi} \qquad i = 1, \ldots, m \qquad (7.126)$$

$$V_{\text{ref}i} = V_i + (V_{Ri}/K_{Ai}) \qquad i = 1, \ldots, m \qquad (7.127)$$

Note that if the machine saturation is included, the calculation for E'_{qi} and E'_{di} may be iterative. The mechanical states ω_i and T_{Mi} are found from (7.3) and (7.4) (after setting the derivatives equal to zero):

$$\omega_i = \omega_s \qquad i = 1, \ldots, m \qquad (7.128)$$

$$T_{Mi} = E'_{di}I_{di} + E'_{qi}I_{qi} + (X'_{qi} - X'_{di})I_{di}I_{qi} \qquad i = 1, \ldots, m \qquad (7.129)$$

This completes the computation of all dynamic-state initial conditions and fixed inputs. Thus, we have computed $x(o)$, $y(o)$, and u from the load-flow data.

For a given disturbance, the inputs remain fixed throughout the simulation. If the disturbance occurs due to a fault or a network change, the algebraic states must change instantaneously. The dynamic states cannot change instantaneously. Thus, it will be necessary to solve all the algebraic equations inclusive of the stator equations with the dynamic states specified at their values just prior to the disturbance as initial conditions to determine the new initial values of the algebraic states.

From the above description, it is clear that once a standard load-flow solution is found, the remaining dynamic states and inputs can be found in a systematic manner. The machine rotor angles δ_i can always be found, provided that:

$$V_i e^{j\theta_i} + (R_{si} + jX_{qi})I_{Gi}e^{j\gamma_i} \neq 0 \qquad i = 1, \ldots, m \qquad (7.130)$$

If control limits are enforced, a solution satisfying these limits may not exist. In this case, the state that is limited would have to be fixed at its limiting value and a corresponding new steady-state solution would have to be found. This would require a new load flow by specifying either different values of generator voltages, different generator real powers, or possibly specifying generator-reactive power injections, thus allowing generator voltage to be a part of the load-flow solution. In fact, the use of reactive power limits in load flow can usually be traced to an attempt to consider excitation system limits or generator capability limits.

Example 7.5

For Example 7.1, compute the initial conditions. The solved load-flow data are given in Table 7.1 (all in pu).

Machine 1

Step 1

$$I_{G1}e^{j\gamma_1} = \frac{P_{G1} - jQ_{G1}}{V_1^*} = \frac{0.716 - j0.27}{1.04\angle 0^\circ}$$

$$= 0.736\angle - 20.66^\circ$$

Step 2

$$\delta_1(0) = \text{Angle of } (V_1 e^{j\theta_1} + (R_{s1} + jX_{q1})I_{G1}e^{j\gamma_1})$$

$$= \text{Angle of } ((1.04\angle 0^\circ + j0.0969)(0.736\angle - 20.66^\circ))$$

$$= 3.58^\circ$$

Step 3

$$I_{d1} + jI_{q1} = I_{G1}e^{j\gamma_1}e^{-j(\delta_1 - \pi/2)}$$

$$= (0.736\angle - 20.66^\circ)(1\angle 86.42^\circ)$$

$$= 0.302 + j0.671$$

$$I_{d1} = 0.302 \quad I_{q1} = 0.671$$

$$V_{d1} + jV_{q1} = V_1 e^{j\theta_1} e^{-j(\delta_1 - \pi/2)}$$
$$= (1.04\angle 0°)(1\angle 86.42°)$$
$$= 0.065 + j1.038$$
$$V_{d1} = 0.065, \quad V_{q1} = 1.038$$

Step 4

$$E'_{d1} = (X_{q1} - X'_{q1})I_{q1}$$
$$= (0.0969 - 0.0969)(0.671)$$
$$= 0$$

It can be verified that this is also equal to $V_{d1} + R_{s1}I_{d1} - X'_{q1}I_{q1}$.

Step 5

$$E'_{q1} = V_{q1} + R_{s1}I_{q1} + X'_{d1}I_{d1}$$
$$= 1.038 + 0 + (0.0608)(0.302)$$
$$= 1.056$$

Step 6

$$E_{fd1} = E'_{q1} + (X_{d1} - X'_{d1})I_{d1}$$
$$= 1.056 + (0.146 - 0.0608)(0.302)$$
$$= 1.082$$

Step 7

$$V_{R1} = (K_{E1} + 0.0039e^{1.555E_{fd1}})E_{fd1} = (1.021)(1.082) = 1.105$$
$$R_{f1} = \frac{K_{F1}}{T_{F1}}E_{fd1} = \left(\frac{0.063}{0.35}\right)1.082 = 0.195$$
$$V_{ref1} = V_1 + \frac{V_{R1}}{K_{A1}} = 1.04 + \frac{1.105}{20}$$
$$= 1.095$$

The mechanical input T_{M1} is computed as follows:

$$T_{M1} = E'_{d1}I_{d1} + E'_{q1}I_{q1} + (X'_{q1} - X'_{d1})I_{d1}I_{q1}$$
$$= (0)(0.302) + (1.056)(0.671) + (0.0969 - 0.0608)(0.302)(0.671)$$
$$= 0.716$$

Similarly, for other machines, we can compute the state and the algebraic variables as

$$
\begin{aligned}
\delta_2(o) &= 61.1^o & \delta_3(o) &= 54.2^o \\
I_{d2} &= 1.29 & I_{d3} &= 0.562 \\
I_{q2} &= 0.931 & I_{q3} &= 0.619 \\
V_{d2} &= 0.805 & V_{d3} &= 0.779 \\
V_{q2} &= 0.634 & V_{q3} &= 0.666 \\
E'_{d2} &= 0.622 & E'_{d3} &= 0.624 \\
E'_{q2} &= 0.788 & E'_{q3} &= 0.768 \\
E_{fd2} &= 1.789 & E_{fd3} &= 1.403 \\
R_{f2} &= 0.322 & R_{f3} &= 0.252 \\
V_{R2} &= 1.902 & V_{R3} &= 1.453 \\
V_{ref2} &= 1.12 & V_{ref3} &= 1.09 \\
T_{M2} &= 1.63 & T_{M3} &= 0.85
\end{aligned}
$$

$\square$

Angle Reference, Infinite Bus, and COI Reference

As explained in Section 6.10, by taking one of the angles as a reference, the order of the dynamic system can be reduced from $7m$ to $7m$-1. Furthermore, if the inertia constant on this reference angle machine is infinity, the order of the system can be reduced to $7m$-2. This is also possible if the machines have zero or uniform damping. Finally, we can also use center-of-angle formulation instead of relative rotor angle formulation. The center-of-angle formulation is discussed in Chapter 9.

7.7   Numerical Solution: Power-Balance Form

The number of algorithms that have been proposed for the numerical solution of the DAE system of equations is very large. There are basically two approaches used in power system simulation packages.

1. Simultaneous-implicit (SI) method

2. Partitioned-explicit (PE) method

The SI is numerically more stable than the PE method. It is also the method used in the EPRI 1208 stability program known as the ETMSP (Extended Transient Midterm Stability Program) program [70]. We illustrate the SI method on the WSCC system with a two-axis model for the machine and with the exciter on all the three machines.

7.7.1 SI Method

We illustrate with the differential-algebraic model in power-balance form.

$$\dot{x} = f_o(x, I_{d-q}, \overline{V}, u) \tag{7.131}$$

$$I_{d-q} = h(x, \overline{V}) \tag{7.132}$$

$$0 = g_o(x, I_{d-q}, \overline{V}) \tag{7.133}$$

All the initial conditions at $t = 0$ have been computed. In the SI method, the differential equations in (7.131) are algebraized using either implicit Euler's method or a trapezoidal integration method. These resulting algebraic equations are then solved simultaneously with the remaining algebraic equations (7.132)–(7.133) using Newton's method at each time step.

Review of Newton's Method

Let $f(x) = 0$ be the set of nonlinear algebraic equations, i.e.,

$$
\begin{aligned}
f_1(x_1, \ldots, x_n) &= 0 \\
f_2(x_1, \ldots, x_n) &= 0 \\
&\vdots \\
f_n(x_1, \ldots, x_n) &= 0
\end{aligned}
\tag{7.134}
$$

Assume an initial guess $x_1^{(o)}, \ldots, x_n^{(o)}$. Expand the equations in a Taylor series and retain only the linear term.

$$f(x^{(o)}) + \left.\frac{\partial f}{\partial x}\right|_{x=x^{(o)}} (x - x^{(o)}) \approx 0 \tag{7.135}$$

Solving for x results in an improved estimate for x:

$$x^{(1)} = x^{(o)} - \left[\left.\frac{\partial f}{\partial x}\right|_{x=x^{(o)}}\right]^{-1} f(x^{(o)}) \tag{7.136}$$

In general,

$$x^{(k+1)} = x^{(k)} - [J^{(k)}]^{-1} f(x^{(k)}) \tag{7.137}$$

where k is the iteration count and $[J] \triangleq \frac{\partial f}{\partial x}$ is called the Jacobian. In an expanded version:

$$[J] = \begin{bmatrix} \frac{\partial f_1}{\partial x_1} & \cdots & \frac{\partial f_1}{\partial x_n} \\ \vdots & & \\ \frac{\partial f_n}{\partial x_1} & \cdots & \frac{\partial f_n}{\partial x_n} \end{bmatrix} \tag{7.138}$$

Define

$$\Delta x^{(k)} = x^{(k+1)} - x^{(k)} \tag{7.139}$$

Equation (7.137) can be recast as follows:

$$[J^{(k)}]\Delta x^{(k)} = -f(x^{(k)}), \quad k = 0, 1, 2, \ldots \tag{7.140}$$

Equation (7.140) is a linear one and has to be solved for $\Delta x^{(k)}$. Then from (7.139)

$$x^{(k+1)} = x^{(k)} + \Delta x^{(k)}, \quad k = 0, 1, 2, \ldots \tag{7.141}$$

With an initial value of $x^{(o)}$, steps corresponding to (7.140) and (7.141) are repeated and at the end of each iteration compute $\overset{\max}{i} \left| f_i(x^{(k+1)}) \right|$. If this is $< \epsilon$ where ϵ is the specified tolerance, the Newton iterates have converged.

Numerical Solution Using SI Method

Let the subscripts n and $n+1$ denote the time instants t_n and t_{n+1}, respectively. Then, integrating the differential equations in (7.131) from t_n to t_{n+1} using the trapezoidal rule and solving the resulting algebraic equations with the remaining algebraic equations at t_{n+1}, obtain

$$x_{n+1} = x_n + \int_{t_n}^{t_{n+1}} f_o(x, I_{d-q}, \overline{V}, u) dt \tag{7.142}$$

i.e.,

$$x_{n+1} = x_n + \frac{\Delta t}{2}[f_o(x_{n+1}, I_{d-q,n+1}, \overline{V}_{n+1}, u_{n+1})$$

$$+ f_o(x_n, I_{d-q,n}, \overline{V}_n, u_n)] \tag{7.143}$$

$$0 = I_{d-q,n+1} - h(x_{n+1}, \overline{V}_{n+1}) \tag{7.144}$$

$$0 = g_o(x_{n+1}, I_{d-q,n+1}, \overline{V}_{n+1}) \tag{7.145}$$

where $\Delta t = t_{n+1} - t_n$ is the integration time step. Rearranging (7.143)–(7.145),

$$[x_{n+1} - \frac{\Delta t}{2}f_o(x_{n+1}, I_{d-q,n+1}, \overline{V}_{n+1}, u_{n+1})] - [x_n + \frac{\Delta t}{2}f_o(x_n, I_{d-q,n}, \overline{V}_n, u_n)]$$

$$\triangleq F_1(x_{n+1}, I_{d-q,n+1}, \overline{V}_{n+1}, u_{n+1}, x_n, I_{d-q,n}, \overline{V}_n, u_n) = 0 \tag{7.146}$$

$$I_{d-q,n+1} - h(x_{n+1}, \overline{V}_{n+1}) \triangleq F_2(x_{n+1}, I_{d-q,n+1}, \overline{V}_{n+1}) = 0 \tag{7.147}$$

$$g_o(x_{n+1}, I_{d-q,n+1}, \overline{V}_{n+1}) \triangleq F_3(x_{n+1}, I_{d-q,n+1}, \overline{V}_{n+1}) = 0 \tag{7.148}$$

At each time step, (7.146)–(7.148) are solved by Newton's method. The Newton iterates are

$$\begin{matrix} & \begin{matrix} 7m & 2m & 2n \end{matrix} \\ \begin{matrix} 7m \\ 2m \\ 2n \end{matrix} & \begin{pmatrix} J_1 & J_2 & J_3 \\ J_4 & J_5 & J_6 \\ J_7 & J_8 & J_9 \end{pmatrix} \end{matrix}^{(k)} \begin{bmatrix} \Delta x_{n+1} \\ \Delta I_{d-q,n+1} \\ \Delta \overline{V}_{n+1} \end{bmatrix}^{(k)} = - \begin{bmatrix} F_1 \\ F_2 \\ F_3 \end{bmatrix}^{(k)} \tag{7.149}$$

$$x_{n+1}^{(k+1)} = x_{n+1}^{(k)} + \Delta x_{n+1}^{(k)} \tag{7.150}$$

$$I_{d-q,n+1}^{(k+1)} = I_{d-q,n+1}^{(k)} + \Delta I_{d-q,n+1}^{(k)} \tag{7.151}$$

$$\overline{V}_{n+1}^{(k+1)} = \overline{V}_{n+1}^{(k)} + \Delta \overline{V}_{n+1}^{(k)} \tag{7.152}$$

where k is the iteration number at time step t_{n+1}. $x_{n+1}^{(o)}$ is the converged value x_n at the previous time step. The iterations are continued until the norm of the mismatch vector $[F_1, F_2, F_3]^t$ is close to zero. This completes the computation at time step t_{n+1}.

If there is a change in reference input $V_{\mathrm{ref}i}$, or T_{Mi}, the DAE (differential algebraic equation) model can be integrated with known values of all variables. But if there is a disturbance in the network, a different procedure has to be adopted, as explained below.

Disturbance Simulation

The typical disturbance corresponds to a network disturbance, such as a fault where the parameters in the algebraic equations change at $t = 0$. The algebraic variables can change instantaneously, whereas the state variables do not. Hence, at $t = 0+$, with the network disturbance reflected in the network equations, we solve the set of algebraic equations for $I_{d-q}(0+)$, $\overline{V}(0+)$ as

$$I_{d-q}(0+) = h^f(x(0), \overline{V}(0+)) \tag{7.153}$$

$$0 = g_o^f(x(0), I_{d-q}(0+), \overline{V}(0+)) \tag{7.154}$$

where the superscript f indicates that the algebraic equations correspond to the faulted state. With the value of $I_{d-q}(0+), \overline{V}(0+)$ so obtained, the trapezoidal method is then applied. Note that the initial guess for the vector $[x^t\ I_{d-q}^t\ \overline{V}^t]^t$ at time instant t_{n+1} is the converged value at the previous time instant, i.e.,

$$\begin{bmatrix} x_{n+1} \\ I_{d-q,n+1} \\ \overline{V}_{n+1} \end{bmatrix}^{(o)} = \begin{bmatrix} x_n \\ I_{d-q,n} \\ \overline{V}_n \end{bmatrix}, \quad n = 0, 1, 2, \ldots \tag{7.155}$$

If there is a change in generation or load, this is simply taken care of by changing P_{Gi} and Q_{Gi} or P_{Li} and Q_{Li} in (7.153) and (7.154), and computing I_{d-q} and $\overline{V}$ at $t = 0+$. If there is a short circuit at bus i, then set $\overline{V}_i \equiv 0$ and delete the P and Q equations at bus i from $g_o(x, I_{d-q}, \overline{V})$ to obtain $g_o^f(x, I_{d-q}, \overline{V})$.

7.7.2   PE method

1. Incorporate the system disturbance and solve for $\overline{V}(0+)$, $I_{d-q}(0+)$ as in (7.153) and (7.154).

2. Using the values of $I_{d-q}(0+), \overline{V}(0+)$ integrate the differential equations

$$\dot{x} = f_o(x, I_{d-q}, \overline{V}) , \quad x(0) = x^o \tag{7.156}$$

to obtain $x(1)$.

3. Go to Step 1 and solve for $I_{d-q}(1)$, $\overline{V}(1)$ again from the algebraic equations.

$$I_{d-q}(1) = h^f(x(1), \overline{V}(1)) \tag{7.157}$$

$$0 = g_o^f(x(1), I_{d-q}(1), \overline{V}(1)) \tag{7.158}$$

4. Integrate the differential equations to obtain $x(2)$, and again solve the algebraic equation to obtain $I_{d-q}(2)$, $\overline{V}(2)$.

5. This procedure is repeated until $t = T_{\text{desired}}$ or there is a change in the configuration. In the second case, a similiar procedure is followed.

The PE scheme, although conceptually simple, has numerical convergence problems such as interface errors, etc. [73].

7.8   Numerical Solution: Current-Balance Form

In this section, we explain the industrial approach to implement the SI method forming the structured approach to the problem. It is the basis of the well-known ETMSP program of EPRI [70]. The current-balance approach is favored, since the bus admittance matrix is easily formed and factorized. The DAE model from (7.36), (7.44), and (7.45) is

$$\dot{x} = f_o(x, I_{d-q}, \overline{V}, u) \tag{7.159}$$

$$I_{d-q} = h(x, \overline{V}) \tag{7.160}$$

$$\overline{I}_o(I_{d-q}, x, \overline{V}) = \overline{Y}_N \overline{V} \tag{7.161}$$

The use of the SI method to solve these equations yields the set of algebraic equations:

$$F_1(x_{n+1}, I_{d-q,n+1}, \overline{V}_{n+1}, u_{n+1}, x_n, I_{d-q,n}, \overline{V}_n, u_n) = 0 \tag{7.162}$$

$$F_2(x_{n+1}, I_{d-q,n+1}, \overline{V}_{n+1}) = 0 \tag{7.163}$$

$$\overline{I}_o(I_{d-q,n+1}, x_{n+1}, \overline{V}_{n+1}) = \overline{Y}_N \overline{V}_{n+1} \tag{7.164}$$

Instead of treating x and I_{d-q} as separate vectors, we form a new vector $X = [X_1 \ldots X_m^t]^t$. Associate $I_{d-q,i}$ with the respective x_i so that $X_i = [x_i^t \ I_{d-qi}^t]^t$. Equations (7.162) and (7.163) are replaced by

$$F_M(X_{n+1}, V_{n+1}^e, u_{n+1}, X_n, V_n^e, u_n) = 0$$

Also, we replace (7.164) by its rectangular equivalent $I^e(X_{n+1}, V_{n+1}^e) = Y_N^e V_{n+1}^e$, where e stands for expanded form, i.e.,

$$I^e = [I_{D1} \ I_{Q1} \ldots I_{Dn} \ I_{Qn}]^t$$

$$V^e = [V_{D1} \ V_{Q1} \ldots V_{Dn} \ V_{Qn}]^t$$

and Y_N^e consists of 2×2 square matrices with real numbers. With this, the application of Newton's method yields the following equations.

$$\begin{bmatrix} A_{GG1} & & & B_{GV1} \\ & \ddots & & \vdots \\ & & A_{GGm} & B_{GVm} \\ C_{VG1} & \cdots & C_{VGm} & Y_N^e + Y_L^e \end{bmatrix} \begin{bmatrix} \Delta X_1 \\ \vdots \\ \Delta X_m \\ \Delta V_1^e \\ \vdots \\ \Delta V_n^e \end{bmatrix} = \begin{bmatrix} R_{G1} \\ \vdots \\ R_{Gm} \\ R_{V1}^e \\ \vdots \\ R_{Vn}^e \end{bmatrix}$$

$$(7.165)$$

The various elements of the Jacobian are defined as

$$A_{GGi} \triangleq \frac{\partial F_M}{\partial X_{n+1,i}}, \quad B_{GVi} \triangleq \frac{\partial F_M}{\partial V_{n+1,i}^e}$$

$$C_{VGi} \triangleq -\frac{\partial I^e}{\partial X_{n+1,i}} \quad i = 1, 2, \ldots, m$$

and

$$Y_L^e = -\frac{\partial I^e}{\partial V_{n+1}^e}$$

For ease of understanding, V_i^e can be considered as a vector $(V_{Di}, V_{Qi})^t$, and $I_i^e = (I_{Di}, I_{Qi})^t$. Y_N^e becomes a $2n \times 2n$ matrix of real elements. The right-hand sides of (7.165) are the residuals. In Newton's method for solving $f(x) = 0$, the residuals at any iteration k are $-f(x^{(k)})$. Y_L^e is computed as follows (the suffix $n+1$ is dropped for ease of notation):

$$Y_L^e \triangleq -\left[\frac{\partial I^e}{\partial V^e}\right] \qquad (7.166)$$

Y_L^e consists of 2×2 blocks of the type

$$-\begin{bmatrix} \frac{\partial I_{Di}}{\partial V_{Di}} & \frac{\partial I_{Di}}{\partial V_{Qi}} \\ \frac{\partial I_{Qi}}{\partial V_{Di}} & \frac{\partial I_{Qi}}{\partial V_{Qi}} \end{bmatrix} \tag{7.167}$$

The solution method to solve the linear equation (7.165) is as follows. We recognize that the nonzero columns in B_{GVi} correspond to ΔV_i^e, and that those in C_{GVi} correspond to ΔX_i. Thus, from (7.165)

$$\Delta X_i = A_{GGi}^{-1}(R_{Gi} - B_{GVi}\Delta V^e) \quad i = 1, \ldots, m \tag{7.168}$$

From (7.165),

$$[Y_N^e + Y_L^e]\Delta V^e + \sum_{i=1}^{m}(C_{VGi}\Delta X_i) = [R_V^e] \tag{7.169}$$

Substituting ΔX_i from (7.168),

$$\left[Y^e + Y_L^e - \sum_{i=1}^{m}(C_{VGi}A_{GGi}^{-1}B_{GVi})\right][\Delta V^e] = [R_V^{e'}] \tag{7.170}$$

where $R_V^{e'} = R_V^e - \sum_{i=1}^{m}C_{VGi}A_{GGi}^{-1}R_{Gi}$.

Thus, the algorithm first solves for ΔV^e in (7.170) and then solves (7.169) iteratively to convergence for ΔX_i ($i = 1, \ldots, m$). The same computations are then repeated at the next time instant and so on.

Some Practical Details [70]

The Jacobian in (7.165) is expensive to compute at each iteration. Let $J_{n+1}^{(k)}$ be the Jacobian evaluated at $t = t_{n+1}$ and k represent the iteration count at that time instant. Solving (7.169) and (7.170) results in

$$\begin{bmatrix} X^{(k+1)} \\ V^{e(k+1)} \end{bmatrix} = \begin{bmatrix} X^{(k)} \\ V^{e(k)} \end{bmatrix} + \begin{bmatrix} \Delta X^{(k)} \\ \Delta V^{e(k)} \end{bmatrix} \quad k = 0, 1, 2, \ldots \tag{7.171}$$

The very dishonest Newton method (VDHN) holds the Jacobian in (7.165) fixed for a period of time. Thus

$$\left[J_{n+1}^{(o)}\right]\begin{bmatrix} \Delta X^{(k)} \\ \Delta V^{e(k)} \end{bmatrix} = \left[R_\ell^{(k)}\right] \tag{7.172}$$

where the time instant $\ell \geq n + 1$.

This means that the initial Jacobian at $t = t_{n+1}$ is held constant for some time steps after t_{n+1}. This reduces the overall cost of computation. The choice of when to reevaluate the Jacobian is based on experience. The Jacobian must be reevaluated at any major system change. Between time steps, it is reevaluated if the previous time-step iteration was considered too slow (took three or more iterations). Within a time step, a maximum of five iterations are taken using the same Jacobian.

Prediction [70]

Whenever a Jacobian is evaluated at the beginning of a time step, instead of taking the converged values of the previous time step, we can use a linear prediction for generator variables and geometric prediction for network voltages. Thus, at time instant t_{n+1}, the initial estimate in the evaluation of the Jacobian would be

$$X_{n+1}^{(o)} = X_n + (X_n - X_{n-1}) \tag{7.173}$$

where X_n, X_{n-1} are the previous converged values at the previous time steps.

For the network variables, each voltage initial guess is

$$V_{n+1}^{(o)} = \frac{(V_n)^2}{V_{n-1}} \tag{7.174}$$

7.9   Reduced-Order Multimachine Models

While many types of reduced-order multimachine models are possible, we discuss three specific ones.

1. Flux-decay model with a fast exciter. The network structure is preserved.

2. Structure-preserving model with a classical machine model.

3. Classical model with network nodes eliminated.

7.9.1 Flux-decay model

This model is widely used in eigenvalue analysis and power-system stabilizer design. If the damper-winding constants are very small, then we can set them to zero (i.e., there is an integral manifold for these states, as discussed in Section 6.5), and we obtain from (7.2):

$$0 = -E'_{di} + (X_{qi} - X'_{qi})I_{qi} \quad i = 1, ..., m \tag{7.175}$$

We eliminate E'_{di} from (7.4) and (7.8) using (7.175). The synchronous machine dynamic circuit is modified as shown in Figure 7.12. It is also common, while using the flux-decay model, to have a simplified exciter with one gain and one time constant, as shown in Figure 7.13. The complete set of differential-algebraic equations (7.1)–(7.10) become (assuming no governor and no damping):

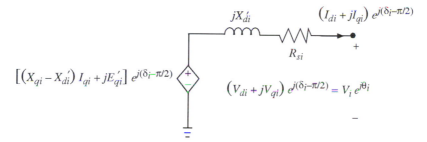

Figure 7.12: *Synchronous machine flux-decay model dynamic circuit*

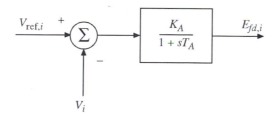

Figure 7.13: *Static exciter (one gain–one time constant)*

Generator Equations

$$T'_{doi}\frac{dE'_{qi}}{dt} = -E'_{qi} - (X_{di} - X'_{di})I_{di} + E_{fdi} \quad i = 1, \ldots, m \tag{7.176}$$

$$\frac{d\delta_i}{dt} = \omega_i - \omega_s \quad i = 1, \ldots, m \tag{7.177}$$

$$\frac{2H_i}{\omega_s}\frac{d\omega_i}{dt} = T_{Mi} - E'_{qi}I_{qi} - (X_{qi} - X'_{di})I_{di}I_{qi} \quad i = 1, \ldots, m$$

$$(7.178)$$

$$T_{Ai}\frac{dE_{fdi}}{dt} = -E_{fdi} + K_{Ai}(V_{\text{ref}i} - V_i) \quad i = 1, \ldots, m \qquad (7.179)$$

Stator Equations

Assuming $R_{si} = 0$, and substituting for E'_{di} from (7.175), we obtain the stator equations (7.13) and (7.14) in polar form as

$$V_i \sin(\delta_i - \theta_i) - X_{qi}I_{qi} = 0 \quad i = 1, \ldots, m \qquad (7.180)$$
$$V_i \cos(\delta_i - \theta_i) + X'_{di}I_{di} - E'_{qi} = 0 \quad i = 1, \ldots, m \qquad (7.181)$$

Network Equations

The network equations can be written in the power-balance form as in Section 7.3.1, or in the current-balance form as in Section 7.3.2.

Initial Conditions

The steps to compute the initial conditions of the multimachine flux-decay model DAE system are given below.

<u>Step 1</u> From the load flow, compute $I_{Gi}e^{j\gamma_i}$ as in (7.113) as $(P_{Gi} - jQ_{Gi})/V_i e^{-j\theta_i}$.

<u>Step 2</u> Compute δ_i as angle of $[V_i e^{j\theta_i} + jX_{qi}I_{Gi}e^{j\gamma_i}]$.

<u>Step 3</u> Compute $\omega_i = \omega_s$ from (7.177).

<u>Step 4</u> Compute I_{di}, I_{qi} from $(I_{di} + jI_{qi}) = I_{Gi}e^{j(\gamma_i - \delta_i + \pi/2)}$.

<u>Step 5</u> Compute E'_{qi} from (7.181) as $E'_{qi} = V_i \cos(\delta_i - \theta_i) + X'_{di}I_{di}$.

<u>Step 6</u> Compute E_{fdi} from (7.176) as $E_{fdi} = E'_{qi} + (X_{di} - X'_{di})I_{di}$.

<u>Step 7</u> Compute $V_{\text{ref}i}$ from (7.179) as $V_{\text{ref},i} = \frac{E_{fdi}}{K_{Ai}} + V_i$.

Step 8 Compute T_{Mi} from (7.178) as $T_{Mi} = E'_{qi}I_{qi} + (X_{qi} - X'_{di})I_{di}I_{qi}$.

7.9.2 Structure-preserving classical model

To obtain the classical model, we set $T'_{doi} = \infty$ and $X_{qi} = X'_{di}$ in (7.176) and (7.178), respectively. This results in E'_{qi} being a constant equal to the initial value E'^{o}_{qi}. Ignoring (7.176) and (7.179), the resulting differential equations are

$$\frac{d\delta_i}{dt} = \omega_i - \omega_s \quad i = 1, \dots, m \tag{7.182}$$

$$\frac{2H_i}{\omega_s}\frac{d\omega_i}{dt} = T_{Mi} - E'^{o}_{qi}I_{qi} \quad i = 1, \dots, m \tag{7.183}$$

These are known as the swing equations in the literature. The stator algebraic equations (7.180) and (7.181) are added after multiplying the former by $-j$. We replace X_{qi} by X'_{di} and E'_{qi} by E'^{o}_{qi}. Thus

$$\begin{aligned} E'^{o}_{qi} &= X'_{di}(I_{di} + jI_{qi}) + V_i(\cos(\delta_i - \theta_i) - j\sin(\delta_i - \theta_i)) \\ &= X'_{di}(I_{di} + jI_{qi}) + V_i e^{-j(\delta_i - \theta_i)} \end{aligned} \tag{7.184}$$

Equation (7.184) can be rearranged as

$$\begin{aligned} E'^{o}_{qi}e^{j\delta_i} &= jX'_{di}(I_{di} + jI_{qi})e^{j(\delta_i - \pi/2)} + V_i e^{j\theta_i} \\ &= jX'_{di}(I_{Di} + jI_{Qi}) + V_i e^{j\theta_i} \end{aligned} \tag{7.185}$$

This represents a voltage $\overline{E}_i = E'^{o}_{qi}e^{j\delta_i}$ of constant magnitude behind a transient reactance X'_{di} (Figure 7.14). Henceforth, we will denote $E'^{o}_{qi} = E_i$.

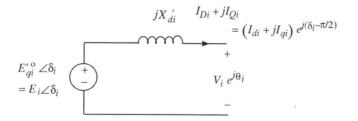

Figure 7.14: *Constant voltage behind transient reactance*

This forms the basis of the structure-preserving model which, is discussed next.

Structure-Preserving Model with Constant Voltage Behind Reactance

This model is widely used in structure-preserving transient energy function and voltage collapse literature that uses energy functions. It consists of the swing equations (7.182) and (7.183) and the network algebraic equations at the nodes $1, \ldots, n$. In Figure 7.15, the n bus transmission network is shown augmented by the constant voltage behind reactances at the generator buses $1, \ldots, m$. The generator internal nodes are denoted as $n+1, \ldots, n+m$. The complex power output at the ith internal node in Figure 7.15 can be expressed as

$$
\begin{aligned}
\overline{E}_i \overline{I}_i^* &= E_i e^{j\delta_i} ((\overline{E}_i - \overline{V}_i)/jX'_{di})^* \\
&= E_i e^{j\delta_i} (E_i e^{-j\delta_i} - V_i e^{-j\theta_i})/(-jX'_{di}) \\
&= j\frac{E_i^2}{X'_{di}} - j\frac{E_i V_i}{X'_{di}}((\cos(\delta_i - \theta_i) + j\sin(\delta_i - \theta_i)) \\
&= \frac{E_i V_i \sin(\delta_i - \theta_i)}{X'_{di}} + j\left(\frac{E_i^2}{X'_{di}} - \frac{E_i V_i \cos(\delta_i - \theta_i)}{X'_{di}}\right) \quad (7.186)
\end{aligned}
$$

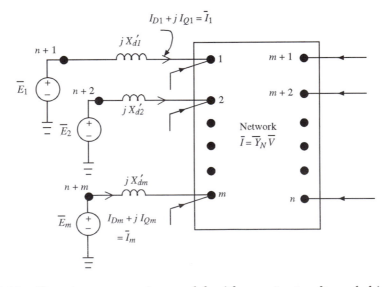

Figure 7.15: *Structure-preserving model with constant voltage behind reactance*

Alternatively,

$$\overline{E}_i\overline{I}_i^* = E_ie^{j\delta_i}(I_{di} - jI_{qi})e^{-j(\delta_i-\pi/2)}$$
$$= E_iI_{qi} + jE_iI_{di} \qquad (7.187)$$

Comparing (7.186) and (7.187), the real and reactive power output at the generator internal nodes are given, respectively, by

$$\text{Real power } E_iI_{qi} = \frac{E_iV_i\sin(\delta_i - \theta_i)}{X'_{di}} \qquad (7.188)$$

$$\text{Reactive power } E_iI_{di} = \frac{E_i^2}{X'_{di}} - \frac{E_iV_i}{X'_{di}}\cos(\delta_i - \theta_i) \qquad (7.189)$$

This shows that real power P is associated with I_q and reactive power is associated with I_d. At the generator buses $1, \ldots, m$, we can express the complex power due to the generators as

$$P_{Gi} + jQ_{Gi} = \overline{V}_i\overline{I}_i^*$$
$$\overline{V}_i\overline{I}_i^* = V_ie^{j\theta_i}\left[\left(\overline{E}_i - \overline{V}_i\right)/jX'_{di}\right]^*$$
$$= V_ie^{j\theta_i}\left(E_ie^{-j\delta_i} - V_ie^{-j\theta_i}\right)/(-jX'_{di})$$
$$= E_iV_i\left(\cos\left(\theta_i - \delta_i\right) + j\sin\left(\theta_i - \delta_i\right)\right)\frac{j}{X'_{di}} - \frac{jV_i^2}{X'_{di}}$$
$$= -\frac{E_iV_i\sin(\theta_i - \delta_i)}{X'_{di}} + j\left(\frac{-V_i^2}{X'_{di}} + \frac{E_iV_i\cos(\theta_i - \delta_i)}{X'_{di}}\right) \qquad (7.190)$$

Therefore

$$P_{Gi} = \frac{-E_iV_i\sin(\theta_i - \delta_i)}{X'_{di}} = \frac{E_iV_i\sin(\delta_i - \theta_i)}{X'_{di}} \qquad (7.191)$$

$$Q_{Gi} = \frac{-V_i^2}{X'_{di}} + \frac{E_iV_i\cos(\theta_i - \delta_i)}{X'_{di}} \qquad (7.192)$$

The structure-preserving model consists of the swing and the network equations (neglecting resistance) as

$$\dot{\delta}_i = \omega_i - \omega_s \qquad i = 1, \ldots, m \qquad (7.193)$$

$$\frac{2H_i}{\omega_s}\dot{\omega}_i = T_{Mi} - \frac{E_iV_i\sin(\delta_i - \theta_i)}{X'_{di}} \qquad i = 1, \ldots, m \qquad (7.194)$$

$$P_{Li}(V_i) + P_{Gi} = \sum_{j=1}^{n} V_i V_j B_{ij} \sin(\theta_i - \theta_j) \quad i = 1, \ldots, n \qquad (7.195)$$

$$Q_{Li}(V_i) + Q_{Gi} = -\sum_{j=1}^{n} V_i V_j B_{ij} \cos(\theta_i - \theta_j) \quad i = 1, \ldots, n \qquad (7.196)$$

where P_{Gi} and Q_{Gi} for $i = 1, \ldots, m$ are given by (7.191) and (7.192), and $P_{Gi} = Q_{Gi} = 0$ for $i = m + 1, \ldots, n$. The right-hand sides of (7.195) and (7.196) are the sums of real and reactive power on the lines emanating from bus i under the assumption of negligible transmission line resistance. This is shown as follows. If we neglect transmission line resistances, then the network admittance matrix is $\overline{Y}_N = [jB_{ij}]$, and the total complex power in the network transmission lines from bus i is given by

$$\overline{V}_i \left(\sum_{j=1}^{n} jB_{ij}\overline{V}_j \right)^* \qquad (7.197)$$

Expanding the expression (7.197) results in

$$V_i e^{j\theta_i} \sum_{j=1}^{n} -jB_{ij}V_j e^{-j\theta_j}$$

$$= \sum_{j=1}^{n} -jV_i V_j B_{ij}(\cos(\theta_i - \theta_j) + j\sin(\theta_i - \theta_j))$$

$$= \sum_{j=1}^{n} V_i V_j B_{ij} \sin(\theta_i - \theta_j) - j \sum_{j=1}^{n} V_i V_j B_{ij} \cos(\theta_i - \theta_j) \qquad (7.198)$$

The real and imaginary parts of (7.198) are the right-hand sides of (7.195) and (7.196), respectively. The model given by (7.193)–(7.196) is called the *structure-preserving classical model*.

An interesting observation from (7.189) and (7.192) regarding reactive power is the following. The reactive power absorbed by X'_{di} is obtained by substracting Q_{Gi} from $E_i I_{di}$, resulting in

$$E_i I_{di} - Q_{Gi} = (E_i^2 - 2E_i V_i \cos(\delta_i - \theta_i) + V_i^2)/X'_{di} \qquad (7.199)$$

An alternative form of the structure-preserving model is to consider the augmented Y matrix obtained by including the admittance corresponding to the transient reactances of the machines. Thus, with proper ordering,

$$\overline{Y}_{\text{aug}} = \begin{array}{c} n+1 \\ \vdots \\ n+m \\ \\ 1 \\ \vdots \\ m \\ \\ m+1 \\ \vdots \\ n \end{array} \begin{pmatrix} \overbrace{\hspace{1cm}}^{n+1\ldots n+m} & \overbrace{\hspace{1cm}}^{1\ldots m} & \overbrace{\hspace{1cm}}^{m+1\ldots n} \\ \overline{y} & -\overline{y} & 0 \\ \hline -\overline{y} & & \\ \hline & \overline{Y}_{N1} & \\ 0 & & \end{pmatrix} \qquad (7.200)$$

where

$$\overline{y} = \text{Diag}\left(\frac{1}{jX'_{di}}\right) \quad i = 1, \ldots, m$$

and

$$\overline{Y}_{N1} = \overline{Y}_N + \begin{bmatrix} \overline{y} & 0 \\ 0 & 0 \end{bmatrix}$$

Considering $\overline{Y}_{\text{aug}} = [jB_{ij}]$, (7.193)–(7.196) can be written more compactly as (denoting, temporarily, δ_i's as θ_i's and E_i's as V_i's).

$$\dot{\theta}_i = \dot{\omega}_i - \omega_s \quad i = n+1, \ldots, n+m \qquad (7.201)$$

$$\frac{2H_i}{\omega_s}\dot{\omega}_i = T_{Mi} - \sum_{j=1}^{n+m} V_i V_j B_{ij} \sin(\theta_i - \theta_j)$$

$$i = n+1, \ldots, n+m \qquad (7.202)$$

$$P_{Li}(V_i) = \sum_{j=1}^{n+m} V_i V_j B_{ij} \sin(\theta_i - \theta_j) \quad i = 1, \ldots, n \qquad (7.203)$$

$$Q_{Li}(V_i) = -\sum_{j=1}^{n+m} V_i V_j B_{ij} \cos(\theta_i - \theta_j) \quad i = 1, \ldots, n \qquad (7.204)$$

It can be verified that, in (7.202), the second term on the right-hand side is only $E_i V_i \sin(\delta_i - \theta_i)/X'_{di}$. It is easy to verify the equivalence of (7.203) and (7.204) with (7.195) and (7.196), respectively. Note that, in this model, we are allowed to have nonlinear load representation. This model is used in voltage stability studies by means of the energy function method [117].

7.9.3   Internal-node model

This is a widely used model in first-swing transient stability analysis. In this model, the loads are assumed to be constant impedances and converted to admittances as

$$\overline{y}_{Li} = \frac{-(P_{Li} - jQ_{Li})}{V_i^2} \quad i = 1, \ldots, n \qquad (7.205)$$

There is a negative sign for $\overline{y}_{Li}$, since loads are assumed as injected quantities. Adding these to the diagonal elements of the $\overline{Y}_{N1}$ matrix in (7.200) makes it $\overline{Y}_{N2} = \overline{Y}_{N1} + \mathrm{Diag}(\overline{y}_{Li})$. The modified augmented Y matrix becomes

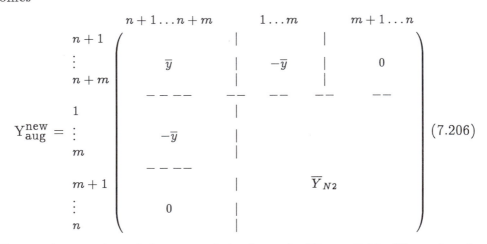

$$\qquad (7.206)$$

The passive portion of the network is shown in Figure 7.16. The network equations for the new augmented network can be rewritten as

$$
\begin{matrix} m \\ n \end{matrix}
\begin{bmatrix} \overline{I}_A \\ 0 \end{bmatrix}
=
\begin{matrix} m \\ n \end{matrix}
\begin{pmatrix} \overline{Y}_A & \overline{Y}_B \\ \overline{Y}_C & \overline{Y}_D \end{pmatrix}
\begin{bmatrix} \overline{E}_A \\ \overline{V}_B \end{bmatrix}
\qquad (7.207)
$$

where $\overline{Y}_A = \overline{y}$, $\overline{Y}_B = [-\overline{y} \mid 0]$, $\overline{Y}_C = \begin{bmatrix} -\overline{y} \\ \hline 0 \end{bmatrix}$, and $\overline{Y}_D = \overline{Y}_{N2}$. The n network buses can be eliminated, since there is no current injection at these buses. Thus

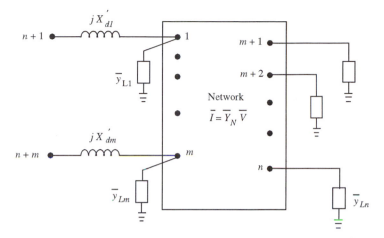

Figure 7.16: *Augmented Y matrix with constant impedance*

$$\overline{I}_A \;=\; \left(\overline{Y}_A - \overline{Y}_B \overline{Y}_D^{-1} \overline{Y}_C\right) \overline{E}_A$$
$$\;=\; \overline{Y}_{\text{int}} \overline{E}_A \tag{7.208}$$

where the elements of $\overline{I}_A$ and $\overline{E}_A$ are, respectively, $\overline{I}_i = (I_{di} + j I_{qi})e^{j(\delta_i - \pi/2)}$ $= I_{Di} + j I_{Qi}$ and $\overline{E}_i = E_i \angle \delta_i$. The elements of $\overline{Y}_{\text{int}}$ are $\overline{Y}_{ij} = G_{ij} + j B_{ij}$. Since the network buses have been eliminated, we may renumber the internal nodes as $1, \ldots, m$ for ease of notation.

$$\overline{I}_i = \sum_{j=1}^{m} \overline{Y}_{ij} \overline{E}_j \qquad i = 1, \ldots, m \tag{7.209}$$

Real electrical power out of the internal node i in Figure 7.16 is given by

$$P_{ei} = Re[\overline{E}_i \overline{I}_i^*]$$
$$= Re\left[E_i e^{j\delta_i} \sum_{j=1}^{m} \overline{Y}_{ij}^* \overline{E}_j^* \right]$$
$$= Re\left[E_i e^{j\delta_i} \sum_{j=1}^{m} (G_{ij} - j B_{ij}) E_j e^{-j\delta_j} \right]$$
$$= Re\left[\sum_{j=1}^{m} (G_{ij} - j B_{ij}) E_i E_j [\cos(\delta_i - \delta_j) + j \sin(\delta_i - \delta_j)] \right] \tag{7.210}$$

Define

$$\delta_i - \delta_j \overset{\Delta}{=} \delta_{ij} \tag{7.211}$$

Then

$$P_{ei} = \sum_{j=1}^{m} E_i E_j (G_{ij} \cos \delta_{ij} + B_{ij} \sin \delta_{ij})$$

$$= E_i^2 G_{ii} + \sum_{\substack{j=1 \\ \neq i}}^{m} (C_{ij} \sin \delta_{ij} + D_{ij} \cos \delta_{ij}) \tag{7.212}$$

where

$$C_{ij} = E_i E_j B_{ij} \tag{7.213}$$
$$D_{ij} = E_i E_j G_{ij} \tag{7.214}$$

Thus, the classical model is

$$\frac{d\delta_i}{dt} = \omega_i - \omega_s \tag{7.215}$$

$$\frac{2H_i}{\omega_s} \frac{d\omega_i}{dt} = T_{Mi} - P_{ei} \quad i = 1, \ldots, m \tag{7.216}$$

where P_{ei} is given by (7.212). Since P_{ei} is a function of the δ_i's, (7.215)–(7.216) can be integrated by any numerical algorithm.

7.10   Initial Conditions

Step 1 From the load flow, compute $\overline{I}_i = I_{Di} + jI_{Qi}$ as $(P_{Gi} - jQ_{Gi})/V_i e^{-j\theta_i}$.

Step 2 Using (7.185), compute E_i, δ_i as $E_i \angle \delta_i = V_i e^{j\theta_i} + jX'_{di}(I_{Di} + jI_{Qi})$.

Step 3 From (7.193) or (7.215), $\omega_i = \omega_s$.

Thus, computation of initial conditions is simple when a classical model is used.

Example 7.6

Compute $E_i \angle \delta_i$ ($i = 1,2,3$) and $\overline{Y}_{\text{int}}$ for Example 7.1 using the classical model (all in pu).

We first compute $\overline{I}_i$ ($i = 1,2,3$)

$$\overline{I}_1 = \frac{0.716 - j0.27}{1.04\angle 0^\circ} = 0.7358\angle -20.66^\circ$$

$$\overline{I}_2 = \frac{1.63 - j0.067}{1.025\angle -9.3^\circ} = 1.5916\angle 6.947^\circ$$

$$\overline{I}_3 = \frac{0.85 + j0.109}{1.02\angle -4.7^\circ} = 0.8361\angle 12.0^\circ$$

$$E_1\angle\delta_1 = 1.04\angle 0^\circ + j0.068\overline{I}_1 = 1.054\angle 2.267^\circ$$

$$E_2\angle\delta_2 = 1.025\angle 9.3^\circ + j0.1198\overline{I}_2 = 1.050\angle 19.75^\circ$$

$$E_3\angle\delta_3 = 1.02\angle 4.7^\circ + j0.1813\overline{I}_3 = 1.017\angle 13.2^\circ$$

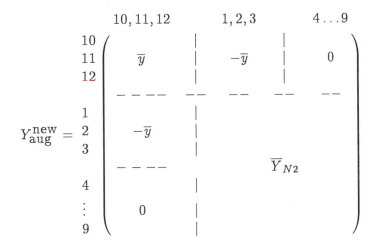

where

$$\overline{y} = \begin{bmatrix} -j16.45 & 0 & 0 \\ 0 & -j8.35 & 0 \\ 0 & 0 & -j5.52 \end{bmatrix}$$

and $\overline{Y}_{N2} = \overline{Y}_{N1} + \begin{bmatrix} \overline{y} & 0 \\ 0 & 0 \end{bmatrix} + \text{Diag}(\overline{y}_{Li})$.

$\overline{Y}_{\text{int}}$ is obtained by eliminating nodes 1 ...9 as in (7.208), and is given by

$$
\overline{Y}_{\text{int}} = \begin{bmatrix}
0.845 - j2.988 & 0.287 + j1.513 & 0.210 + j1.226 \\
0.287 + j1.513 & 0.420 - j2.724 & 0.213 + j1.088 \\
0.210 + j1.226 & 0.213 + j1.088 & 0.277 - j2.368
\end{bmatrix}
$$

□

7.11   Conclusion

This chapter has discussed the formulation of a multimachine system model with the two-axis model, as well as the reduced-order flux-decay model and the classical model. The computation of initial conditions and solution methodology for the two-axis model using the simultaneous-implicit method has been discussed. The methodology of the ETMSP program of EPRI has also been discussed. The reduced-order flux-decay model and the classical model, as well as the computation of initial conditions, are discussed. We also discussed the structure-preserving classical model.

7.12   Problems

7.1 Perform a load flow for the 3-machine system of Example 7.1 by varying the load at bus 5 in increments of 0.5 until $P_L = 4.5$ pu. Plot the voltage magnitude at bus 5 as a function of the load. This is called a *PV* curve.

7.2 In Problem 7.1, the increased load at each increment is allocated to slack bus 1. In practice, the AGC system allocates it through the area control error, etc. As an alternative, consider allocating the increased load in proportion to the inertias of the machine, i.e., ΔP_{L5} is allocated to buses 2 and 3 as $P_{G2}^{k+1} = P_{G2}^k + \frac{H_2}{H_T}\Delta P_{L5}$ and $P_{G3}^{k+1} = P_{G3}^k + \frac{H_3}{H_T}\Delta P_{L5}$, where P_{G2}^k and P_{G3}^k are the generator powers before the load is increased and $H_T = H_1 + H_2 + H_3$. Again draw the *PV* curve.

7.3 Repeat Problems 7.1 and 7.2 for a contingency of lines 5 to 7 being outaged. Draw the *PV* curves for the pre-contingency state and the contingent case on the same graph.

7.4 This problem and Problem 7.5 can be done using symbolic software. As explained at the end of Example 7.1, I_{di}, I_{qi} can be eliminated from (7.47) as well as (7.54) and (7.55) using (7.53). Express the resulting equations in the form

$$\dot{x} = f_1(x, \overline{V}, u)$$
$$0 = g_1(x, \overline{V})$$

This is the DAE model with the algebraic equations in the power-balance form ((7.39) and (7.40)).

7.5 Using (7.53), substitute for I_{di}, I_{qi} ($i = 1,2,3$) in (7.57) to express the DAE in the form

$$\dot{x} = f(x, \overline{V}, u)$$
$$\overline{I}_1(x, \overline{V}) = \overline{Y}_N \overline{V}$$

7.6 Derive $\overline{I}_1(x, \overline{V}) = \overline{Y}_N \overline{V}$ directly in Problem 7.5 from $0 = g_1(x, \overline{V})$ computed in Problem 7.4. (Hint: See Section 7.3.2.)

7.7 Express Example 7.1 in the form

$$\dot{x} = A(x) + BW + Cu$$
$$W = G(E, \overline{V})$$
$$0 = g_1(x, \overline{V})$$

and Example 7.2 in the form

$$\dot{x} = A(x) + BW + Cu$$
$$W = G(E, \overline{V})$$
$$\overline{I}_1(x, \overline{V}) = \overline{Y}_N \overline{V}$$

7.8 In Example 7.1, loads (both real and reactive) at buses 5 and 6 are increased by 50 percent. With the new load flow compute the initial conditions for the variables, I_{di}, I_{qi}, V_{di}, V_{qi}, V_{ref_i} and T_{Mi} ($i = 1,2,3$).

7.9 Under nominal loading conditions of Example 7.1, there is a three-phase ground fault at bus 5 that is self-clearing in six cycles. Do the dynamic simulation using the SI method or any available commercial software for 0–2 sec. Use the two-axis model and the IEEE-Type 1 exciter data in Example 7.1. You may use the model obtained in Problems 7.4, 7.5, or 7.7.

7.10 Repeat Problem 7.8 for the flux-decay model and a fast exciter with $K_A = 25$ and $T_A = 0.2$ sec. Do the simulation as in Problem 7.9.

7.11 In Problems 7.9 and 7.10, find the gain K_A at which the system will become unstable. (Instability occurs when relative rotor angles diverge.)

7.12 In Problems 7.9 and 7.10, find by repetitive simulation the value t_{cr}, the critical-clearing time at which the system becomes unstable.

7.13 Repeat Problems 7.9 and 7.10 for the fault at bus 5 followed by clearing of lines 5 to 7. Assume t_{cl} = six cycles. Also find the critical-clearing time t_{cr}.

7.14 For Example 7.1, write the structure-preserving model in the form of (7.201)–(7.204).

7.15 Obtain $\overline{Y}_{\text{int}}$ for the system in Example 7.1 for both the faulted state and post-fault states for (a) self-clearing fault at bus 5 and (b) fault at bus 5 followed by switching of the lines 5 to 7. Write the equations in the form (7.215)–(7.216).

7.16 Find t_{cr} in Problem 7.15 for both cases by repetitive simulation, and compare the results with Problems 7.12 and 7.13, respectively.

Chapter 8

SMALL-SIGNAL STABILITY

8.1 Background

In this chapter, we consider linearized analysis of multimachine power systems that is necessary for the study of both steady-state and voltage stability. In many cases, instability and eventual loss of synchronism are initiated by some spurious disturbance in the system resulting in oscillatory behavior that, if not damped, may eventually build up. This is very much a function of the operating condition of the power system. Oscillations, even if undamped at low frequencies, are undesirable because they limit power transfers on transmission lines and, in some cases, induce stress in the mechanical shaft. The source of inter-area oscillations is difficult to diagnose. Extensive research has been done in both of these areas. In recent years, there has been considerable interest in dynamic voltage collapse. As regional transfers vary over a wide range due to restructuring and open transmission access, certain parts of the system may face increased loading conditions. Earlier, this phenomenon was analyzed purely on the basis of static considerations, i.e., load-flow equations. In this chapter, we develop a comprehensive dynamic model to study both low-frequency oscillations and voltage stability using a two-axis model with IEEE-Type I exciter, as well as the flux-decay model with a high-gain fast exciter. Both the electromechanical oscillations and their damping, as well as dynamic voltage stability, are discussed. The electromechanical oscillation is of two types:

1. Local mode, typically in the 1 to 3-Hz range between a remotely located power station and the rest of the system.

2. Inter-area oscillations in the range of less than 1 Hz.

Two kinds of analysis are possible: (1) A multimachine linearized analysis that computes the eigenvalues and also finds those machines that contribute to a particular eigenvalue (both local and inter-area oscillations can be studied in such a framework); (2) a single-machine infinite-bus system case that investigates only local oscillations.

Dynamic voltage stability is analyzed by monitoring the eigenvalues of the linearized system as a power system is progressively loaded. Instability occurs when a pair of complex eigenvalues cross to the right-half plane. This is referred to as dynamic voltage instability. Mathematically, it is called Hopf bifurcation.

Also discussed in this chapter is the role of a power system stabilizer that stabilizes a machine with respect to the local mode of oscillation. A brief review of the approaches to the design of the stabilizers is given. For detailed design procedures, it is necessary to refer to the literature. References [77] and [78] are the basic works in this area.

8.2 Basic Linearization Technique

A unified framework is presented in this section for the linear analysis of multimachine systems. The nonlinear model derived in Chapter 7, with a two-axis model with IEEE-Type I exciter or flux-decay model with a static exciter, is of the form

$$\dot{x} = f(x, y, u) \tag{8.1}$$

$$0 = g(x, y) \tag{8.2}$$

where the vector y includes both the I_{d-q} and $\overline{V}$ vectors. Thus, (8.1) is of dimension $7m$, and (8.2) is of dimension $2(n + m)$.

Equation (8.2) consists of the stator algebraic equations and the network equations in the power-balance form. To show explicitly the traditional load-flow equations and the other algebraic equations, we partition y as

$$y = [I_{d-q}^t \; \theta_1 \; V_1 \ldots V_m \mid \theta_2 \ldots \theta_n \; V_{m+1} \ldots V_n]^t$$

$$= [y_a^t \mid y_b^t]^t \tag{8.3}$$

Here, the vector y_b corresponds to the load-flow variables, and the vector y_a corresponds to the other algebraic variables. Bus 1 is taken as the slack bus and buses $2, \ldots, m$ are the PV buses with the buses $m+1, \ldots, n$ being the PQ buses. The dimension of x is $7m$. Linearizing (8.1) and (8.2) around an operating point gives

$$\begin{bmatrix} \frac{d}{dt}\Delta x \\ 0 \\ 0 \end{bmatrix} = \begin{bmatrix} A & B \\ \hline C & \begin{array}{c|c} D_{11} & D_{12} \\ \hline D_{21} & J_{LF} \end{array} \end{bmatrix} \begin{bmatrix} \Delta x \\ \Delta y_a \\ \Delta y_b \end{bmatrix} + E[\Delta u] \qquad (8.4)$$

Define

$$J_{AE} = \begin{bmatrix} D_{11} & D_{12} \\ D_{21} & J_{LF} \end{bmatrix} \qquad (8.5)$$

Eliminating Δy_a, Δy_b we get $\Delta \dot{x} = A_{\text{sys}} \Delta x$ where $A_{\text{sys}} = (A - BJ_{AE}^{-1}C)$. J_{LF} is the load-flow Jacobian. The model represented by (8.4) is useful in both small-signal stability, analysis and voltage stability, since J_{LF} is explicitly shown as part of the system differential-algebraic Jacobian.

We use the above formulation for a multimachine system with a two-axis machine model and the IEEE-Type I exciter (model A) and indicate a similiar extension for the flux-decay model with a fast exciter (model B). The methodology is based on [79] and [80].

8.2.1   Linearization of Model A

The differential equations of the machine and the exciter are the same as in Chapter 7, except that the state variables are reordered. The synchronous machine dynamic circuit and the IEEE-Type I exciter are shown in Figures 8.1 and 8.2, respectively. The differential and algebraic equations follow.

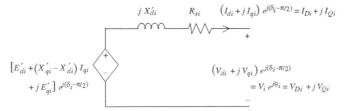

Figure 8.1: *Synchronous machine two-axis model dynamic circuit (i = 1, . . . , m)*

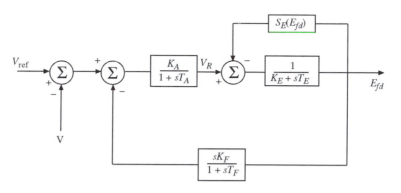

Figure 8.2: *IEEE-Type I exciter model*

Differential Equations

$$\frac{d\delta_i}{dt} = \omega_i - \omega_s \tag{8.6}$$

$$\frac{d\omega_i}{dt} = \frac{T_{Mi}}{M_i} - \frac{[E'_{qi} - X'_{di}I_{di}]I_{qi}}{M_i} - \frac{[E'_{di} + X'_{qi}I_{qi}]I_{di}}{M_i}$$
$$- \frac{D_i(\omega_i - \omega_s)}{M_i} \tag{8.7}$$

$$\frac{dE'_{qi}}{dt} = -\frac{E'_{qi}}{T'_{doi}} - \frac{(X_{di} - X'_{di})I_{di}}{T'_{doi}} + \frac{E_{fdi}}{T'_{doi}} \tag{8.8}$$

$$\frac{dE'_{di}}{dt} = -\frac{E'_{di}}{T'_{qoi}} + \frac{I_{qi}}{T'_{qoi}}(X_{qi} - X'_{qi}) \tag{8.9}$$

$$\frac{dE_{fdi}}{dt} = -\frac{K_{Ei} + S_E(E_{fdi})}{T_{Ei}}E_{fdi} + \frac{V_{Ri}}{T_{Ei}} \tag{8.10}$$

$$\frac{dV_{Ri}}{dt} = -\frac{V_{Ri}}{T_{Ai}} + \frac{K_{Ai}}{T_{Ai}}R_{fi} - \frac{K_{Ai}K_{Fi}}{T_{Ai}T_{Fi}}E_{fdi} + \frac{K_{Ai}}{T_{Ai}}(V_{\text{ref}i} - V_i) \tag{8.11}$$

$$\frac{dR_{Fi}}{dt} = -\frac{R_{Fi}}{T_{Fi}} + \frac{K_{Fi}}{(T_{Fi})^2}E_{fdi} \tag{8.12}$$

$$\text{for } i = 1, \ldots, m$$

Stator Algebraic Equations

The stator algebraic equations in polar form are

$$E'_{di} - V_i \sin(\delta_i - \theta_i) - R_{si}I_{di} + X'_{qi}I_{qi} = 0 \tag{8.13}$$

$$E'_{qi} - V_i \cos(\delta_i - \theta_i) - R_{si}I_{qi} - X'_{di}I_{di} = 0 \qquad (8.14)$$
$$\text{for } i = 1, \ldots, m$$

Network Equations

The network equations are

$$I_{di}V_i \sin(\delta_i - \theta_i) + I_{qi}V_i \cos(\delta_i - \theta_i) + P_{Li}(V_i)$$
$$- \sum_{k=1}^{n} V_i V_k Y_{ik} \cos(\theta_i - \theta_k - \alpha_{ik}) = 0 \qquad (8.15)$$
$$I_{di}V_i \cos(\delta_i - \theta_i) - I_{qi}V_i \sin(\delta_i - \theta_i) + Q_{Li}(V_i)$$
$$- \sum_{k=1}^{n} V_i V_k Y_{ik} \sin(\theta_i - \theta_k - \alpha_{ik}) = 0 \quad i = 1, 2, \ldots, m \ (8.16)$$

$$P_{Li}(V_i) - \sum_{k=1}^{n} V_i V_k Y_{ik} \cos(\theta_i - \theta_k - \alpha_{ik}) = 0 \qquad (8.17)$$
$$Q_{Li}(V_i) - \sum_{k=1}^{n} V_i V_k Y_{ik} \sin(\theta_i - \theta_k - \alpha_{ik}) = 0 \qquad (8.18)$$
$$\text{for } i = m + 1, \ldots, n$$

Equations (8.6)–(8.18) are linearized analytically, as explained below. The linearization of the differential equations (8.6)–(8.12) yields

$$\frac{d\Delta \delta_i}{dt} = \Delta \omega_i \qquad (8.19)$$

$$\frac{d\Delta \omega_i}{dt} = \frac{1}{M_i}\Delta T_{Mi} - \frac{E'_{qio}}{M_i}\Delta I_{qi} + \frac{X'_{di}I_{dio}}{M_i}\Delta I_{qi} + \frac{X'_{di}I_{qio}}{M_i}\Delta I_{di} - \frac{I_{qio}}{M_i}\Delta E'_{qi}$$
$$- \frac{E'_{dio}}{M_i}\Delta I_{di} - \frac{I_{dio}}{M_i}\Delta E_{di} - \frac{X'_{qi}I_{dio}}{M_i}\Delta I_{qi} - \frac{X'_{qi}I_{qio}}{M_i}\Delta I_{di} - \frac{D_i}{M_i}\Delta \omega_i$$

$$\frac{d\Delta E'_{qi}}{dt} = -\frac{\Delta E'_{qi}}{T'_{doi}} - \frac{(X_{di} - X'_{di})\Delta I_{di}}{T'_{doi}} + \frac{\Delta E_{fdi}}{T'_{doi}} \qquad (8.20)$$

$$\frac{d\Delta E'_{di}}{dt} = -\frac{\Delta E'_{di}}{T'_{qoi}} + \frac{(X_{qi} - X'_{qi})}{T'_{qoi}}\Delta I_{qi} \qquad (8.21)$$

$$\frac{d\Delta E_{fdi}}{dt} = f_{si}(E_{fdio})\Delta E_{fdi} + \frac{\Delta V_{Ri}}{T_{Ei}} \qquad (8.22)$$

$$\frac{d\Delta V_{Ri}}{dt} = -\frac{\Delta V_{Ri}}{T_{Ai}} + \frac{K_{Ai}}{T_{Ai}}\Delta R_{Fi} - \frac{K_{Ai}K_{Fi}}{T_{Ai}T_{Fi}}\Delta E_{fdi} - \frac{K_{Ai}}{T_{Ai}}\Delta V_i$$

$$+ \frac{K_{Ai}}{T_{Ai}}\Delta V_{\mathrm{refi}} \tag{8.23}$$

$$\frac{d\Delta R_{Fi}}{dt} = -\frac{\Delta R_{Fi}}{T_{Fi}} + \frac{K_{Fi}}{(T_{Fi})^2}\Delta E_{fdi} \tag{8.24}$$

$$\text{for } i = 1, \ldots, m$$

where $f_{si}(E_{fdio}) = -\frac{K_{Bi}+E_{fdio}\partial S_B(E_{fdio})+S_B(E_{fdio})}{T_{Bi}}$ where the symbol ∂ stands for partial derivative. Writing (8.19) through (8.24) in matrix notation, we obtain

$$
\begin{bmatrix}
\Delta\dot{\delta}_i \\
\Delta\dot{\omega}_i \\
\Delta\dot{E}'_{qi} \\
\Delta\dot{E}'_{di} \\
\Delta\dot{E}'_{fdi} \\
\Delta\dot{V}_{Ri} \\
\Delta\dot{R}_{Fi}
\end{bmatrix}
=
\begin{bmatrix}
0 & 1 & 0 & 0 & 0 & 0 & 0 \\
0 & -\frac{D_i}{M_i} & -\frac{I_{qio}}{M_i} & -\frac{I_{dio}}{M_i} & 0 & 0 & 0 \\
0 & 0 & -\frac{1}{T'_{doi}} & 0 & \frac{1}{T'_{doi}} & 0 & 0 \\
0 & 0 & 0 & -\frac{1}{T'_{qoi}} & 0 & 0 & 0 \\
0 & 0 & 0 & 0 & f_{si}(E_{fdio}) & \frac{1}{T_{Bi}} & 0 \\
0 & 0 & 0 & 0 & -\frac{K_{Ai}K_{Fi}}{T_{Ai}T_{Fi}} & -\frac{1}{T_{Ai}} & \frac{K_{Ai}}{T_{Ai}} \\
0 & 0 & 0 & 0 & \frac{K_{Fi}}{(T_{Fi})^2} & 0 & -\frac{1}{T_{Fi}}
\end{bmatrix}
\begin{bmatrix}
\Delta\delta_i \\
\Delta\omega_i \\
\Delta E'_{qi} \\
\Delta E'_{di} \\
\Delta E_{fdi} \\
\Delta V_{Ri} \\
\Delta R_{Fi}
\end{bmatrix}
$$

$$
+
\begin{bmatrix}
0 & 0 \\
\frac{I_{qio}(X'_{di}-X'_{qi})-E'_{dio}}{M_i} & \frac{I_{dio}(X'_{di}-X'_{qi})-E'_{qio}}{M_i} \\
-\frac{(X_{di}-X'_{di})}{T'_{doi}} & 0 \\
0 & \frac{X_{qi}-X'_{qi}}{T'_{qoi}} \\
0 & 0 \\
0 & 0 \\
0 & 0
\end{bmatrix}
\begin{bmatrix}
\Delta I_{di} \\
\Delta I_{qi}
\end{bmatrix}
$$

$$
+
\begin{bmatrix}
0 & 0 \\
0 & 0 \\
0 & 0 \\
0 & 0 \\
0 & 0 \\
0 & -\frac{K_{Ai}}{T_{Ai}} \\
0 & 0
\end{bmatrix}
\begin{bmatrix}
\Delta\theta_i \\
\Delta V_i
\end{bmatrix}
+
\begin{bmatrix}
0 & 0 \\
\frac{1}{M_i} & 0 \\
0 & 0 \\
0 & 0 \\
0 & 0 \\
0 & \frac{K_{Ai}}{T_{Ai}} \\
0 & 0
\end{bmatrix}
\begin{bmatrix}
\Delta T_{Mi} \\
\Delta V_{\mathrm{refi}}
\end{bmatrix}
$$

$$i = 1, \ldots, m$$

$$\tag{8.25}$$

Denoting $\begin{bmatrix} \Delta I_{di} \\ \Delta I_{qi} \end{bmatrix} = \Delta I_{gi}$, $\begin{bmatrix} \Delta\theta_i \\ \Delta V_i \end{bmatrix} = \Delta V_{gi}$, and $\begin{bmatrix} \Delta T_{Mi} \\ \Delta V_{\mathrm{refi}} \end{bmatrix} = \Delta u_i$, (8.25) can be written as

$$\Delta \dot{x}_i = A_{1i}\Delta x_i + B_{1i}\Delta I_{gi} + B_{2i}\Delta V_{gi} + E_i\Delta u_i \text{ for } i = 1,\ldots,m \quad (8.26)$$

For the m-machine system, (8.26) can be expressed in matrix form as

$$\Delta \dot{x} = A_1\Delta x + B_1\Delta I_g + B_2\Delta V_g + E_1\Delta U \quad (8.27)$$

where A_1, B_1, B_2, and E_1 are block diagonal matrices.

We now linearize the stator algebraic equations (8.13) and (8.14):

$$\Delta E'_{di} - \sin(\delta_{io} - \theta_{io})\Delta V_i - V_{io}\cos(\delta_{io} - \theta_{io})\Delta\delta_i + V_{io}\cos(\delta_{io} - \theta_{io})\Delta\theta_i$$
$$- R_{si}\Delta I_{di} + X'_{qi}\Delta I_{qi} = 0 \quad (8.28)$$
$$\Delta E'_{qi} - \cos(\delta_{io} - \theta_{io})\Delta V_i + V_{io}\sin(\delta_{io} - \theta_{io})\Delta\delta_i - V_{io}\sin(\delta_{io} - \theta_{io})\Delta\theta_i$$
$$- R_{si}\Delta I_{qi} - X'_{di}\Delta I_{di} = 0 \qquad i = 1,\ldots,m \quad (8.29)$$

Writing (8.28) and (8.29) in matrix form, we have

$$\begin{bmatrix} -V_{io}\cos(\delta_{io} - \theta_{io}) & 0 & 0 & 1 & 0 & 0 & 0 \\ V_{io}\sin(\delta_{io} - \theta_{io}) & 0 & 1 & 0 & 0 & 0 & 0 \end{bmatrix} \begin{bmatrix} \Delta\delta_i \\ \Delta\omega_i \\ \Delta E'_{qi} \\ \Delta E'_{di} \\ \Delta E_{fdi} \\ \Delta V_{Ri} \\ \Delta R_{Fi} \end{bmatrix}$$

$$+ \begin{bmatrix} -R_{si} & X'_{qi} \\ -X'_{di} & -R_{si} \end{bmatrix} \begin{bmatrix} \Delta I_{di} \\ \Delta I_{qi} \end{bmatrix}$$

$$+ \begin{bmatrix} V_{io}\cos(\delta_{io} - \theta_{io}) & -\sin(\delta_{io} - \theta_{io}) \\ -V_{io}\sin(\delta_{io} - \theta_{io}) & -\cos(\delta_{io} - \theta_{io}) \end{bmatrix} \begin{bmatrix} \Delta\theta_i \\ \Delta V_i \end{bmatrix} = 0$$

$$i = 1,\ldots,m \quad (8.30)$$

Rewriting (8.30), we obtain

$$0 = C_{1i}\Delta x_i + D_{1i}\Delta I_{gi} + D_{2i}\Delta V_{gi} \quad i = 1,\ldots,m \quad (8.31)$$

In matrix notation, (8.31) can be written as

$$0 = C_1\Delta x + D_1\Delta I_g + D_2\Delta V_g \quad (8.32)$$

where C_1, D_1, and D_2 are block diagonal. Linearizing the network equations (8.15) and (8.16), which pertain to generators, we obtain

$$
\begin{aligned}
&V_{io}\sin(\delta_{io}-\theta_{io})\Delta I_{di} + I_{dio}\sin(\delta_{io}-\theta_{io})\Delta V_i + I_{dio}V_{io}\cos(\delta_{io}-\theta_{io})\Delta\delta_i \\
&\quad -I_{dio}V_{io}\cos(\delta_{io}-\theta_{io})\Delta\theta_i + V_{io}\cos(\delta_{io}-\theta_{io})\Delta I_{qi} \\
&\quad +I_{qio}\cos(\delta_{io}-\theta_{io})\Delta V_i - I_{qio}V_{io}\sin(\delta_{io}-\theta_{io})\Delta\delta_i \\
&\quad +I_{qio}V_{io}\sin(\delta_{io}-\theta_{io})\Delta\theta_i - \left[\sum_{k=1}^{n}V_{ko}Y_{ik}\cos(\theta_{io}-\theta_{ko}-\alpha_{ik})\right]\Delta V_i \\
&\quad -V_{io}\sum_{k=1}^{n}\left[Y_{ik}\cos(\theta_{io}-\theta_{ko}-\alpha_{ik})\right]\Delta V_k \\
&\quad +\left[V_{io}\sum_{\substack{k=1\\\neq i}}^{n}V_{ko}Y_{ik}\sin(\theta_{io}-\theta_{ko}-\alpha_{ik})\right]\Delta\theta_i \\
&\quad -V_{io}\sum_{\substack{k=1\\\neq i}}^{n}\left[V_{ko}Y_{ik}\sin(\theta_{io}-\theta_{ko}-\alpha_{ik})\right]\Delta\theta_k + \frac{\partial P_{Li}(V_i)}{\partial V_i}\Delta V_i = 0 \qquad (8.33)
\end{aligned}
$$

$$
\begin{aligned}
&V_{io}\cos(\delta_{io}-\theta_{io})\Delta I_{di} + I_{dio}\cos(\delta_{io}-\theta_{io})\Delta V_i - I_{dio}V_{io}\sin(\delta_{io}-\theta_{io})\Delta\delta_i \\
&\quad +I_{dio}V_{io}\sin(\delta_{io}-\theta_{io})\Delta\theta_i - V_{io}\sin(\delta_{io}-\theta_{io})\Delta I_{qi} \\
&\quad -I_{qio}\sin(\delta_{io}-\theta_{io})\Delta V_i - I_{qio}V_{io}\cos(\delta_{io}-\theta_{io})\Delta\delta_i \\
&\quad +I_{qio}V_{io}\cos(\delta_{io}-\theta_{io})\Delta\theta_i - \left[\sum_{k=1}^{n}V_{ko}Y_{ik}\sin(\theta_{io}-\theta_{ko}-\alpha_{ik})\right]\Delta V_i \\
&\quad -V_{io}\sum_{k=1}^{n}\left[Y_{ik}\sin(\theta_{io}-\theta_{ko}-\alpha_{ik})\right]\Delta V_k \\
&\quad -\left[V_{io}\sum_{\substack{k=1\\\neq i}}^{n}V_{ko}Y_{ik}\cos(\theta_{io}-\theta_{ko}-\alpha_{ik})\right]\Delta\theta_i \\
&\quad +V_{io}\sum_{\substack{k=1\\\neq i}}^{n}\left[V_{ko}Y_{ik}\cos(\theta_{io}-\theta_{ko}-\alpha_{ik})\right]\Delta\theta_k + \frac{\partial Q_{Li}(V_i)}{\partial V_i}\Delta V_i = 0 \\
&i = 1, 2, \ldots, m \qquad\qquad\qquad\qquad\qquad\qquad\qquad\qquad\qquad\qquad (8.34)
\end{aligned}
$$

Rewriting (8.33) and (8.34) in matrix form, we obtain

$$
0 = \begin{bmatrix} C_{21} & & \\ & \ddots & \\ & & C_{2m} \end{bmatrix} \begin{bmatrix} \Delta x_1 \\ \vdots \\ \Delta x_m \end{bmatrix} + \begin{bmatrix} D_{31} & & \\ & \ddots & \\ & & D_{3m} \end{bmatrix}
$$

$$
\begin{bmatrix} \Delta I_{g1} \\ \vdots \\ \Delta I_{gm} \end{bmatrix} + \begin{bmatrix} D_{41,1} & \cdots & D_{41,m} \\ \vdots & \vdots & \vdots \\ D_{4m,1} & \cdots & D_{4m,m} \end{bmatrix} \begin{bmatrix} \Delta V_{g1} \\ \vdots \\ \Delta V_{gm} \end{bmatrix}
$$

$$
+ \begin{bmatrix} D_{51,m+1} & \cdots & D_{51,n} \\ \vdots & \vdots & \vdots \\ D_{5m,m+1} & \cdots & D_{5m,n} \end{bmatrix} \begin{bmatrix} \Delta V_{\ell m+1} \\ \vdots \\ \Delta V_{\ell n} \end{bmatrix} \tag{8.35}
$$

where the various submatrices of (8.35) can be easily identified. In matrix notation, (8.35) is

$$
0 = C_2 \Delta x + D_3 \Delta I_g + D_4 \Delta V_g + D_5 \Delta V_\ell \tag{8.36}
$$

where

$$
\Delta V_{\ell i} = \begin{bmatrix} \Delta \theta_i \\ \Delta V_i \end{bmatrix}
$$

for the non-generator buses $i = m+1, \ldots, n$.

Note that C_2, D_3 are block diagonal, whereas D_4, D_5 are full matrices. Linearizing network equations (8.17) and (8.18) for the load buses (PQ buses), we obtain

$$
0 = \frac{\partial P_{Li}(V_i)}{\partial V_i} \Delta V_i - \left[\sum_{k=1}^{n} V_{ko} Y_{ik} \cos(\theta_{io} - \theta_{ko} - \alpha_{ik}) \right] \Delta V_i
$$

$$
+ \left[\sum_{\substack{k=1 \\ \neq i}}^{n} V_{io} V_{ko} Y_{ik} \sin(\theta_{io} - \theta_{ko} - \alpha_{ik}) \right] \Delta \theta_i
$$

$$
- V_{io} \sum_{k=1}^{n} [Y_{ik} \cos(\theta_{io} - \theta_{ko} - \alpha_{ik})] \Delta V_k
$$

$$
- V_{io} \sum_{\substack{k=1 \\ \neq i}}^{n} [V_{ko} Y_{ik} \sin(\theta_{io} - \theta_{ko} - \alpha_{ik})] \Delta \theta_k \tag{8.37}
$$

$$0 = \frac{\partial Q_{Li}(V_i)}{\partial V_i} \Delta V_i - \left[\sum_{k=1}^{n} V_{ko} Y_{ik} \sin(\theta_{io} - \theta_{ko} - \alpha_{ik}) \right] \Delta V_i$$

$$- \left[\sum_{\substack{k=1 \\ \neq i}}^{n} V_{io} V_{ko} Y_{ik} \cos(\theta_{io} - \theta_{ko} - \alpha_{ik}) \right] \Delta \theta_i$$

$$-V_{io} \sum_{k=1}^{n} \left[Y_{ik} \sin(\theta_{io} - \theta_{ko} - \alpha_{ik}) \right] \Delta V_k$$

$$+V_{io} \sum_{\substack{k=1 \\ \neq i}}^{n} \left[V_{ko} Y_{ik} \sin(\theta_{io} - \theta_{ko} - \alpha_{ik}) \right] \Delta \theta_k$$

$$i = m + 1, ..., n \tag{8.38}$$

Rewriting (8.37) and (8.38) in matrix form gives

$$0 = \begin{bmatrix} D_{6m+1,1} & \cdots & D_{6m+1,m} \\ \vdots & \vdots & \vdots \\ D_{6n,1} & \cdots & D_{6n,m} \end{bmatrix} \begin{bmatrix} \Delta V_{g1} \\ \vdots \\ \Delta V_{gm} \end{bmatrix}$$

$$+ \begin{bmatrix} D_{7m+1,m+1} & \cdots & D_{7m+1,n} \\ \vdots & \vdots & \vdots \\ D_{7n,m+1} & \cdots & D_{7n,n} \end{bmatrix} \begin{bmatrix} \Delta V_{\ell m+1} \\ \vdots \\ \Delta V_{\ell n} \end{bmatrix} \tag{8.39}$$

where, again, the various submatrices in (8.39) can be identified from (8.37) and (8.38). Rewriting (8.39) in a compact notation,

$$0 = D_6 \Delta V_g + D_7 \Delta V_\ell \tag{8.40}$$

where D_6, D_7 are full matrices. Rewriting (8.27), (8.32), (8.36), and (8.40) together,

$$\Delta \dot{x} = A_1 \Delta x + B_1 \Delta I_g + B_2 \Delta V_g + E_1 \Delta u \tag{8.41}$$

$$0 = C_1 \Delta x + D_1 \Delta I_g + D_2 \Delta V_g \tag{8.42}$$

$$0 = C_2 \Delta x + D_3 \Delta I_g + D_4 \Delta V_g + D_5 \Delta V_\ell \tag{8.43}$$

$$0 = D_6 \Delta V_g + D_7 \Delta V_\ell \tag{8.44}$$

where

$$x = [x_1^t \ldots x_m^t]^t$$

$$x_i = [\delta_i \; \omega_i \; E'_{qi} \; E'_{di} \; E_{fdi} \; V_{Ri} \; R_{fi}]^t$$
$$I_g = [I_{d1} \; I_{q1} \ldots I_{dm} \; I_{qm}]^t$$
$$V_g = [\theta_1 \; V_1 \ldots \theta_m \; V_m]^t$$
$$V_\ell = [\theta_{m+1} \; V_{m+1} \ldots \theta_n \; V_n]^t$$
$$u = [u_1^t \ldots u_m^t]^t$$
$$u_i = [T_{Mi} \; V_{\text{ref}i}]^t$$

This is the linearized DAE model for the multimachine system. Equations (8.41)–(8.44) are equivalent to the linear model of (8.4), except that in these equations the dependence of loads on the voltages has not been specified, whereas in (8.4) the loads were assumed to be of the constant power type. This model is quite general and can easily be expanded to include frequency or $\dot{V}$ dependence at the load buses. The power system stabilizer (PSS) dynamics can also be included easily. In the above model, ΔI_g is not of interest and, hence, is eliminated from (8.41) and (8.43) using (8.42). Thus, from (8.42),

$$\Delta I_g = -D_1^{-1}C_1\Delta x - D_1^{-1}D_2\Delta V_g \qquad (8.45)$$

Substituting (8.45) into (8.43), we get

$$C_2\Delta x + D_3(-D_1^{-1}C_1\Delta x - D_1^{-1}D_2\Delta V_g) + D_4\Delta V_g + D_5\Delta V_\ell = 0 \qquad (8.46)$$

Let

$$\left[D_4 - D_3D_1^{-1}D_2\right] \triangleq K_1$$

and

$$\left[C_2 - D_3D_1^{-1}C_1\right] \triangleq K_2 \qquad (8.47)$$

Note that D_1^{-1} involves taking a series of 2×2 inverses of $Z_{d-q,i}(i = 1, \ldots, m)$. Equation (8.46) is now expressed as

$$K_2\Delta x + K_1\Delta V_g + D_5\Delta V_\ell = 0 \qquad (8.48)$$

If (8.45) is substituted in (8.41) to eliminate ΔI_g, the new overall differential–algebraic model is given by

$$\Delta \dot{x} = (A_1 - B_1D_1^{-1}C_1)\Delta x + (B_2 - B_1D_1^{-1}D_2)\Delta V_g + E_1\Delta u \qquad (8.49)$$
$$0 = K_2\Delta x + K_1\Delta V_g + D_5\Delta V_\ell \qquad (8.50)$$
$$0 = D_6\Delta V_g + D_7\Delta V_\ell \qquad (8.51)$$

Equations (8.49)–(8.51) can be put in the more compact form

$$
\begin{bmatrix} \Delta \dot{x} \\ 0 \end{bmatrix} = \begin{bmatrix} A' & B' \\ C' & D' \end{bmatrix} \begin{bmatrix} \Delta x \\ \Delta V_N \end{bmatrix} + \begin{bmatrix} E_1 \\ 0 \end{bmatrix} \Delta u \qquad (8.52)
$$

where $\Delta V_N = \begin{bmatrix} \Delta V_g \\ \Delta V_\ell \end{bmatrix}$. Reorder the variables in the voltage vector and de-

fine $\Delta V_p = [\Delta y_c^t \ \Delta y_b^t]^t = [\Delta \theta_1 \ \Delta V_1 \ \dots \ \Delta V_m \mid \Delta \theta_2 \dots \Delta \theta_n \ \Delta V_{m+1} \dots \Delta V_n]^t$.
Δy_b is the set of load-flow variables, and Δy_c is the set of other algebraic
variables in the network equations. Note that we have eliminated the stator
algebraic variables.

In any rotational system, the reference for angles is arbitrary. The order
of the dynamical system in (8.52) is 7m, and can be reduced to $(7m\text{ -}1)$ by
introducing relative rotor angles. Selecting δ_1 as the reference, we have

$$
\begin{aligned}
\delta_i' &= \delta_i - \delta_1 & i &= 2, \dots, m \\
\delta_1' &= 0 \\
\dot{\delta}_i' &= \omega_i - \omega_1 & i &= 2, \dots, m \\
\dot{\delta}_1' &= 0 \\
\theta_i' &= \theta_i - \delta_1 & i &= 1, \dots, n
\end{aligned}
$$

Since the angles δ_i $(i = 1, \dots, m)$ and θ_i $(i = 1, \dots, n)$ always appear as differences, we can retain the notation with the understanding that these angles
are referred to δ_1. This implies that the differential equation corresponding
to δ_1 can be deleted from (8.52) and also from the column corresponding to
$\Delta \delta_1$ in A' and B'. The differential equations $\Delta \dot{\delta}_i = \omega_i$ $(i = 2, \dots, m)$ are
replaced by $\Delta \dot{\delta}_i' = \Delta \omega_i - \Delta \omega_1$. This means that in the A' matrix we place
minus 1 in the intersections of the rows corresponding to $\Delta \dot{\delta}_i(i = 2, \dots, m)$
and the column corresponding to $\Delta \omega_1$. This process is analytically neat,
but is not carried out in linear analysis packages. Since angles appear as
differences, they are computed to within a constant. Hence, a zero eigenvalue is always present. Recognizing this fact, we retain the formulation
corresponding to (8.52). We rewrite the differential-algebraic (DAE) system
(8.52) after the reordering of algebraic variables as

$$
\begin{bmatrix} \Delta \dot{x} \\ 0 \\ 0 \end{bmatrix} = \begin{bmatrix} A' & B_1' & B_2' \\ C_1' & D_{11}' & D_{12}' \\ C_2' & D_{21}' & D_{22}' \end{bmatrix} \begin{bmatrix} \Delta x \\ \Delta y_c \\ \Delta y_b \end{bmatrix} + \begin{bmatrix} E_1 \\ 0 \\ 0 \end{bmatrix} \Delta u \qquad (8.53)
$$

For voltage-dependent loads, only the appropriate diagonal elements of D'_{11} and D'_{22} will be affected. Now, D'_{22} is the load-flow Jacobian J_{LF} modified by the load representation and

$$\begin{bmatrix} D'_{11} & D'_{12} \\ D'_{21} & D'_{22} \end{bmatrix} = J'_{AE} \tag{8.54}$$

the network algebraic Jacobian with voltage dependencies of the load included. Note that, compared to the algebraic Jacobian J_{AE} of (8.4), the stator algebraic variables have been eliminated to obtain J'_{AE}. The system A_{sys} matrix is obtained as

$$\Delta \dot{x} = A_{\text{sys}} \Delta x + E \Delta u \tag{8.55}$$

where

$$A_{\text{sys}} = A' - [B'_1 \, B'_2][J'_{AE}]^{-1} \begin{bmatrix} C'_1 \\ C'_2 \end{bmatrix} \tag{8.56}$$

This model is used later to examine the effects of increased loading on the eigenvalues of A_{sys} and the determinants of J'_{AE} and J_{LF} for (1) constant power case, (2) constant current case, and (3) constant impedance case for the models A and B.

We now show that the model in (8.53) is consistent with the development from the nonlinear model in power balance form from Chapter 7. These are (7.36)-(7.38):

$$\dot{x} = f_o(x, I_{d-q}, \overline{V}, u) \tag{8.57}$$
$$I_{d-q} = h(x, \overline{V}) \tag{8.58}$$
$$0 = g_o(x, I_{d-q}, \overline{V}) \tag{8.59}$$

By substituting I_{d-q} from (8.58) in (8.57) and (8.59), we obtain

$$\dot{x} = f_1(x, \overline{V}, u) \tag{8.60}$$
$$0 = g_1(x, \overline{V}) \tag{8.61}$$

Linearizing (8.60) and (8.61) results in

$$\Delta \dot{x} = \frac{\partial f_1}{\partial x} \Delta x + \frac{\partial f_1}{\partial V_p} \Delta V_p + \frac{\partial f_1}{\partial u} \Delta u$$

$$0 = \frac{\partial g_1}{\partial x} \Delta x + \frac{\partial g_1}{\partial V_p} \Delta V_p$$

where $V_p = [\theta_1 \, V_1 \ldots V_m \mid \theta_2 \ldots \theta_n \, V_{m+1} \ldots V_n]^t$, then we can identify

$$\frac{\partial f_1}{\partial x} = A' \quad , \quad \frac{\partial f_1}{\partial V_p} = [B'_1 \, B'_2]$$

$$\frac{\partial g_1}{\partial x} = \begin{bmatrix} C'_1 \\ C'_2 \end{bmatrix}, \quad \frac{\partial g_1}{\partial V_p} = \begin{bmatrix} D'_{11} & D'_{12} \\ D'_{21} & D'_{22} \end{bmatrix}, \quad \frac{\partial f_1}{\partial u} = \begin{bmatrix} E_1 \\ 0 \\ 0 \end{bmatrix}$$

8.2.2 Linearization of Model B

As discussed in Chapter 7, the multimachine model with the flux–decay model and fast exciter (Figures 8.3 and 8.4) is given by the following set of differential-algebraic equations:

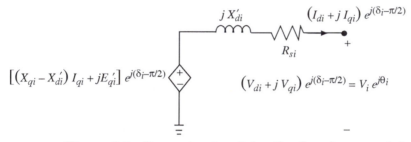

Figure 8.3: *Dynamic circuit for the flux-decay model*

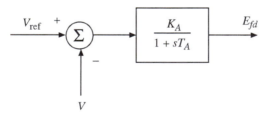

Figure 8.4: *Static exciter model*

Differential Equations

$$\frac{d\delta_i}{dt} = \omega_i - \omega_s \tag{8.62}$$

$$\frac{d\omega_i}{dt} = \frac{T_{Mi}}{M_i} - \frac{E'_{qi}I_{qi}}{M_i} - \frac{(X_{qi} - X'_{di})}{M_i}I_{di}I_{qi} - \frac{D_i(\omega_i - \omega_s)}{M_i} \tag{8.63}$$

$$\frac{dE'_{qi}}{dt} = -\frac{E'_{qi}}{T'_{doi}} - \frac{(X_{di} - X'_{di})}{T'_{doi}}I_{di} + \frac{E_{fdi}}{T'_{doi}} \tag{8.64}$$

$$\frac{dE_{fdi}}{dt} = -\frac{E_{fdi}}{T_{Ai}} + \frac{K_{Ai}}{T_{Ai}}(V_{\text{ref},i} - V_i) \tag{8.65}$$
$$\text{for } i = 1, \ldots, m$$

Stator Algebraic Equations

The stator algebraic equations are

$$V_i \sin(\delta_i - \theta_i) + R_{si}I_{di} - X_{qi}I_{qi} = 0 \tag{8.66}$$
$$E'_{qi} - V_i \cos(\delta_i - \theta_i) - R_{si}I_{qi} - X'_{di}I_{di} = 0 \tag{8.67}$$
$$\text{for } i = 1, \ldots, m$$

Network Equations

The network equations are the same as (8.15)–(8.18). The linearization of this model is done in the same manner as in model A and, hence, is not discussed.

8.3 Participation Factors

Due to the large size of the power system, it is often necessary to construct reduced-order models for dynamic stability studies by retaining only a few modes. The appropriate definition and determination as to which state variables significantly participate in the selected modes become very important. This requires a tool for identifying the state variables that have significant participation in a selected mode. It is natural to suggest that the significant state variables for an eigenvalue λ_i are those that correspond to large entries in the corresponding eigenvector v_i. But the entries in the eigenvector are dependent on the dimensions of the state variables which are, in general, incommensurable (for example, angle, velocities, and flux). Verghese et al. [81] have suggested a related but dimensionless measure of state variable participation called participation factors.

Participation factor analysis aids in the identification of how each dynamic variable affects a given mode or eigenvalue. Specifically, given a linear system

$$\dot{x} = Ax \tag{8.68}$$

a participation factor is a sensitivity measure of an eigenvalue to a diagonal entry of the system A matrix. This is defined as

$$p_{ki} = \frac{\partial \lambda_i}{\partial a_{kk}} \tag{8.69}$$

where λ_i is the ith system eigenvalue, a_{kk} is a diagonal entry in the system A matrix, and p_{ki} is the participation factor relating the kth state variable to the ith eigenvalue. The participation factor may also be defined by

$$p_{ki} = \frac{w_{ki} v_{ik}}{w_i^t v_i} \tag{8.70}$$

where w_{ki} and v_{ki} are the kth entries in the left and right eigenvector associated with the ith eigenvalue. The right eigenvector v_i and the left eigenvector w_i associated with the ith eigenvalue λ_i satisfy

$$A v_i = \lambda_i v_i \tag{8.71}$$

$$w_i^t A = w_i^t \lambda_i \tag{8.72}$$

It is not obvious that the definitions given in (8.69) and (8.70) are equivalent. We establish the equivalence as follows. Consider the system

$$[A - \lambda_i I] v_i = 0 \tag{8.73}$$

$$w_i^t [A - \lambda_i I] = 0 \tag{8.74}$$

where v_i and w_i need not be normalized eigenvectors. It is our goal to examine the sensitivity of the eigenvalue to a diagonal element of the A matrix. From (8.68), assuming that the eigenvalues and eigenvectors vary continuously with respect to the elements of the A matrix, we write the perturbation equation as

$$(A + \Delta A)(v_i + \Delta v_i) = (\lambda_i + \Delta \lambda_i)(v_i + \Delta v_i) \tag{8.75}$$

Expansion yields

$$[A v_i] + [\Delta A v_i + A \Delta v_i] + [\Delta A \Delta v_i] = [\lambda_i v_i] + [\Delta \lambda_i v_i$$
$$+ \lambda_i \Delta v_i] + [\Delta \lambda_i \Delta v_i] \tag{8.76}$$

Neglecting the second-order terms $\Delta A \Delta v_i$ and $\Delta \lambda_i \Delta v_i$ and using (8.71), we obtain

$$[A - \lambda_i I] \Delta v_i + \Delta A v_i = \Delta \lambda_i v_i \tag{8.77}$$

Multiply (8.77) by the left eigenvector w_i^t to give

$$w_i^t[A - \lambda_i I]\Delta v_i + w_i^t \Delta A v_i = w_i^t \Delta \lambda_i v_i \qquad (8.78)$$

The first term in the left-hand side of (8.78) is identically zero in view of (8.74), leaving

$$w_i^t \Delta A v_i = w_i^t \Delta \lambda_i v_i \qquad (8.79)$$

Now, the sensitivities of λ_i with respect to diagonal entries of A are related to the participation factors, as follows. Assume only that the kth diagonal entry of A is perturbed so that

$$\Delta A = \begin{bmatrix} 0 & \cdots & 0 & 0 \\ 0 & & \Delta a_{kk} & 0 \\ 0 & \cdots & 0 & 0 \end{bmatrix} \qquad (8.80)$$

Then, in (8.79), the left-hand side can be simplified, resulting in

$$w_i^t \Delta A v_i = w_{ki} \Delta a_{kk} v_{ik} = w_i^t \Delta \lambda_i v_i \qquad (8.81)$$

Solving for the sensitivity gives the participation factor as

$$\frac{\Delta \lambda_i}{\Delta a_{kk}} = \frac{w_{ki} v_{ik}}{w_i^t v_i} = p_{ki} \qquad (8.82)$$

Equation (8.82) thus shows that (8.69) and (8.70) are equivalent.

An eigenvector may be scaled by any value resulting in a new vector, which is also an eigenvector. We can use this property to choose a scaling that simplifies the use of participation factors, for instance, choosing the eigenvectors such that $w_i^t v_i = 1$ simplifies the definition of the participation factor. In any case, since $\sum_{k=1}^n w_{ki} v_{ik} = w_i^t v_i$, it follows from (8.82) that the sum of all the participation factors associated with a given eigenvalue is equal to 1, i.e.,

$$\sum_{k=1}^n p_{ki} = 1 \qquad (8.83)$$

This property is useful, since all participation factors lie on a scale from zero to one. To handle participation factors corresponding to complex eigenvalues, we introduce some modifications as follows. The eigenvectors corresponding to a complex eigenvalue will have complex elements. Hence, p_{ki} is defined as

$$p_{ki} = \frac{|v_{ik}||w_{ki}|}{\sum_{k=1}^n |v_{ik}||w_{ki}|} \qquad (8.84)$$

A further normalization can be done by making the largest of the participation factors equal to unity.

Example 8.1

Compute the participation factors of the 2×2 matrix $\dot{x} = Ax$, where

$$A = \begin{bmatrix} 1 & 4 \\ 3 & 2 \end{bmatrix}$$

The eigenvalues are $\lambda_1 = 5$, $\lambda_2 = $ -2. The right eigenvectors are

$$v_1 = \begin{bmatrix} 1 \\ 1 \end{bmatrix}$$

and

$$v_2 = \begin{bmatrix} 4 \\ -3 \end{bmatrix}$$

The left eigenvectors are computed as $w_1^t = [3 \quad 4]$ and $w_2^t = [1 - 1]$. Verify that $w_2^t v_1 = w_1^t v_2 = 0$. Also, $w_1^t v_1 = 7 = w_1^t v_1$. Letting $i = 1$ and $k = 1,2$, successively, we obtain the participation of the state variables x_1, x_2 in the mode $\lambda_1 = 5$ as

$$p_{11} = \frac{w_{11} v_{11}}{w_1^t v_1} = \left(\frac{(3)(1)}{7}\right) = \frac{3}{7}$$

$$p_{21} = \frac{w_{21} v_{12}}{w_2^t v_2} = \left(\frac{(4)(1)}{7}\right) = \frac{4}{7}$$

Letting $i = 2$ and $k = 1,2$, we obtain the participation of the state variables x_1, x_2 in the mode $\lambda_2 = $ -2 as

$$p_{12} = \frac{w_{12} v_{21}}{w_2^t v_2} = \frac{4}{7}$$

$$p_{22} = \frac{w_{22} v_{22}}{w_2^t v_2} = \frac{3}{7}$$

The participation matrix is therefore

$$P = \begin{bmatrix} p_{11} & p_{12} \\ p_{21} & p_{22} \end{bmatrix} = \begin{bmatrix} \frac{3}{7} & \frac{4}{7} \\ \frac{4}{7} & \frac{3}{7} \end{bmatrix}$$

Normalizing the largest participation factor as equal to 1 in each column results in

$$P_{\text{norm}} = \begin{bmatrix} 0.75 & 1 \\ 1 & 0.75 \end{bmatrix}$$

□

The ith column entries in the P or P_{norm} matrix are the sensitivities of the ith eigenvalue with respect to the states.

Example 8.2

Compute the participation factors corresponding to the complex eigenvalue of

$$A = \begin{bmatrix} -0.4 & 0 & -0.01 \\ 1 & 0 & 0 \\ -1.4 & 9.8 & -0.02 \end{bmatrix}$$

The eigenvalues are $\lambda_1 = $ -0.6565 and $\lambda_{2,3} = 0.1183 \pm j0.3678$. The right and left eigenvectors corresponding to the complex eigenvalue $\lambda_2 = 0.1183 + j0.3678$ are

$$v_2 = \begin{bmatrix} 0.0138 - j0.0075 \\ -0.0075 - j0.04 \\ -0.9918 - j0.1203 \end{bmatrix}, \; w_2 = \begin{bmatrix} 0.838 - j0.0577 \\ 0.4469 + j0.307 \\ -0.0061 + j0.0205 \end{bmatrix}$$

Using the formula in the previous section, we obtain

$$p_{21} = 0.2332, \; p_{22} = 0.3896, \; p_{23} = 0.3772$$

Note $p_{21} + p_{22} + p_{23} = 1$. We can normalize with respect to p_{22} by making it unity, in which case $p_{21(\text{norm})} = 0.598$, $p_{22(\text{norm})} = 1$ and $p_{23(\text{norm})} = 0.968$.

□

Example 8.3

The numerical Example 7.1 is used to illustrate the eigenvalue computation. Compute the eigenvalues, as well as the participation factors, for the eigenvalues for the nominal loading of Example 7.1. The damping $D_i \equiv 0$ (i = 1,2,3). The machine and exciter data are given in Table 7.3. Loads are assumed as constant power type.

Solution

Following the linearization procedure results in a 21×21 sized A_{sys} matrix. Because of zero-damping, two zero eigenvalues are obtained. The eigenvalues are shown in Table 8.1. The participation factors associated with the eigenvalues are given in Table 8.2. Only the participation factors greater than 0.2 are listed. Also shown are the state variables and the machines associated with these state variables. From a practical point of view, this information is very useful.

Table 8.1: Eigenvalues of the 3-Machine System

-0.7209	$\pm j12.7486$
-0.1908	$\pm j8.3672$
-5.4875	$\pm j7.9487$
-5.3236	$\pm j7.9220$
-5.2218	$\pm j7.8161$
-5.1761	
-3.3995	
-0.4445	$\pm j1.2104$
-0.4394	$\pm j0.7392$
-0.4260	$\pm j0.4960$
-0.0000	
-0.0000	
-3.2258	

Table 8.2: Eigenvalues and Their Participation Factors

Eigenvalue	Machine Number	Machine Variable	PF
$-0.7209 \pm j12.7486$	3	δ, ω	1.0, 1.0
	2	δ, ω	0.22, 0.22
$-0.1908 \pm j8.3672$	2	δ, ω	1.0, 1.0
	1	δ, ω	0.42, 0.42
$-5.4875 \pm j7.79487$	2	V_R, E_{fd}	1.0, 0.98
	2	R_f	0.29
$-5.3236 \pm j7.9220$	3	V_R, E_{fd}	1.0, 0.98
	3	R_f	0.29
$-5.2218 \pm j7.8161$	1	V_R, E_{fd}	1.0, 0.97
	1	R_f	0.31
-5.1761	2	E_d'	1.0
	3	E_d'	0.92
-3.3995	3	E_d'	1.0
-3.2258	2	E_d'	0.89
$-0.4445 \pm j1.2104$	1	E_q', R_f	1.0, 0.74
	2	E_q', R_f	0.67, 0.48
	3	E_q', R_f	0.38, 0.28
$-0.4394 \pm j0.7392$	1	E_q', R_f	1.0, 0.78
	2	E_q', R_f	0.78, 0.60
	3	E_q'	0.22
$-0.4260 \pm j0.4960$	3	E_q', R_f	1.0, 0.83
	2	E_q', R_f	0.43, 0.33
0.0000	1	δ, ω	1.0, 1.0
	2	δ, ω	0.26, 0.26
0.0000	1	δ, ω	1.0, 1.0
	2	δ, ω	0.26, 0.26

□

8.4 Studies on Parametric Effects

In this section, the effect of various parameters on the small-signal stability of the system is studied.

8.4.1   Effect of loading

The WSCC 3-machine, 9-bus system of Chapter 7 is considered. The real or reactive loads at a particular bus/buses are increased continuously. At each step, the initial conditions of the state variables are computed, after running the load flow, and linearization of the equations is done. Ideally, the increase in load is picked up by the generators through the economic load dispatch scheme. To simplify matters, the load increase is allocated among the generators (real power) in proportion to their inertias. In the case of increase of reactive power, it is picked up by the PV buses. The A_{sys} matrix is formed, and its eigenvalues are checked for stability. Also $\det J_{LF}$ and $\det J'_{AE}$ are computed. The step-by-step algorithm is as follows:

1. Increase the load at bus/buses for a particular generating unit model.

2. If the real load is increased, then distribute the load among the various generators in proportion to their inertias.

3. Run the load flow.

4. Stop, if the load flow fails to converge.

5. Compute the initial conditions of the state variables, as discussed in Chapter 7.

6. From the linearized DAE model, compute the various matrices.

7. Compute $\det J_{LF}$, $\det J'_{AE}$, and the eigenvalues of A_{sys}.

8. If A_{sys} is stable, then go to step (1).

9. If unstable, identify the states associated with the unstable eigenvalue(s) of A_{sys} using the participation factor method, and go to step (1).

The above algorithm is implemented for models A and B. Nonuniform damping is assumed by choosing $D_1 = 0.0254$, $D_2 = 0.0066$, and $D_3 = 0.0026$. The results are summarized in Tables 8.3 and 8.4. It is observed that for constant power load, with model A and the IEEE-Type I slow exciter, it is the voltage control mode that goes unstable at $P_{L5} = 4.5$ pu. Examination of the participation factor indicates that the pair of state variables E'_q, R_f of machine 1 in the excitation system is responsible for this model. In the case of model B, the mode that goes unstable is due to the electromechanical variables δ, ω of machine 2 at a load of 4.6 pu. A value of $K_A = 45$ is assumed. The point at which the eigenvalues cross over to the right-half plane is called the Hopf bifurcation point, and the point at which the $\det J'_{AE}$ changes sign is the singularity-induced bifurcation. These are discussed in the literature in detail [82, 85].

Table 8.3: Eigenvalues with Model A, $K_A = 20$

Load at Bus	$sgn(\det J_{LF})$	$sgn(\det J'_{AE})5$	Critical Eigenvalue(s)	Associated States
4.3	+	+	$-0.1618 \pm j1.9769$	E'_{q1}, R_{f1}
4.4	+	+	$-0.0522 \pm j2.1102$	E'_{q1}, R_{f1}
4.5 (A)	+	+	$0.1268 \pm j2.2798$	E'_{q1}, R_{f1}
4.6	+	+	$0.4446 \pm j2.4911$	E'_{q1}, E'_{q2}
4.7	+	+	$1.0825 \pm j2.7064$	$E'_{q1}, E'_{q2}, E'_{d2}$
4.8	+	+	$2.6051 \pm j2.4392$	$E'_{d2}, E'_{q1}, E'_{q2}, E'_{d3}$
4.9	+	+	$17.568, 1.7849$	E'_{d2}, E'_{q1}
5.0 (B)	+	$-$	1.0526	E'_{q1}
5.1	+	$-$	0.6553	E'_{q1}
5.2	+	$-$	0.3505	E'_{q1}
5.3	+	$-$	0.0496	$E'_{q1}, \delta_1, \delta_2$
5.35	$-$	$-$	-0.1454	ω_1
5.45			*LF does not converge*	

Table 8.4: Eigenvalues with Model B, $K_A = 45$

Load at Bus 5	$sgn(\det J_{LF})$	$sgn(\det J'_{AE})$	Critical Eigenvalue(s)	Associated States
4.3	+	+	$-0.1119 \pm j8.8738$	δ_2, ω_2
4.4	+	+	$-0.0729 \pm j8.8401$	δ_2, ω_2
4.5	+	+	$-0.0035 \pm j8.8183$	δ_2, ω_2
4.6 (A)	+	+	$0.0901 \pm j8.8421$	δ_2, ω_2
4.7	+	+	$0.1587 \pm j8.9371$	δ_2, ω_2
4.8	+	+	$0.1292 \pm j9.0538$	δ_2, ω_2
4.9	+	+	$0.7565 \pm j20.1162$	E'_{q1}, E'_{d1}
			$0.0471 \pm j9.0902$	δ_2, ω_2
5.0 (B)	+	$-$	14.7308	E'_{q1}, E'_{d1}
5.1	+	$-$	7.1144	E'_{q1}
5.2	+	$-$	4.2567	E'_{q1}
5.3	+	$-$	2.3120	E'_{q1}
5.4	$-$	$-$	$-0.0597 \pm j8.7819$	δ_2, ω_2
5.5			*LF does not converge*	

8.4.2   Effect of K_A

It was found that, for model A, the increase in K_A alone did not lead to any instability. The stabilizing feedback in the IEEE-Type I exciter was removed, and then an increase in K_A led to instability for this model, as well. For model B, a sufficient increase in K_A led to instability even for a nominal load.

8.4.3   Effect of type of load

Appropriate voltage-dependent load modeling can be incorporated into the dynamic model by specifying the load functions. The load at any bus i is given by

$$P_{Li} = P_{Lio}\left(\frac{V_i}{V_{io}}\right)^{n_{pi}} \qquad i = 1, \ldots, n \qquad (8.85)$$

$$Q_{Li} = Q_{Lio}\left(\frac{V_i}{V_{io}}\right)^{n_{qi}} \qquad i = 1, \ldots, n \qquad (8.86)$$

where P_{Lio} and Q_{Lio} are the nominal real and reactive powers, respectively, at bus i, with the corresponding voltage magnitude V_{io}, and n_{pi}, n_{qi} are the load indices. Three types of load are considered.

1. Constant power type $(n_p = n_q = 0)$

2. Constant current type $(n_p = n_q = 1)$

3. Constant impedance type $(n_p = n_q = 2)$

The step-by-step procedure of analysis for a given generating-unit model is as follows:

1. Select the type of the load at various buses (i.e., choose values of n_p and n_q at each bus).

2. Compute the system matrix.

3. Compute the eigenvalues of A_{sys} for stability analysis.

   For the three types of loads mentioned earlier, the eigenvalues of model A for increased values of load P_{Lo} at bus 5 are listed in Tables 8.5 to 8.7. First of all, the relative stability of constant power, constant current, and constant impedance-type load has been shown for a nominal operating point

$P_{Lo} = 1.5$ pu and $Q_{Lo} = 0.5$ pu (Table 8.5). We observe that the system is dynamically stable for all types of loads. For an increased value of load at bus 5 ($P_{Lo} = 4.5$ pu, $Q_{Lo} = 0.5$ pu), the eigenvalues are listed in Table 8.6. From Table 8.6 we observe that, for this increased load at bus 5, the system becomes dynamically unstable if the load is treated as a constant power type, whereas for the other two types of loads the system remains stable. Finally, we take another case ($P_{Lo} = 4.6$ pu, $Q_{Lo} = 0.5$ pu), in which we show that the constant impedance type load is more stable than the constant current type. To demonstrate this condition, we take model B with a high gain of the exciter ($K_A = 175$). The eigenvalues for various kind of loads are listed in Table 8.7. Both the constant power and constant current cases are unstable, whereas the constant impedance type is stable. These results corroborate the observation in the literature that constant power gives poor results as far as network loadability is concerned [84].

Table 8.5: Eigenvalues for Different Types of Load at Bus 5 for model A ($P_{Lo} = 1.5$ pu; $Q_{Lo} = 0.5$ pu): (a) constant power; (b) constant current; (c) constant impedance

Constant Power (a)	Constant Current (b)	Constant Impedance (c)
$-0.7927 \pm j12.7660$	$-0.7904 \pm j12.7686$	$-0.7887 \pm j12.7706$
$-0.2849 \pm j8.3675$	$-0.2768 \pm j8.3447$	$-0.2703 \pm j8.3271$
$-5.5187 \pm j7.9508$	$-5.5214 \pm j7.9516$	$-5.5236 \pm j7.9523$
$-5.3325 \pm j7.9240$	$-5.3335 \pm j7.9247$	$-5.3344 \pm j7.9253$
$-5.2238 \pm j7.8156$	$-5.2273 \pm j7.8259$	$-5.2301 \pm j7.8337$
-5.2019	-5.2030	-5.2039
-3.4040	-3.4462	-3.4801
$-0.4427 \pm j1.2241$	$-0.4537 \pm j1.1822$	$-0.4617 \pm j1.1489$
$-0.4404 \pm j0.7413$	$-0.4412 \pm j0.7416$	$-0.4419 \pm j0.7418$
-0.0000	-0.0000	-0.0000
-0.1975	-0.1974	-0.1973
$-0.4276 \pm j0.4980$	$-0.4276 \pm j0.4980$	$-0.4277 \pm j0.4980$
-3.2258	-3.2258	-3.2258

Table 8.6: Eigenvalues for Different Types of Load at Bus 5 for Model A
($P_{Lo} = 4.5$ pu; $Q_{Lo} = 0.5$ pu): (a) constant power; (b) constant current; (c)
constant impedance

Constant Power (a)	Constant Current (b)	Constant Impedance (c)
$-0.7751 \pm j12.7373$	$-0.7335 \pm j12.7842$	$-0.7285 \pm j12.7936$
$-0.2845 \pm j8.0723$	$-0.2497 \pm j8.0650$	$-0.2444 \pm j8.0659$
$-6.7291 \pm j7.8883$	$-6.7669 \pm j7.9730$	$-6.7760 \pm j7.9895$
$-5.6034 \pm j7.9238$	$-5.6287 \pm j7.9557$	$-5.6338 \pm j7.9639$
$-5.2935 \pm j7.6433$	$-5.2812 \pm j7.8419$	$-5.2938 \pm j7.8712$
-5.2541	-5.2715	-5.2790
$0.1268 \pm j2.2798$	-3.5296	-3.8105
-2.5529	$-0.5020 \pm j1.2531$	$-0.5303 \pm j1.0434$
$-0.4858 \pm j0.7475$	-0.0000	$-0.4950 \pm j0.7653$
-0.0000	$-0.4910 \pm j0.7561$	$-0.5371 \pm j0.5336$
$-0.5341 \pm j0.5306$	$-0.5360 \pm j0.7561$	-0.0000
-0.1976	-0.1972	-0.1970
-3.2258	-3.2258	-3.2258

Table 8.7: Eigenvalues for Different Types of Load at Bus 5 for Model B with
$K_A = 175$ ($P_{Lo} = 4.6$ pu, $Q_{Lo} = 0.5$ pu): (a) constant power; (b) constant
current; (c) constant impedance

Constant Power (a)	Constant Current (b)	Constant Impedance (c)
$-1.9610 \pm j19.3137$	$-0.2039 \pm j15.5128$	$-0.2054 \pm j15.5285$
$-0.1237 \pm j15.5812$	$-2.2586 \pm j12.4273$	$-2.1834 \pm j11.1128$
$0.4495 \pm j9.1844$	$0.1441 \pm j9.0636$	$-0.0607 \pm j9.1218$
$-3.1711 \pm j8.2119$	$-3.1359 \pm j8.1099$	$-3.0930 \pm j8.0364$
$-2.7621 \pm j7.1753$	$-2.7599 \pm j7.1594$	$-2.7575 \pm j7.1462$
-0.0000	-0.0000	-0.0000
-0.1987	-0.1989	-0.1990

8.4.4 Hopf bifurcation

For model A, when the load is increased at bus 5, it is observed that the critical modes for the unstable eigenvalues are the electrical ones associated with the exciter, and are complex (Table 8.3). At a load of 4.5 pu, the eigenvalues cross the $j\omega$ axis. This is known as Hopf bifurcation (point A). When the load is increased from 4.8 pu to 4.9 pu, the complex pair of unstable eigenvalues splits into real ones that move in opposite directions along the real axis. The one moving along the positive real axis eventually comes back to the left-half plane via $+\infty$ when the load at bus 5 is increased from 4.9 pu to 5.0 pu (point B). This is the point at which $\det J'_{AE}$ changes sign. This is also known as singularity-induced bifurcation in the literature [85]. The other unstable real eigenvalue moves to the left, and is sensitive to the variable E'_{q1} of the exciter. This eigenvalue returns to the left-half plane at a loading of approximately 5.4 pu, and the system is again dynamically stable (point C). For the load at bus 5 = 5.5 pu, the load flow does not converge. It is possible through other algebraic techniques to reach the nose of the PV curve or the saddle node bifurcation. This phenomenon is pictorially indicated in the PV curve for model A and also is the locus of critical eigenvalue(s) in the s-plane (Figures 8.5 and 8.6).

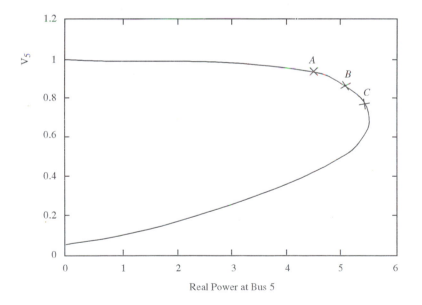

Figure 8.5: *PV curve for bus 5 with model A*

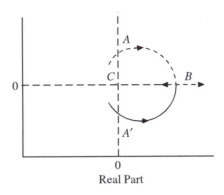

Figure 8.6: *The qualitative behavior of the critical modes of A_{sys} as a function of the load at bus 5 (model A)*

At point A, Hopf bifurcation occurs, and it has been shown to be subcritical, i.e., the limit cycle corresponding to the $E_q' - R_f$ pair is unstable [86]. However, load-flow solution still exists. In this region, E_q' and R_f state variables are clearly dominant initially. As the eigenvalues become real and positive, other state variables start participating substantially in the unstable eigenvalues, as indicated in Table 8.3. For model B, which has the fast static exciter with a single time constant, the modes that go unstable are the electromechanical ones (Table 8.4).

When the Hopf bifurcation phenomenon in power systems was first discussed in the literature for a single-machine case, the electromechanical mode was considered as the critical one [78, 87]. It was called low-frequency oscillatory instability. In studies relating to voltage collapse, it was shown that the exciter mode may go unstable first [82]. From Tables 8.3 and 8.4, it is seen that both the exciter modes and the electromechanical modes are critical in steady-state stability and voltage collapse, and that they both participate in the dynamic instability depending on the machine and exciter models. Hence, decoupling the QV dynamics from the $P\delta$ dynamics as suggested in the literature may not always hold. It may be true for special system configuration/operating conditions. Load dynamics, if included, can be considered as fast dynamics, and the phenomenon of $\det J_{AE}'$ changing sign will still exist. In conventional bifurcation-theory terms, one can think of solving $g(x, y) = 0$ for $y = h(x)$ in (8.2) and substituting this in the differential equation (8.1) to get $\dot{x} = f(x, h(x))$. The change in sign of $\det J_{AE}'$ (which generally agrees with the sign of J_{AE} in (8.5)) is the instant at which the solution of y is no longer possible. This is also tied in with the concept of the implicit function theorem in singular perturbation theory.

8.5 Electromechanical Oscillatory Modes

These are the modes associated with the rotor angles of the machines. These can be identified through a participation factor analysis in the detailed model. In the classical model with internal node description, we have only the rotor angle modes. We now discuss the computation of these modes as a special case. The equations have been derived in Chapter 7 and are reproduced below.

$$\frac{d\delta_i}{dt} = \omega_i - \omega_s \quad i = 1, \ldots, m \tag{8.87}$$

$$\frac{2H_i}{\omega_s} \frac{d\omega_i}{dt} = T_{Mi} - \left[E_i^2 G_{ii} + \sum_{\substack{j=1 \\ \neq i}}^{m} (C_{ij} \sin \delta_{ij} + D_{ij} \cos \delta_{ij}) \right]$$

$$= T_{Mi} - P_{ei} \quad i = 1, \ldots, m \tag{8.88}$$

Linearization around an operating point o gives

$$\frac{d}{dt} \Delta \delta_i = \Delta \omega_i \tag{8.89}$$

$$\frac{d}{dt} \Delta \omega_i = \frac{\omega_s}{2H_i} [\Delta T_{Mi} - \Delta P_{ei}] \tag{8.90}$$

Because $T_{Mi} = \text{constant}$, and $\Delta P_{ei} = \sum_{j=1}^{m} \frac{\partial P_{ei}}{\partial \delta_j} \Delta \delta_j$, we get

$$\frac{d}{dt} \Delta \delta_i = \Delta \omega_i \tag{8.91}$$

$$\frac{d}{dt} \Delta \omega_i = -\frac{\omega_s}{2H_i} \sum_{j=1}^{m} \left(\frac{\partial P_{ei}}{\partial \delta_j} \Delta \delta_j \right) \tag{8.92}$$

where

$$-\frac{\partial P_{ei}}{\partial \delta_j} \begin{cases} = -\sum_{j=1 \neq i}^{m} (C_{ij} \cos \delta_{ij}^o - D_{ij} \sin \delta_{ij}^o) & j = i \\ = C_{ij} \cos \delta_{ij}^o - D_{ij} \sin \delta_{ij}^o & j \neq i \end{cases} \tag{8.93}$$

In matrix form we have

$$
\begin{bmatrix} \Delta\dot{\delta}_1 \\ \vdots \\ \Delta\dot{\delta}_m \\ \hline \Delta\dot{\omega}_1 \\ \vdots \\ \Delta\dot{\omega}_m \end{bmatrix} = \left[\begin{array}{c|c} 0 & I \\ \hline A_\omega & 0 \end{array} \right] \begin{bmatrix} \Delta\delta_1 \\ \vdots \\ \Delta\delta_m \\ \hline \Delta\omega_1 \\ \vdots \\ \Delta\omega_m \end{bmatrix} \tag{8.94}
$$

The elements of A_ω are given by $\frac{-\omega_s}{2H_i}\frac{\partial P_{ei}}{\partial\delta_j}$. The matrix A_ω can be written as

$$
A_\omega = \begin{bmatrix} \frac{-\omega_s}{2H_1}\frac{\partial P_{e1}}{\partial\delta_1} & \cdots & \frac{-\omega_s}{2H_1}\frac{\partial P_{e1}}{\partial\delta_m} \\ \vdots & & \vdots \\ \frac{-\omega_s}{2H_m}\frac{\partial P_{em}}{\partial\delta_1} & \cdots & \frac{\omega_s}{2H_m}\frac{\partial P_{em}}{\partial\delta_m} \end{bmatrix}_o \tag{8.95}
$$

The elements of A_ω can also be expressed in a polar notation by noting that

$$
C_{ij} = E_i E_j B_{ij} = E_i E_j Y_{ij} \sin\alpha_{ij}
$$

and

$$
D_{ij} = E_i E_j G_{ij} = E_i E_j Y_{ij} \cos\alpha_{ij}
$$

Hence, in (8.93)

$$
C_{ij} \cos\delta_{ij}^o - D_{ij} \sin\delta_{ij}^o = E_i E_j Y_{ij} \sin(\alpha_{ij} - \delta_{ij}^o) \tag{8.96}
$$

Therefore

$$
-\frac{\partial P_{ei}}{\partial\delta_j} = \begin{cases} -\sum_{\substack{j=1 \\ \neq i}}^{m} E_i E_j Y_{ij} \sin(\alpha_{ij} - \delta_{ij}^o) & j = i \\ E_i E_j Y_{ij} \sin(\alpha_{ij} - \delta_{ij}^o) & j \neq i \end{cases} \tag{8.97}
$$

For the 3-machine case

$$A_\omega = \omega_s \begin{bmatrix} \frac{-1}{2H_1}[E_1 E_2 Y_{12}\sin(\alpha_{12}-\delta^o_{12}) \\ +E_1 E_3 Y_{13}\sin(\alpha_{13}-\delta^o_{13})] & \frac{1}{2H_1}E_1 E_2 Y_{12}\sin(\alpha_{12}-\delta^o_{12}) & \frac{1}{2H_1}[E_1 E_3 Y_{13}\sin(\alpha_{13}-\delta^o_{13})] \\ \frac{1}{2H_2}[E_1 E_2 Y_{21}\sin(\alpha_{21}-\delta^o_{21})] & \frac{-1}{2H_2}[E_2 E_1 Y_{21}\sin(\alpha_{21}-\delta^o_{21}) \\ +E_2 E_3 Y_{23}\sin(\alpha_{23}-\delta^o_{23})] & \frac{1}{2H_2}[E_2 E_3 Y_{23}\sin(\alpha_{23}-\delta^o_{23})] \\ \frac{1}{2H_3}[E_3 E_1 Y_{13}\sin(\alpha_{31}-\delta^o_{31})] & \frac{1}{2H_3}[E_3 E_2 Y_{23}\sin(\alpha_{23}-\delta^o_{32})] & \frac{-1}{2H_3}[E_3 E_1 Y_{13}\sin(\alpha_{31}-\delta^o_{31}) \\ +E_2 E_2 Y_{23}\sin(\alpha_{32}-\delta^o_{32})] \end{bmatrix}$$

$$(8.98)$$

A_ω is not symmetric, since $\delta_{ij} = -\delta_{ij}$, although $\alpha_{ij} = \alpha_{ji}$. However, the sum of the three columns $= 0$, hence, the rank of $A_\omega = 2$. This means that one eigenvalue of $A_\omega = 0$. A_ω can be expressed as $A_\omega = M^{-1}K$, where $M^1 = [\text{diag}(\frac{\omega_s}{2H_i})]$ and $K = [K_{ij}]$:

$$K_{ij} = -\sum_{j=1 \neq i}^{m} E_i E_j Y_{ij}\sin(\alpha_{ij}-\delta^o_{ij}) \qquad j = i \qquad (8.99)$$

$$= E_i E_j Y_{ij}\sin(\alpha_{ij}-\delta^o_{ij}) \qquad j \neq i \qquad (8.100)$$

K_{ij}'s are called the synchronizing coefficients, and are equivalent to those in (8.93).

Eigenvalues of A and A_ω

Case (a): Transfer conductances are neglected, i.e., $G_{ij} = 0$.

This implies $\alpha_{ij} = \frac{\pi}{2}$ and $K_{ij} = K_{ji}$ by inspection from (8.99) and (8.100). Hence, the matrix K is symmetric. It can be shown mathematically that (1) the eigenvalues of

$$A = \left[\begin{array}{c|c} 0 & U \\ \hline M^{-1}K & 0 \end{array} \right] \qquad (8.101)$$

are the square roots of the eigenvalues of $A_\omega = M^{-1}K$. (2) The eigenvalues of A_ω are real, including one zero eigenvalue. The zero eigenvalue is due to the fact that a reference angle is necessary, and the angles appear only as differences. Hence, A has a pair of zero eigenvalues. If the real eigenvalues of A_ω are negative, then A has complex pairs of eigenvalues on the imaginary axis. The nonzero eigenvalues $\pm j\omega_k$ of $A(k = 1, 2, \ldots, m-1)$ define the *electromechanical modes*. If A_ω has a positive real eigenvalue, then A has at least one positive real eigenvalue, and, hence, the system is unstable.

Case (b): $G_{ij} \neq 0 (i \neq j)$.

K is not symmetric. However, A_ω will still have a zero eigenvalue. Hence, A will also have two zero eigenvalues. It appears that the eigenvalues of A are also the square roots of the eigenvalues of A_ω.

Example 8.4

Compute the electromechanical modes with and without transfer conductances for the 3-machine system. In Example 7.6, the voltages behind transient reactances and the reduced-order model at the internal buses around the operating point have been computed. They are reproduced below.

$$\overline{E}_1 = 1.054\angle 2.27° \quad \overline{E}_2 = 1.05\angle 19.75° \quad \overline{E}_3 = 1.02\angle 13.2°$$

The $\overline{Y}_{int}$ matrix is given by

$$\overline{Y}_{int} = \begin{bmatrix} 0.845 - j2.988 & 0.287 + j1.513 & 0.210 + j1.226 \\ 0.287 + j1.513 & 0.420 - j2.724 & 0.213 + j1.088 \\ 0.210 + j1.226 & 0.213 + j1.088 & 0.277 - j2.368 \end{bmatrix}$$

The elements of A_ω are calculated as follows.

$$\frac{\omega_s}{2H_1} = 7.974 , \quad \frac{\omega_s}{2H_2} = 29.453 , \quad \frac{\omega_s}{2H_3} = 62.625$$

$$C_{12} = 1.678 , \ C_{13} = 1.321 , \ G_{11} = 0.845 , \ |\ \overline{E}_1 \ | = 1.054$$

$$D_{12} = 0.318 , \ D_{13} = 0.226$$

$$C_{21} = 1.678 , \ C_{23} = 1.165 , \ G_{22} = 0.420 , \ |\ \overline{E}_2 \ | = 1.05$$

$$D_{21} = 0.318 , \ D_{23} = 0.228$$

$$C_{31} = 1.313 , \ C_{32} = 1.165 , \ C_{33} = 0.277 , \ |\ \overline{E}_3 \ | = 1.02$$

$$D_{31} = 0.226 , \ D_{32} = 0.228$$

$$\delta_{12}^o = -17.46° , \ \delta_{13}^o = -10.91° , \ \delta_{23}^o = 6.55°$$

We next obtain the following quantities for use in computing elements of A_ω.

$$C_{12} \cos \delta_{12}^o - D_{12} \sin \delta_{12}^o = 1.696$$

$$C_{13} \cos \delta_{13}^o - D_{13} \sin \delta_{13}^o = 1.332$$

$$C_{21} \cos \delta_{21}^o - D_{21} \sin \delta_{21}^o = 1.506$$

$$C_{23} \cos \delta_{23}^o - D_{23} \sin \delta_{23}^o = 1.139$$

$$C_{31} \cos \delta_{31}^o - D_{31} \sin \delta_{31}^o = 1.246$$

$$C_{32} \cos \delta_{32}^o - D_{32} \sin \delta_{32}^o = 1.191$$

From (8.93), we can now calculate the elements of A_ω.

Case (a):

A_ω for $G_{ij} \neq 0$ is

$$A_\omega = \begin{bmatrix} -24.209 & 13.521 & 10.688 \\ 44.356 & -77.902 & 33.546 \\ 78.031 & 74.586 & -152.617 \end{bmatrix}$$

Case (b):

With $G_{ij} = 0$

$$A_\omega = \begin{bmatrix} -23.83 & 13.15 & 10.68 \\ 45.64 & -79.04 & 33.40 \\ 78.48 & 74.28 & -152.76 \end{bmatrix}$$

The eigenvalues of A are obtained as

Case (a):

$$0, \ 0, \ \pm j8.7326, \ \pm j13.393$$

Case (b):

$$0, \ 0, \ \pm j9.807, \ \pm j13.435$$

Notice the difference by neglecting G_{ij}.

□

8.6   Power System Stabilizers

So far we have discussed the linearized model, eigenvalues, and its application to voltage stability in the context of a multimachine power system. We now discuss a stabilizing device, called the power system stabilizer (PSS), used to damp out the low-frequency oscillations. Although considerable research is being done in designing PSS for a multimachine system, no definitive results have been applied in the field. The design is still done on the basis of a single machine infinite bus (SMIB) system. The parameters are then tuned on-line to suppress the modes, both the local and inter-area modes. We explain the basic approach based on control theory, and illustrate it with an example.

8.6.1   Basic approach

The differential-algebraic nonlinear model of a single machine connected to an infinite bus is given by

$$\dot{x} = f(x, y, u) \tag{8.102}$$

$$0 = g(x, y) \tag{8.103}$$

where x is the state vector, y is the vector of algebraic variables, and g consists of the stator algebraic and the network equations. Let the operating point be x^o, y^o, u^o. The perturbed variables are $x = x^o + \Delta x$, $y = y^0 + \Delta y$, and $u = u^o + \Delta u$. Linearization of (8.102) and (8.103) leads to

$$\Delta \dot{x} = A\Delta x + B\Delta y + E\Delta u \tag{8.104}$$

$$0 = C\Delta x + D\Delta y \tag{8.105}$$

If D is invertible,

$$\Delta y = -D^{-1}C\Delta x \tag{8.106}$$

Therefore

$$\Delta \dot{x} = (A - BD^{-1}C)\Delta x + E\Delta u = A_{\text{sys}}\Delta x + E\Delta u \tag{8.107}$$

A, B, C, D, and E are appropriate Jacobians of (8.102) and (8.103) evaluated at the operating point. We illustrate the procedure for a single-machine system. Equation (8.103) consists of the two stator algebraic equations and the network equations in either the power-balance or the current-balance form.

8.6.2 Derivation of $K1$-$K6$ constants [78,88]

Historically, when low-frequency oscillations were first investigated analytically, they took a two-step approach. A single machine connected to an infinite bus is chosen to analyze the local (plant) mode of oscillation in the 1- to 3-Hz range. A flux-decay model is linearized with E_{fd} as an input, and the model so obtained is put in a block diagram form. Then a fast-acting exciter between ΔV_t and ΔE_{fd} is introduced in the block diagram. In the resulting state-space model, certain constants called the $K1$-$K6$ are identified. These constants are functions of the operating point. The state-space model is then used to examine the eigenvalues, as well as to design supplementary controllers to ensure adequate damping of the dominant modes. The real and imaginary parts of the electromechanical mode are associated with the damping and synchronizing torques, respectively. We now outline this approach, which is a special case of the general development discussed in the earlier sections of this chapter.

The single machine connected to an infinite bus through an external reactance X_e and resistance R_e is the widely used configuration with a flux-decay model and stator resistance equal to zero. No local load is assumed at the generator bus. Figure 8.7 shows the system. The flux-decay model of the machine only is given by (8.62)-(8.64), with E_{fd} being treated as an input; the equations are given below.

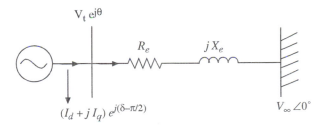

Figure 8.7: *Single-machine infinite-bus system*

$$\dot{E}'_q = -\frac{1}{T'_{do}}(E'_q + (X_d - X'_d)I_d - E_{fd}) \tag{8.108}$$

$$\dot{\delta} = \omega - \omega_s \tag{8.109}$$

$$\dot{\omega} = \frac{\omega_s}{2H}[T_M - (E'_q I_q + (X_q - X'_d)I_d I_q + D(\omega - \omega_s))] \tag{8.110}$$

The stator algebraic equations are given by (8.66) and (8.67), and we assume $R_s = 0$. We use V_t to denote the magnitude of the generator terminal voltage. The equations are

$$X_q I_q - V_t \sin (\delta - \theta) = 0 \tag{8.111}$$

$$E_q' - V_t \cos (\delta - \theta) - X_d' I_d = 0 \tag{8.112}$$

Now

$$(V_d + jV_q)e^{j(\delta - \pi/2)} = V_t e^{j\theta}$$

Hence

$$V_d + jV_q = V_t e^{j\theta} e^{-j(\delta - \pi/2)} \tag{8.113}$$

Expansion of the right-hand side results in

$$V_d + jV_q = V_t \sin (\delta - \theta) + jV_t \cos (\delta - \theta) \tag{8.114}$$

Hence, $V_d = V_t \sin(\delta - \theta)$ and $V_q = V_t \cos(\delta - \theta)$. Substituting for V_d and V_q in (8.111) and (8.112), we get the stator algebraic equations

$$X_q I_q - V_d = 0 \tag{8.115}$$

$$E_q' - V_q - X_d' I_d = 0 \tag{8.116}$$

The network equation is (assuming zero phase angle at the infinite bus)

$$(I_d + jI_q)e^{j(\delta - \pi/2)} = \frac{(V_d + jV_q)e^{j(\delta - \pi/2)} - V_\infty \angle 0^\circ}{R_e + jX_e} \tag{8.117}$$

$$(I_d + jI_q) = \frac{(V_d + jV_q) - V_\infty e^{-j(\delta - \pi/2)}}{R_e + jX_e} \tag{8.118}$$

Cross-multiplying and separating into real and imaginary parts,

$$R_e I_d - X_e I_q = V_d - V_\infty \sin \delta \tag{8.119}$$

$$X_e I_d + R_e I_q = V_q - V_\infty \cos \delta \tag{8.120}$$

We thus have, for the single-machine case, the differential equations (8.108)–(8.110) and the algebraic equations (8.115), (8.116), (8.119), and (8.120). We linearize them around an operating point, and eliminate the algebraic variables I_d, I_q, θ, V_d, V_q as follows.

Linearization

Step 1

Linearize the algebraic equations (8.115) and (8.116):

$$\begin{bmatrix} \Delta V_d \\ \Delta V_q \end{bmatrix} = \begin{bmatrix} 0 & X_q \\ -X'_d & 0 \end{bmatrix} \begin{bmatrix} \Delta I_d \\ \Delta I_q \end{bmatrix} + \begin{bmatrix} 0 \\ \Delta E'_q \end{bmatrix} \qquad (8.121)$$

Step 2

Linearize the load-flow equations (8.119) and (8.120):

$$\begin{bmatrix} \Delta V_d \\ \Delta V_q \end{bmatrix} = \begin{bmatrix} R_e & -X_e \\ X_e & R_e \end{bmatrix} \begin{bmatrix} \Delta I_d \\ \Delta I_q \end{bmatrix} + \begin{bmatrix} V_\infty \cos \delta^o \\ -V_\infty \sin \delta^o \end{bmatrix} \Delta\delta \qquad (8.122)$$

Step 3

Equate the right-hand sides of (8.121) and (8.122), and simplify

$$\begin{bmatrix} R_e & -(X_e + X_q) \\ (X_e + X'_d) & R_e \end{bmatrix} \begin{bmatrix} \Delta I_d \\ \Delta I_q \end{bmatrix} = \begin{bmatrix} 0 \\ \Delta E'_q \end{bmatrix} + \begin{bmatrix} -V_\infty \cos \delta^o \\ V_\infty \sin \delta^o \end{bmatrix} \Delta\delta$$

$$(8.123)$$

Now

$$\begin{bmatrix} R_e & -(X_e + X_q) \\ (X_e + X'_d) & R_e \end{bmatrix}^{-1} = \frac{1}{\Delta} \begin{bmatrix} R_e & (X_e + X_q) \\ -(X_e + X'_d) & R_e \end{bmatrix}$$

$$(8.124)$$

where the determinant Δ is given by

$$\Delta = R_e^2 + (X_e + X_q)(X_e + X'_d)$$

Solve for ΔI_d, ΔI_q in (8.123), to get (after simplification)

$$\begin{bmatrix} \Delta I_d \\ \Delta I_q \end{bmatrix} = \frac{1}{\Delta} \begin{bmatrix} (X_e + X_q) & | & -R_e V_\infty \cos \delta^o + V_\infty \sin \delta^o (X_q + X_e) \\ R_e & | & R_e V_\infty \sin \delta^o + V_\infty \cos \delta^o (X'_d + X_e) \end{bmatrix}$$

$$\begin{bmatrix} \Delta E'_q \\ \Delta\delta \end{bmatrix} \qquad (8.125)$$

Step 4

Linearize the differential equations (8.108)–(8.110). We introduce the normalized frequency $\nu = \frac{\omega}{\omega_s}$ so that the linearized differential equations become

$$
\begin{bmatrix} \Delta \dot{E}'_q \\ \Delta \dot{\delta} \\ \Delta \dot{\nu} \end{bmatrix} = \begin{bmatrix} \frac{-1}{T'_{do}} & 0 & 0 \\ 0 & 0 & \omega_s \\ \frac{-I^o_q}{2H} & 0 & -\frac{D\omega_s}{2H} \end{bmatrix} \begin{bmatrix} \Delta E'_q \\ \Delta \delta \\ \Delta \nu \end{bmatrix}
$$

$$
+ \begin{bmatrix} \frac{-1}{T'_{do}}(X_d - X'_d) & 0 \\ 0 & 0 \\ \frac{1}{2H}I^o_q(X'_d - X_q) & \frac{1}{2H}(X'_d - X_q)I^o_d - \frac{1}{2H}E'^o_q \end{bmatrix} \begin{bmatrix} \Delta I_d \\ \Delta I_q \end{bmatrix}
$$

$$
+ \begin{bmatrix} \frac{1}{T'_{do}} & 0 \\ 0 & 0 \\ 0 & \frac{1}{2H} \end{bmatrix} \begin{bmatrix} \Delta E_{fd} \\ \Delta T_M \end{bmatrix} \qquad (8.126)
$$

Step 5

Substitute for ΔI_d, ΔI_q from (8.125) into (8.126) to obtain

$$
\Delta \dot{E}'_q = -\frac{1}{K_3 T'_{do}} \Delta E'_q - \frac{K_4}{T'_{do}} \Delta \delta + \frac{1}{T'_{do}} \Delta E_{fd} \qquad (8.127)
$$

$$
\Delta \dot{\delta} = \omega_s \Delta \nu \qquad (8.128)
$$

$$
\Delta \dot{\nu} = -\frac{K_2}{2H} \Delta E'_q - \frac{K_1}{2H} \Delta \delta - \frac{D\omega_s}{2H} \Delta \nu + \frac{1}{2H} \Delta T_M \qquad (8.129)
$$

where

$$
\frac{1}{K_3} = 1 + \frac{(X_d - X'_d)(X_q + X_e)}{\Delta} \qquad (8.130)
$$

$$
K_4 = \frac{V_\infty (X_d - X'_d)}{\Delta} [(X_q + X_e) \sin \delta^o - R_e \cos \delta^o] \qquad (8.131)
$$

$$
K_2 = \frac{1}{\Delta}[I^o_q \Delta - I^o_q(X'_d - X_q)(X_q + X_e) - R_e(X'_d - X_q)I^o_d + R_e E'^o_q] \qquad (8.132)
$$

$$
K_1 = -\frac{1}{\Delta}[I^o_q V_\infty (X'_d - X_q)\{(X_q + X_e) \sin \delta^o - R_e \cos \delta^o\}
$$

$$+V_\infty\{(X_d'-X_q)I_d^o-E_q'^o\}\{(X_d'+X_e)\cos\ \delta^o+R_e\ \sin\ \delta^o\}]$$

$$(8.133)$$

Since

$$V_t = \sqrt{V_d^2+V_q^2}$$

$$V_t^2 = V_d^2+V_q^2$$

$$2V_t\Delta V_t = 2V_d^o\Delta V_d+2V_q^o\Delta V_q$$

$$\Delta V_t = \frac{V_d^o}{V_t}\Delta V_d+\frac{V_q^o}{V_t}\Delta V_q \qquad (8.134)$$

Substituting (8.125) in (8.121) results in

$$\begin{bmatrix} \Delta V_d \\ \Delta V_q \end{bmatrix} = \frac{1}{\Delta}\begin{bmatrix} 0 & X_q \\ -X_d' & 0 \end{bmatrix}$$

$$\begin{bmatrix} X_q+X_e & -R_eV_\infty\cos\ \delta^o+V_\infty(X_q+X_e)\sin\ \delta^o) \\ R_e & R_eV_\infty\sin\ \delta^o+V_\infty\cos\ \delta^o(X_d'+X_e) \end{bmatrix}$$

$$\left\{ \begin{bmatrix} \Delta E_q' \\ \Delta\delta \end{bmatrix} + \begin{bmatrix} 0 \\ \Delta E_q' \end{bmatrix} \right\}$$

$$= \frac{1}{\Delta}\begin{bmatrix} X_qR_e \\ -X_d'(X_q+X_e) \end{bmatrix}$$

$$\begin{matrix} X_q(R_eV_\infty\sin\ \delta^o+V_\infty\cos\ \delta^o(X_d'+X_e) \\ -X_d'(-R_eV_\infty\cos\ \delta^o+V_\infty(X_q+X_e)\sin\ \delta^o) \end{matrix} \Bigg]$$

$$\left\{ \begin{bmatrix} \Delta E_q' \\ \Delta\delta \end{bmatrix} + \begin{bmatrix} 0 \\ \Delta E_q' \end{bmatrix} \right\} \qquad (8.135)$$

Substituting (8.135) in (8.134) gives

$$\Delta V_t = K_5\Delta\delta + K_6\Delta E_q' \qquad (8.136)$$

where

$$K_5 = \frac{1}{\Delta}\left\{ \frac{V_d^o}{V_t}X_q\left[R_eV_\infty\sin\ \delta^o+V_\infty\cos\ \delta^o(X_d'+X_e)\right] \right.$$

$$\left. +\frac{V_q^o}{V_t}\left[X_d'(R_eV_\infty\cos\ \delta^o-V_\infty(X_q+X_e)\sin\ \delta^o)\right] \right\} \qquad (8.137)$$

$$K_6 = \frac{1}{\Delta}\left\{ \frac{V_d^o}{V_t}X_qR_e - \frac{V_q^o}{V_t}X_d'(X_q+X_e) \right\} + \frac{V_q^o}{V_t} \qquad (8.138)$$

The constants that we have derived are called the $K1$-$K6$, developed by Heffron-Phillips [88], and later by DeMello-Concordia [78], for the study of local low-frequency oscillations.

Example 8.5

In Figure 8.7, assume that $R_e = 0$, $X_e = 0.5$ pu, $V_t \angle \theta = 1 \angle 15°$ pu, and $V_\infty \angle 0° = 1.05 \angle 0°$ pu. The machine data are $H = 3.2$ sec, $T'_{do} = 9.6$ sec, $K_A = 400$, $T_A = 0.2$ sec, $R_s = 0.0$ pu, $X_q = 2.1$ pu, $X_d = 2.5$ pu, $X'_d = 0.39$ pu, $D = 0$, and $\omega_s = 377$. Using the flux-decay model, find (1) the initial values of state and algebraic variables, as well as V_{ref}, T_M, and (2) $K1$-$K6$ constants.

1. Computation of Initial Conditions

The technique discussed in Section 7.6 is followed. The superscript o on the algebraic and state variables is omitted.

$$I_G e^{j\gamma} = (I_d + jI_q)e^{j(\delta - \pi/2)} = \frac{1\angle 15° - 1.05\angle 0°}{j0.5} = 0.5443\angle 18°$$

$\delta(0) = $ angle of $\overline{E}$ where $\overline{E} = V_t e^{j\theta} + (R_s + jX_q)I_G e^{j\gamma}$.

$$\overline{E} = 1\angle 15° + (j2.1)(0.5443\angle 18°)$$

$$= 1.4788\angle 65.52°$$

Therefore $\delta(0) = 65.52°$. $I_d + jI_q = I_G e^{j\gamma}e^{-j(\delta - \pi/2)} = 0.5443\angle 42.48°$, $I_d = 0.4014$, and $I_q = 0.3676$.

$$V_d + jV_q = V e^{j\theta}e^{-j(\delta - \pi/2)}$$

$$= 1\angle 39.48°$$

Hence

$$V_d = 0.77185, \ V_q = 0.63581$$

From (8.116):

$$E'_q = V_q + X'_d I_d$$

$$= 0.63581 + (0.39)(0.4014) = 0.7924$$

From (8.62)–(8.65), setting derivatives $= 0$,

$$E_{fd} = E_q' + (X_d - X_d')I_d$$

$$= 0.7924 + (2.5 - 0.39)0.4014 = 1.6394$$

$$V_{\text{ref}} = V_t + \frac{E_{fd}}{K_A} = 1 + \frac{1.6394}{400} = 1.0041$$

$$\omega_s = 377, \quad T_M = E_q'I_q + (X_q - X_d')I_dI_q$$

$$= (0.7924)(0.3676) + (2.1 - 0.39)(0.4014)(0.3676)$$

$$= 0.5436$$

This completes the calculation of the initial values.

2. Computation of *K1-K6* Constants

The formulas given in (8.130)–(8.134) and (8.137)–(8.138) are used.

$$\Delta = R_e^2 + (X_e + X_q)(X_e + X_d')$$

$$= 2.314$$

$$\frac{1}{K_3} = 1 + \frac{(X_d - X_d')(X_q + X_e)}{\Delta}$$

$$= 3.3707$$

$$K_3 = 0.296667$$

$$K_4 = \frac{V_\infty(X_d - X_d')}{\Delta}[(X_q + X_e)\sin\delta° - R_e\cos\delta°]$$

$$= 2.26555$$

$$K_2 = \frac{1}{\Delta}[I_q°\Delta - I_q°(X_d' - X_q)(X_q + X_e) - R_e(X_d' - X_q)I_d° + R_eE_q'°]$$

$$= 1.0739$$

Similarly, K_1, K_5, and K_6 are calculated as

$$K_1 = 0.9224$$

$$K_5 = 0.005$$

$$K_6 = 0.3572$$

□

8.6.3 Synchronizing and damping torques

To the single-machine infinite-bus system of Figure 8.7, we add a fast exciter whose state-space equation is

$$T_A \dot{E}_{fd} = -E_{fd} + K_A(V_{ref} - V_t) \tag{8.139}$$

The linearized form of (8.139) is

$$T_A \Delta \dot{E}_{fd} = -\Delta E_{fd} + K_A(\Delta V_{ref} - \Delta V_t) \tag{8.140}$$

Then the machine differential equations (8.127)–(8.129), the exciter equation (8.140), and the algebraic equation (8.136) can be put in the block diagram form shown in Figure 8.8. Both the normalized frequency ν and ω in rad/sec are shown.

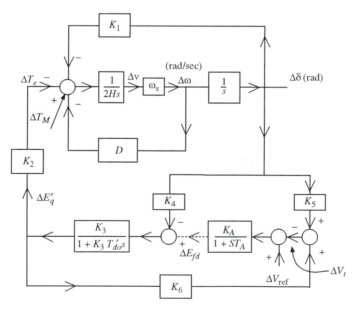

Figure 8.8: *Block diagram of the incremental flux-decay model with fast exciter (dotted portion represents the exciter)*

System loading, as well as the external network parameter X_e, affect the parameters *K1-K6*. Generally these are > 0, but under heavy loading, K_5 might become negative, contributing to negative damping and instability, as we explain below.

Damping of Electromechanical Modes

There are two ways to explain the damping phenomena:

1. State-space analysis [89]

2. Frequency-domain analysis [78]

1. State-space analysis (assume that $\Delta T_M \equiv 0$)

Equations (8.127)–(8.129) are rewritten in matrix form as

$$
\begin{bmatrix} \Delta \dot{E}'_q \\ \Delta \dot{\delta} \\ \Delta \dot{\nu} \end{bmatrix} = \begin{bmatrix} -\dfrac{1}{K_3 T'_{do}} & \dfrac{-K_4}{T'_{do}} & 0 \\ 0 & 0 & \omega_s \\ \dfrac{-K_2}{2H} & \dfrac{-K_1}{2H} & -\dfrac{D\omega_s}{2H} \end{bmatrix} \begin{bmatrix} \Delta E'_q \\ \Delta \delta \\ \Delta \nu \end{bmatrix} + \begin{bmatrix} \dfrac{1}{T'_{do}} \\ 0 \\ 0 \end{bmatrix} \Delta E_{fd}
$$

$$(8.141)$$

Note that, instead of $\Delta \omega$, we have used the normalized frequency deviation $\Delta \nu = \Delta \omega / \omega_s$. Hence, the last row in (8.141) is

$$
\Delta \dot{\nu} = -\frac{K_2}{2H} \Delta E'_q - \frac{K_1}{2H} \Delta \delta - \frac{D\omega_s}{2H} \Delta \nu \qquad (8.142)
$$

ΔE_{fd} is the perturbation in the field voltage. Without the exciter, the machine is said to be on "manual control." The matrix generally has a pair of complex eigenvalues and a negative real eigenvalue. The former corresponds to the electromechanical mode (1 to 3-Hz range), and the latter the flux-decay mode. Without the exciter (i.e., $K_A = 0$), there are three loops in the block diagram (Figure 8.8), the top two loops corresponding to the complex pair of eigenvalues, and the bottom loop due to $\Delta E'_q$ through K_4, resulting in the real eigenvalue. Note that the bottom loop contributes to positive feedback. Hence, the torque-angle eigenvalues tend to move to the left-half plane, and the negative real eigenvalue to the right. Thus, with constant E_{fd}, there is "natural" damping. With enough gain, the real pole may go to the right-half plane. This is referred to as monotonic instability. In the power literature, this twofold effect is described in more graphic physical terms. The effect of the lag associated with the time constant T'_{do} is to increase the damping torque but to decrease the synchronizing torque. Now, if we add the exciter through the simplified representation, the state-space equation will now be modified by making ΔE_{fd} a state variable. The equation for ΔE_{fd} is given by (8.140) as

$$\Delta \dot{E}_{fd} = -\frac{1}{T_A}\Delta E_{fd} + \frac{K_A}{T_A}(\Delta V_{ref} - \Delta V_t)$$

$$= -\frac{1}{T_A}\Delta E_{fd} - \frac{K_A K_5}{T_A}\Delta\delta - \frac{K_A K_6}{T_A}\Delta E_q' + \frac{K_A}{T_A}\Delta V_{ref} \quad (8.143)$$

Ignoring the dynamics of the exciter for the moment, if $K_5 < 0$ and K_A is large enough, then the gain through T_{do}' is approximately $-(K_4 + K_A K_5)K_3$. This gain may become positive, resulting in negative feedback for the torque-angle loop and pushing the complex pair to the right-half plane. Hence, this complicated action should be studied carefully. The overall state-space model for Figure 8.8 becomes

$$\begin{bmatrix} \Delta \dot{E}_q' \\ \Delta \dot{\delta} \\ \Delta \dot{\nu} \\ \Delta \dot{E}_{fd} \end{bmatrix} = \begin{bmatrix} \frac{-1}{K_3 T_{do}'} & \frac{-K_4}{T_{do}'} & 0 & \frac{1}{T_{do}'} \\ 0 & 0 & \omega_s & 0 \\ \frac{-K_2}{2H} & \frac{-K_1}{2H} & -\frac{D\omega_s}{2H} & 0 \\ \frac{-K_A K_6}{T_A} & \frac{-K_A K_5}{T_A} & 0 & \frac{-1}{T_A} \end{bmatrix} \begin{bmatrix} \Delta E_q' \\ \Delta\delta \\ \Delta\nu \\ \Delta E_{fd} \end{bmatrix} + \begin{bmatrix} 0 \\ 0 \\ 0 \\ \frac{K_A}{T_A} \end{bmatrix} \Delta V_{ref}$$

$$(8.144)$$

The exciter introduces an additional negative real eigenvalue.

Example 8.6

For the following two test systems whose $K_1 - K_6$ constants and other parameters are given, find the eigenvalues for $K_A = 50$. Plot the root locus for varying K_A. Note that in system 1 $K_5 > 0$, and in system 2 $K_5 < 0$.

Test System 1

$$K_1 = 3.7585 \quad K_2 = 3.6816$$
$$K_3 = 0.2162 \quad K_4 = 2.6582$$
$$K_5 = 0.0544 \quad K_6 = 0.3616$$
$$T_{do}' = 5 \text{ sec} \quad H = 6 \text{ sec}$$
$$T_A = 0.2 \text{ sec}$$

Test System 2

$$K_1 = 0.9831 \quad K_2 = 1.0923$$
$$K_3 = 0.3864 \quad K_4 = 1.4746$$
$$K_5 = -0.1103 \quad K_6 = 0.4477$$
$$T'_{do} = 5 \text{ sec} \quad H = 6 \text{ sec}$$
$$T_A = 0.2 \text{ sec}$$

The eigenvalues for $K_A = 50$ are shown below using (8.144).

Test System 1	Test System 2
-0.353 ± j10.946	0.015 ± j5.38
-2.61 ± j3.22	-2.77 ± j2.88

Notice that test system 2 is unstable for this value of gain. The root loci for the two systems can be drawn using MATLAB. An alternative way to draw the root locus is to remove the exciter and compute the transfer function $\frac{\Delta V_t(s)}{\Delta E_{fd}(s)} = H(s)$ in Figure 8.8. Note that $H(s)$ includes all the dynamics except that of the exciter. With the $G(s) = \frac{K_A}{1+0.2s}$, we can view $H(s)$ as a feedback transfer function, as in Figure 8.9.

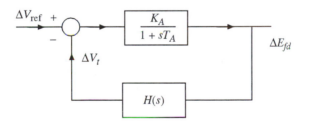

Figure 8.9: *Small-signal model viewed as a feedback system*

$H(s)$ can be computed for each of the two systems as

System 1

$$H(s) = \frac{0.0723s^2 + 7.2811}{s^3 + 0.9251s^2 + 118.0795s + 47.74}$$

System 2

$$H(s) = \frac{0.0895s^2 + 3.5225}{s^3 + 0.5176s^2 + 30.886s^2 + 5.866}$$

The closed-loop characteristic equation is given by $1 + G(s)H(s) = 0$, where $G(s) = \frac{K_A}{1+0.2s}$. The root locus for each of the two test systems is shown in Figure 8.10. System 1 is stable for all values of gain, whereas system 2 becomes unstable for $K_A = 22.108$.

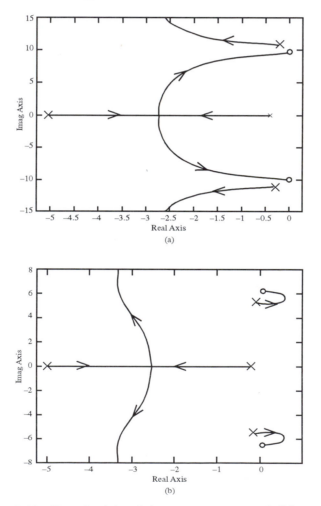

Figure 8.10: *Root loci for (a) test system 1 and (b) test system 2*

□

2. Frequency-domain analysis through block diagram

For simplicity, assume that the exciter is simply a high constant gain K_A, i.e., assume $T_A = 0$ in Figure 8.8. Now, we can compute the transfer function $\Delta E_q'(s)/\Delta\delta(s)$ as

$$\frac{\Delta E_q'(s)}{\Delta\delta(s)} = \frac{-K_3(K_4 + K_A K_5)}{1 + K_A K_3 K_6 + s K_3 T_{do}'} \qquad (8.145)$$

This assumes that $\Delta V_{\text{ref}} = 0$. The effect of the feedback around T_{do}' is to reduce the time constant. If $K_5 > 0$, the overall situation does not differ qualitatively from the case without the exciter, i.e., the system has three open loop poles, with one of them being complex and positive feedback. Thus, the real pole tends to move into the right-half plane. If $K_5 < 0$ and, consequently, $K_4 + K_A K_5 < 0$, the feedback from $\Delta\delta$ to ΔT_e changes from positive to negative, and, with a large enough gain K_A, the electromechanical modes may move to the right-half plane and the real eigenvalue to the left on the real axis. The situation is changed in detail, but not in its general features, if a more detailed exciter model is considered.

Thus, a fast-acting exciter is bad for damping, but it has beneficial effects also. It minimizes voltage fluctuations, increases the synchronizing torque, and improves transient stability. With the time constant T_A present,

$$\frac{\Delta E_q'(s)}{\Delta\delta(s)} = \frac{-[(K_4(1 + sT_A) + K_A K_5)]K_3}{K_A K_6 K_3 + (1 + K_3 T_{do}' s)(1 + sT_A)} \qquad (8.146)$$

The contribution of this expression to the torque-angle loop is given by

$$\frac{\Delta T_e(s)}{\Delta\delta(s)} = K_2 \frac{\Delta E_q'(s)}{\Delta\delta(s)} \triangleq H(s) \qquad (8.147)$$

Torque-Angle Loop

Letting $\Delta T_M = 0$, the torque-angle loop is given by Figure 8.11. The undamped frequency of the torque-angle loop ($D \equiv 0$) is given by the roots of the characteristic equation

$$\frac{2H}{\omega_s}s^2 + K_1 = 0 \qquad (8.148)$$

$$s_{1,2} = \pm j\sqrt{\frac{K_1\omega_s}{2H}} \text{ rad/sec} \qquad (8.149)$$

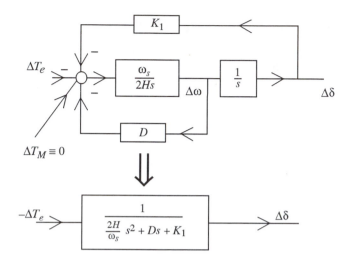

Figure 8.11: *Torque-angle loop*

With a higher synchronizing torque coefficient K_1 and lower H, $s_{1,2}$ is higher. K_1 is a complicated expression involving loading conditions and external reactances. The value of D is generally small and, hence, neglected.

We wish to compute the damping due to E'_q. The overall block diagram neglecting damping is shown in Figure 8.12. From this diagram and the closed-loop transfer function, it can be verified that the characteristic equation is given by

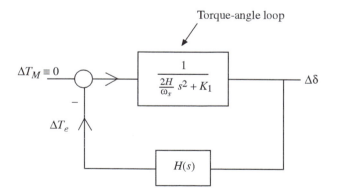

Figure 8.12: *Torque-angle loop with other dynamics added*

$$\frac{2H}{\omega_s} s^2 \Delta\delta + K_1\Delta\delta + H(s)\Delta\delta = 0 \qquad (8.150)$$

$H(s)\Delta\delta$ therefore contributes to both the synchronizing torque and the damping torque. The contributions are now computed approximately. At oscillation frequencies of 1 to 3 Hz, it can be shown that K_4 has negligible effect.

Neglecting the effect of K_4 in Figure 8.8, we get from (8.147)

$$H(s) = \frac{-K_2 K_A K_5}{\frac{1}{K_3} + K_A K_6 + s\left(\frac{T_A}{K_3} + T'_{do}\right) + s^2 T'_{do} T_A} \qquad (8.151)$$

Let $s = j\omega$. Then

$$H(j\omega) = \frac{-K_2 K_A K_5}{\left(\frac{1}{K_3} + K_A K_6 - \omega^2 T'_{do} K_A\right) + j\omega\left(\frac{T_A}{K_3} + T'_{do}\right)}$$

$$= \frac{-K_2 K_A K_5 (x - jy)}{x^2 + y^2} \qquad (8.152)$$

where

$$x = \frac{1}{K_3} + K_A K_6 - \omega^2 T'_{do} T_A \qquad (8.153)$$

$$y = \omega\left(\frac{T_A}{K_3} + T'_{do}\right) \qquad (8.154)$$

From (8.150), it is clear that, at the oscillation frequency, if $Im[H(j\omega)] > 0$, positive damping is implied, i.e., the roots move to the left-half plane. If $Im[H(j\omega)] < 0$, it tends to make the system unstable, i.e., negative damping results. Thus

$$Re[H(j\omega)] = \frac{-K_2 K_A K_5 x}{x^2 + y^2} \triangleq \text{ contribution to the synchronizing}$$

torque component due to $H(s)$ \qquad (8.155)

$$Im[H(j\omega)] = \frac{+K_2 K_A K_5 y}{x^2 + y^2} \triangleq \text{ contribution to damping}$$

torque component due to $H(s)$ \qquad (8.156)

Synchronizing Torque

For low frequencies, we set $\omega \approx 0$. Thus, from (8.155):

$$Re[H(j\omega)] = \frac{-K_2 K_A K_5}{\frac{1}{K_3} + K_A K_6} \approx \frac{-K_2 K_5}{K_6} \text{ for high } K_A \qquad (8.157)$$

Thus, the total synchronizing component is $K_1 - \frac{K_2 K_5}{K_6} > 0$. K_1 is usually high, so that even with $K_5 > 0$ (low to medium external impedance and low-to-medium loadings), $\frac{K_1 - K_2 K_5}{K_6} > 0$. With $K_5 < 0$ (moderate to high external impedance and heavy loadings), the synchronizing torque is enhanced positively.

Damping torque

$$Im[H(j\omega)] = \frac{K_2 K_A K_5 \left(\frac{T_A}{K_3} + T'_{do}\right) \omega}{x^2 + y^2} \qquad (8.158)$$

This expression contributes to positive damping for $K_5 > 0$ but negative damping for $K_5 < 0$, which is a cause for concern. Further, with $K_5 < 0$, a higher K_A spells trouble (see Figure 8.10). This may offset the inherent machine damping torque D. To introduce damping, a power system stabilizer (PSS) is therefore introduced. The stabilizing signal may be $\Delta\nu$, ΔP_{acc}, or a combination of both. We discuss this briefly next. For an extensive discussion of PSS design, the reader is referred to the literature [77, 78].

8.6.4   Power system stabilizer design

Speed Input PSS

Stabilizing signals derived from machine speed, terminal frequency, or power are processed through a device called the power system stabilizer (PSS) with a transfer function $G(s)$ and its output connected to the input of the exciter. Figure 8.13 shows the PSS with speed input and the signal path from $\Delta\nu$ to the torque-angle loop.

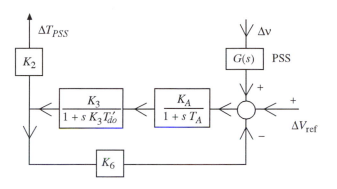

Figure 8.13: *Speed input PSS*

Frequency-Domain Approach [78]

From Figure 8.13, the contribution of the PSS to the torque-angle loop is (assuming $\Delta V_{\text{ref}} \equiv 0$ and $\Delta \delta \equiv 0$)

$$\frac{\Delta T_{PSS}}{\Delta \nu} = \frac{G(s) K_2 K_A K_3}{K_A K_3 K_6 + (1 + s K_3 T'_{do})(1 + sT_A)}$$

$$= \frac{G(s) K_2 K_A}{\left(\frac{1}{K_3} + K_A K_6\right) + s\left(\frac{T_A}{K_3} + T'_{do}\right) + s^2 T'_{do} T_A}$$

$$= G(s) GEP(s) \tag{8.159}$$

For the usual range of constants [78], the above expression can be approximated as

$$\approx \frac{G(s) K_2 K_A}{\left(\frac{1}{K_3} + K_A K_6\right) \left[1 + s\left(T'_{do}/K_A K_6\right)\right] \left(1 + sT_A\right)} \tag{8.160}$$

For large values of K_A (high gain exciter), this is further approximated by

$$\frac{\Delta T_{PSS}}{\Delta \nu} = \frac{K_2}{K_6} \frac{G(s)}{\left[1 + s(T'_{do}/K_A K_6)\right]\left[1 + sT_A\right]} \tag{8.161}$$

If this were to provide pure damping throughout the frequency range, then $G(s)$ should be a pure lead function with zeros, i.e., $G(s) = K_{PSS}[1 + s(T'_{do}/K_A K_6)](1 + sT_A)$ where K_{PSS} = gain of the PSS. Such a function is not physically realizable. Hence, we have a compromise resulting in what is called a lead-lag type transfer function such that it provides enough phase lead over the expected range of frequencies. For design purposes, $G(s)$ is of the form

$$G(s) = K_{PSS} \frac{(1+sT_1)}{(1+sT_2)} \frac{(1+sT_3)}{(1+sT_4)} \frac{sT_W}{(1+sT_W)} = K_{PSS}G_1(s) \qquad (8.162)$$

The time constants T_1, T_2, T_3, T_4 should be set to provide damping over the range of frequencies at which oscillations are likely to occur. Over this range they should compensate for the phase lag introduced by the machine and the regulator. A typical technique [77] is to compensate for the phase lag in the absence of PSS such that the net phase lag is:

1. Between 0 to 45° from 0.3 to 1 Hz

2. Less than 90° up to 3 Hz

Typical values of the parameters are:

> K_{PSS} is in the range of 0.1 to 50
> T_1 is the lead time constant, 0.2 to 1.5 sec
> T_2 is the lag time constant, 0.02 to 0.15 sec
> T_3 is the lead time constant, 0.2 to 1.5 sec
> T_4 is the lag time constant, 0.02 to 0.15 sec

The desired stabilizer gain is obtained by first finding the gain at which the system becomes unstable. This may be obtained by actual test or by root locus study. T_W, called the washout time constant, is set at 10 sec. The purpose of this constant is to ensure that there is no steady-state error of voltage reference due to speed deviation. K_{PSS} is set at $\frac{1}{3} K_{PSS}^*$, where K_{PSS}^* is the gain at which the system becomes unstable [77].

It is important to avoid interaction between the PSS and the torsional modes of vibration. Analysis has revealed that such interaction can occur on nearly all modern excitation systems, as they have relatively high gain at high frequencies. A stabilizer-torsional instability with a high-response excitation system may result in shaft damage, particularly at light generator loads where the inherent mechanical damping is small. Even if shaft damage does not occur, such an instability can cause saturation of the stabilizer output, causing it to be ineffective, and possibly causing saturation of the voltage regulator, resulting in loss of synchronism and tripping the unit. It is imperative that stabilizers do not induce torsional instabilities. Hence, the PSS is put in series with another transfer function FILT(s) [77]. A typical value of FILT(s) $\approx \frac{570}{570+35s+s^2}$. The overall transfer function of PSS is $G(s)$FILT(s).

Design Procedure Using the Frequency-Domain Method

The following procedure is adapted from [90]. In Figure 8.13, let

$$\frac{\Delta T_{PSS}}{\Delta \nu} = GEP(s)G(s) \tag{8.163}$$

where $GEP(s)$ from (8.159) is given by

$$GEP(s) = \frac{K_2 K_A K_3}{K_A K_3 K_6 + (1 + sT'_{do}K_3)(1 + sT_A)} \tag{8.164}$$

Step 1

Neglecting the damping due to all other sources, find the undamped natural frequency ω_n in rad/sec of the torque-angle loop from

$$\frac{2H}{\omega_s}s^2 + K_1 = 0 \ , \ \text{i.e.} \ s_{1,2} = \pm j\omega_n, \ \text{where} \ \omega_n = \sqrt{\frac{K_1\omega_s}{2H}} \tag{8.165}$$

Step 2

Find the phase lag of $GEP(s)$ at $s = j\omega_n$ in (8.164).

Step 3

Adjust the phase lead of $G(s)$ in (8.163) such that

$$\angle G(s) \, |_{s=j\omega_n} + \angle GEP(s) \, |_{s=j\omega_n} = 0 \tag{8.166}$$

Let

$$G(s) = K_{PSS} \left(\frac{1 + sT_1}{1 + sT_2} \right)^k \tag{8.167}$$

ignoring the washout filter whose net phase contribution is approximately zero. $k = 1$ or 2 with $T_1 > T_2$. Thus, if $k = 1$:

$$\angle 1 + j\omega_n T_1 = \angle 1 + j\omega_n T_2 - \angle GEP(j\omega_n) \tag{8.168}$$

Knowing ω_n and $\angle GEP(j\omega_n)$, we can select T_1. T_2 can be chosen as some value between 0.02 to 0.15 sec.

Step 4

To compute K_{PSS}, we can compute K^*_{PSS}, i.e., the gain at which the system becomes unstable using the root locus, and then have $K_{PSS} = \frac{1}{3}K^*_{PSS}$. An alternative procedure that avoids having to do the root locus is to design for a damping ratio ξ due to PSS alone. In a second-order system whose characteristic equation is

$$\frac{2H}{\omega_s}s^2 + Ds + K_1 = 0 \tag{8.169}$$

The damping ratio is $\xi = \frac{1}{2}D/\sqrt{MK_1}$ where $M = 2H/\omega_s$. This is shown in Figure 8.14.

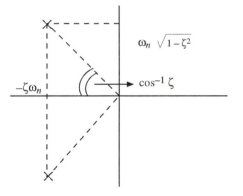

Figure 8.14: *Damping ratio*

The characteristic roots of (8.169) are

$$s_{1,2} = -\frac{-\frac{D}{M} \pm \sqrt{\left(\frac{D}{M}\right)^2 - \frac{4K_1}{M}}}{2}$$

$$= -\frac{D}{2M} \pm j\sqrt{\frac{K_1}{M} - \left(\frac{D}{2M}\right)^2} \text{ if } \left(\frac{D}{M}\right)^2 < \frac{4K_1}{M}$$

$$= -\xi\omega_n \pm j\omega_n\sqrt{1 - \xi^2}$$

We note that $\omega_n = \sqrt{\frac{K_1}{M}}$. Therefore

$$\xi = \frac{D}{2M\omega_n} = \frac{D}{2M}\sqrt{\frac{M}{K_1}} = \frac{D}{2\sqrt{K_1 M}} \qquad (8.170)$$

Verify that

$$\omega_n^2 \xi^2 = \frac{K_1}{M} \cdot \frac{D^2}{4K_1 M} = \left(\frac{D}{2M}\right)^2$$

To revert to step 4, since the phase lead of $G(s)$ cancels phase lag due to $GEP(s)$ at the oscillation frequency, the contribution of the PSS through $GEP(s)$ is a pure damping torque with a damping coefficient D_{PSS}. Thus, again ignoring the phase contribution of the washout filter,

$$D_{PSS} = K_{PSS} \mid GEP(s) \mid_{s=j\omega_n} \mid\mid G_1(s) \mid_{s=j\omega_n} \mid \qquad (8.171)$$

Therefore, the characteristic equation is

$$s^2 + \frac{D_{PSS}}{M} s + \frac{K_1}{M} = 0 \qquad (8.172)$$

i.e., $s^2 + 2\xi\omega_n s + \omega_n^2 = 0$. As a result,

$$D_{PSS} = 2\xi\omega_n M = K_{PSS} \mid GEP(j\omega_n) \mid\mid G_1(j\omega_n) \mid \qquad (8.173)$$

We can thus find K_{PSS}, knowing ω_n and the desired ξ. A reasonable choice for ξ is between 0.1 and 0.3.

Step 5

Design of the washout time constant is now discussed. The PSS should be activated only when low-frequency oscillations develop and should be automatically terminated when the system oscillation ceases. It should not interfere with the regular function of the excitation system during steady-state operation of the system frequency. The washout stage has the transfer function

$$G_W(s) = \frac{sT_W}{1 + sT_W} \qquad (8.174)$$

Since the washout filter should not have any effect on phase shift or gain at the oscillating frequency, it can be achieved by choosing a large value of T_W so that sT_W is much larger than unity.

$$G_W(j\omega_n) \approx 1 \tag{8.175}$$

Hence, its phase contribution is close to zero. The PSS will not have any effect on the steady state of the system since, in steady state,

$$\Delta\nu = 0 \tag{8.176}$$

Example 8.7

The purpose of this example is to show that the introduction of the PSS will improve the damping of the electromechanical mode. Without the PSS, the A matrix, for example, 8.5, is calculated as

$$\begin{bmatrix} -0.3511 & -0.236 & 0 & 0.104 \\ 0 & 0 & 377 & 0 \\ -0.1678 & -0.144 & 0 & 0 \\ -714.4 & -10 & 0 & -5 \end{bmatrix}$$

The eigenvalues are $\lambda_{1,2}$ = -0.0875 $\pm$ $j7.11$, $\lambda_{3,4}$ = -2.588 $\pm$ $j8.495$. The electromechanical mode $\lambda_{1,2}$ is poorly damped. Instead of a two-stage lag lead compensator, we will have a single-stage lag-lead PSS. Assume that the damping D in the torque-angle loop is zero. The input to the stabilizer is $\Delta\nu$. An extra state equation will be added. The washout stage is omitted, since its objective is to offset only the dc steady-state error. Hence, it does not play any role in the design. The block diagram in Figure 8.15 shows a single lag-lead stage of the PSS. The added state equation due to the PSS is

$$\Delta\dot{y} = -\frac{1}{T_2}\Delta y + \frac{K_{PSS}}{T_2}\Delta\nu + K_{PSS}\frac{T_1}{T_2}\Delta\dot{\nu}$$

$$= \frac{-1}{T_2}\Delta y + \frac{K_{PSS}}{T_2}\Delta\nu + K_{PSS}\frac{T_1}{T_2}\left(\frac{-K_2}{2H}\Delta E_q' - \frac{K_1}{2H}\Delta\delta\right) \tag{8.177}$$

The new A matrix is given as

$$\begin{bmatrix} \frac{-1}{K_3 T_{do}'} & \frac{-K_4}{T_{do}'} & 0 & \frac{1}{T_{do}'} & 0 \\ 0 & 0 & 377 & 0 & 0 \\ \frac{-K_2}{2H} & \frac{-K_1}{2H} & 0 & 0 & 0 \\ \frac{K_A K_6}{T_A} & \frac{-K_A K_5}{T_A} & 0 & \frac{-1}{T_A} & \frac{K_A}{T_A} \\ \frac{-K_2 T_1}{T_2}\left(\frac{K_{PSS}}{2H}\right) & \frac{-K_1 T_1}{T_2}\left(\frac{K_{PSS}}{2H}\right) & \frac{K_{PSS}}{T_2} & 0 & \frac{-1}{T_2} \end{bmatrix}$$

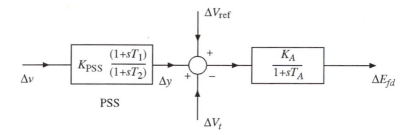

Figure 8.15: *Exciter with PSS*

With a choice of $K_{PSS} = 0.5$, $T_1 = 0.5$, $T_2 = 0.1$, the new A matrix is

$$\begin{bmatrix} -0.3511 & -0.236 & 0 & 0.104 & 0 \\ 0 & 0 & 377 & 0 & 0 \\ -0.1678 & -0.144 & 0 & 0 & 0 \\ -714.4 & -10 & 0 & -5 & 2000 \\ -0.42 & -0.36 & 5 & 0 & -10 \end{bmatrix}$$

and the eigenvalues are $\lambda_{1,2}$ = -0.8612 $\pm$ $j7.7042$ $\lambda_{3,4} = -1.6314 \pm j8.5504$, λ_5 = -10.3661. Note the improvement in damping of the electromechanical mode $\lambda_{1,2}$.

□

8.7 Conclusion

In this chapter, we have discussed linear models of single and multimachine systems with different degrees of machine and load modeling. The effect of different types of loading on the steady-state stability was discussed for the multimachine case; Hopf bifurcation in the context of voltage collapse was discussed. The design of a power-system stabilizer for damping the local mode of oscillation was discussed for the single-machine infinite-bus case.

8.8 Problems

8.1 The single line diagram for the two-area system is given in Figure 8.16. The transmission line data, machine data, excitation system data, and load-flow results are given in Tables 8.8 to 8.11.

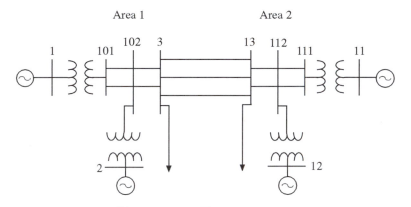

Figure 8.16: *Two-area system*

Table 8.8: Transmission Line Data on 100 MVA Base

From Bus Number	To Bus Number	Series Resistance (R_s) pu	Series Reactance (X_s) pu	Shunt Susceptance (B) pu
1	101	0.001	0.012	0.00
2	102	0.001	0.012	0.00
3	13	0.022	0.22	0.33
3	13	0.022	0.22	0.33
3	13	0.022	0.22	0.33
3	102	0.002	0.02	0.03
3	102	0.002	0.02	0.03
11	111	0.001	0.012	0.00
12	112	0.001	0.012	0.00
13	112	0.002	0.02	0.03
13	112	0.002	0.02	0.03
101	102	0.005	0.05	0.075
101	102	0.005	0.05	0.075
111	112	0.005	0.05	0.075
111	112	0.005	0.05	0.075

Table 8.9: Machine Data

Variable	Machine at Bus 1	Machine at Bus 2	Machine at Bus 11	Machine at Bus 12
X_1 (pu)	0.022	0.022	0.022	0.022
R_s (pu)	0.00028	0.00028	0.00028	0.00028
X_d (pu)	0.2	0.2	0.2	0.2
X'_d (pu)	0.033	0.033	0.033	0.033
T'_{do} (sec)	8.0	8.0	8.0	8.0
X_q (pu)	0.19	0.19	0.19	0.19
X'_q (pu)	0.061	0.061	0.061	0.061
T'_{qo} (pu)	0.4	0.4	0.4	0.4
H (sec)	54.0	54.0	63.0	63.0
D (pu)	0.0	0.0	0.0	0.0

Table 8.10: Excitation System Data

Variable	Machine at Bus 1	Machine at Bus 2	Machine at Bus 11	Machine at Bus 12
K_A (pu)	200	200	200	200
T_A (pu)	0.0001	0.0001	0.0001	0.0001

Table 8.11: Load-Flow Results for the System

Bus Number	Bus Type	Voltage Magnitude (pu)	Angle (degrees)	Real Power Gen. (pu)	Reactive Power Gen. (pu)	Real Power Load (pu)	Reactive Power Load (pu)
1	PV	1.03	8.2154	7.0	1.3386	0.0	0.0
2	PV	1.01	-1.5040	7.0	1.5920	0.0	0.0
11	Swing	1.03	0.0	7.2172	1.4466	0.0	0.0
12	PV	1.01	-10.2051	7.0	1.8083	0.0	0.0
101	PQ	1.0108	3.6615	0.0	0.0	0.0	0.0
102	PQ	0.9875	-6.2433	0.0	0.0	0.0	0.0
111	PQ	1.0095	-4.6977	0.0	0.0	0.0	0.0
112	PQ	0.9850	-14.9443	0.0	0.0	0.0	0.0
3	PQ	0.9761	-14.4194	0.0	0.0	11.59	2.12
13	PQ	0.9716	-23.2922	0.0	0.0	15.75	2.88

(a) Using the two-axis model for the generator and constant power load representation, obtain eigenvalues of the linearized system.

(b) Repeat (a) with one tie line out of service. Any comments?

8.2 With three tie lines in service, add a PSS at bus 12 with the following parameters. $K_{PSS} = 25$, $T_W = 10$ sec, $T_1 = 0.047$ sec, $T_2 = 0.021$ sec, $T_3 = 3.0$ sec, and $T_4 = 5.4$ sec. What are the new eigenvalues?

8.3 Repeat Problem 8.2 with two tie lines in service.

8.4 Consider the single machine connected to an infinite bus in Figure 8.7. Assume that $V_\infty = 1.0$. The parameters are as follows:

Line: $R_e = 0.0$, $X_e = 0.4$ pu

Generator: $X_d = 1.6$ sec, $X_q = 1.55$ pu, $X_d' = 0.32$ pu, $T_{do}' = 6.0$ sec, $H = 3.0$ sec

Injected power into the bus: $P = 0.8$ pu, $Q = 0.4$ pu

Exciter: $K_A = 50$, $T_A = 0.05$ sec

(a) Compute the *K1-K6* constants.

(b) Compute the eigenvalues. (ans: -14.5662, -5.7351, -.0808 $\pm$ j8.55)

8.5 In a single-machine-infinite-bus system (Figure 8.17), there is a local load at the generator bus. The parameters are

Line: $R_E = -.034$ pu, $X_E = 0.977$ pu

Generator: $X_d = 0.973$ pu, $X_q = 0.550$ pu, $X_d' = 0.230$ pu, $T_{do}' = 7.76$ sec, $H = 4.63$ sec

Injected power: $P = 1.0$ pu, $Q = 0.015$ pu

Generator terminal voltage: $V_t = 1.05$ pu

Local load (constant impedance): $G = 0.249$ pu, $B = 0.262$ pu

Exciter: $K_A = 50$, $T_A = 0.05$ sec

(a) Assuming $\overline{V}_t = V_t \angle 0°$, compute $\overline{V}_\infty = V_\infty \angle \beta$.

(b) Compute the Thevenin equivalent looking into the external network from the generator bus as $\overline{V}_{th} = V_\infty' \angle \beta'$ in series with an impedance Z_{th}. Show that

$$R_e + jX_e = \frac{R_E + jX_E}{1 + (R_E + jX_E)(G + jB)}$$

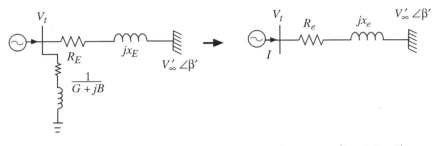

Figure 8.17: *Single-machine infinite-bus case (local load)*

and

$$\overline{V}_{\text{th}} = V_\infty \angle \beta \frac{1}{1 + (R_E + jX_E)(G + jB)} = V'_\infty \angle \beta'$$

(c) Compute $\delta(0)$.

(d) To apply the results of Section 8.6.2, we set $\beta' = 0$ and replace $\delta(0)$ by $\delta(0) - \beta'$. This makes the infinite bus a reference bus with phase-angle zero.

(e) Compute the *K1-K6* constants and the eigenvalues. (ans: -10.316 $\pm$ j3.2644, -0.2838 $\pm$ j4.9496)

8.6 For each of the two test systems below (Example 8.6), whose *K1-K6* constants and other parameters are given:

(a) Write the state space model in the form (8.144). Assume $D \equiv 0$.

(b) Plot the root locus as K_A is varied from a small to a high value. At what value of K_A does instability occur and what are the unstable eigenvalues.

(c) Find the eigenvalues at $K_A = 50$.

Test System 1

$K_1 = 3.7585$, $K_2 = 3.6816$, $K_3 = 0.2162$, $K_4 = 2.6582$, $K_5 = 0.0544$, $K_6 = 0.3616$, $T'_{do} = 5$ sec, $H = 6$ sec, $T_A = 0.2$ sec.

Test System 2

$K_1 = 0.9831$, $K_2 = 1.0923$, $K_3 = 0.3864$, $K_4 = 1.4746$, $K_5 = -0.1103$, $K_6 = 0.4477$, $T'_{do} = 5$ sec, $H = 6$ sec, $T_A = 0.2$ sec.

8.7 A single machine with a flux-decay model and a fast exciter is connected to an infinite bus through a reactance of $j0.5$ pu. The generator terminal voltage is $1\angle15°$ and the infinite bus voltage is $1.05\angle0°$. The parameters and initial conditions of the state variables are given below.

Parameters

$$H = 3.2 \text{ sec}, T'_{do} = 9.6 \text{ sec}, K_A = 400, T_A = 0.2 \text{ sec}$$
$$R_s = 0.0017 \text{ pu}, \ X_q = 2.1 \text{ pu}, \ X_d = 2.5 \text{ pu}, \ X'_d = 0.39 \text{ pu}$$
$$D \equiv 0, \omega_s = 377 \text{ rad/sec}$$

Initial conditions using the flux-decay model and the fast exciter

$$
\begin{aligned}
\delta(0) &= 65.52°, V_d(0) = 0.7719, V_q(0) = 0.6358 \\
I_d(0) &= 0.3999, I_q(0) = 0.3662 \\
E'_q(0) &= 0.7949, E_{fd}(0) = 1.6387, \omega(0) = 377 \text{ rad/sec} \\
V_{\text{ref}} &= 1.0041, T_M = 0.542
\end{aligned}
$$

(a) Compute the *K1-K6* constants and the undamped natural frequency of the torque-angle loop.

(b) Compute the eigenvalues.

8.8 Find the participation factors of the eigenvalues for the following systems, where $\dot{x} = Ax$.

(a)

$$A = \begin{bmatrix} 3 & 8 \\ 2 & 3 \end{bmatrix}$$

(b)

$$A = \begin{bmatrix} 1 & 2 & 1 \\ 0 & 3 & 1 \\ 0 & 5 & -1 \end{bmatrix}$$

Chapter 9

ENERGY FUNCTION METHODS

9.1 Background

In this chapter, we discuss energy function methods for transient stability analysis. In transient stability, we are interested in computing the critical clearing time of circuit breakers to clear a fault when the system is subjected to large disturbances. In real-world applications, the critical clearing time can be interpreted in terms of meaningful quantities such as maximum power transfer in the prefault state. The energy function methods have proved to be reliable after many decades of research [93, 96]. It is now considered a promising tool in dynamic security assessment.

9.2 Physical and Mathematical Aspects of the Problem

The ultimate objective of nonlinear dynamic simulation of power systems is to see whether synchronism is preserved in the event of a disturbance. This is judged by the variation of rotor angles as a function of time. If the rotor angle δ_i of a machine or a group of machines continues to increase with respect to the rest of the system, the system is unstable. The rotor angle δ_i of each machine is measured with respect to a fixed rotating reference frame that is the synchronous network reference frame. Hence, instability of a machine means that the rotor angle of machine i pulls away from the rest

of the system. Thus, relative rotor angles rather than absolute rotor angles
must be monitored to test stability/instability. Figure 9.1 shows the rotor
angles for the cases of stability and instability. Figure 9.1(a) shows that all
relative rotor angles are finite, as $t \to \infty$. In Figure 9.1(b), the rotor angle
of one machine is increasing with respect to the rest of the system; hence,
it is a single-machine instability. Figure 9.1(c) is a group of two machines
going unstable with respect to the rest of the system.

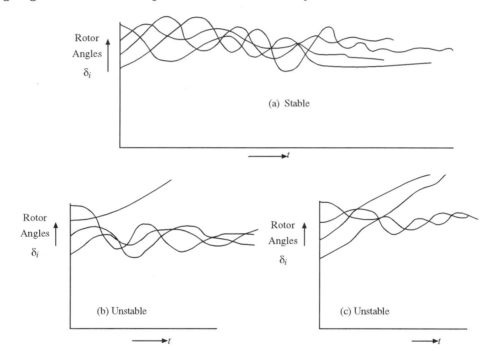

Figure 9.1: *Behavior of rotor angles for the (a) stable, and (b), (c) the
unstable cases*

    In its simplest form, a power system undergoing a disturbance can be
described by a set of three differential equations:

$$\dot{x}(t) \;=\; f^I(x(t)) \quad -\infty < t \le 0 \tag{9.1}$$

$$\dot{x}(t) \;=\; f^F(x(t)) \quad 0 < t \le t_{cl} \tag{9.2}$$

$$\dot{x}(t) \;=\; f(x(t)) \quad t_{cl} < t < \infty \tag{9.3}$$

$x(t)$ is the vector of state variables of the system at time t. At $t = 0$, a fault occurs in the system and the dynamics change from f^I to f^F. During $0 < t \leq t_{cl}$, called the faulted period, the system is governed by the fault-on dynamics f^F. Actually, before the fault is cleared at $t = t_{cl}$, we may have several switchings in the network, each giving rise to a different f^F. For simplicity, we have taken a single f^F, indicating that there are no structural changes between $t = 0$ and $t = t_{cl}$. When the fault is cleared at $t = t_{cl}$, we have the postfault system with its dynamics $f(x(t))$. In the prefault period $-\infty < t \leq 0$, the system would have settled down to a steady state, so that $x(o) = x_o$ is known. Therefore, we need not discuss (9.1). We then have only

$$\dot{x}(t) = f^F(x(t)) \quad 0 < t \leq t_{cl}$$
$$x(o) = x_o \tag{9.4}$$

and

$$\dot{x}(t) = f(x(t)) \quad t > t_{cl} \tag{9.5}$$

with the initial condition $x(t_{cl})$ for (9.5) provided by the solution of the faulted system (9.4) evaluated at $t = t_{cl}$. Viewed in another manner, the solution of (9.4) provides at each instant of time the possible initial conditions for (9.5). Let us assume that (9.5) has a stable equilibrium point x_s. The question is whether the trajectory $x(t)$ for (9.5) with the initial condition $x(t_{cl})$ will converge to x_s as $t \to \infty$. The largest value of t_{cl} for which this holds true is called the critical clearing time t_{cr}.

From this discussion, it is clear that if we have an accurate estimate of the region of attraction of the postfault stable equilibrium point (s.e.p) x_s, then t_{cr} is obtained when the trajectory of (9.4) exits the region of attraction of (9.5) at $x = x^*$. Figure 9.2 illustrates this concept for a two-dimensional system. The computation of the region of attraction for a general nonlinear dynamical system is far from easy. It is not, in general, a closed region. In the case of power systems with simple-machine models, the characterization of this region has been discussed theoretically in the literature. The stability region consists of surfaces passing through the unstable equilibrium points (u.e.p's) of (9.5). For each fault, the mode of instability (i.e., one or more machines going unstable) may be different if the fault is not cleared in time. We may describe the interior of the region of attraction of the postfault system (9.5) through an inequality of the type $V(x) < V_{cr}$, where $V(x)$ is the Lyapunov or energy function for (9.5). $V(x)$ is generally the sum of the kinetic and potential energies of the postfault system. The computation of V_{cr}, called the critical energy, is different for each fault and is a difficult step. There are currently three basic methods, with a number of variations on each method.

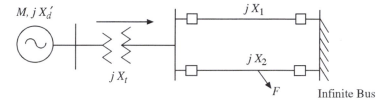

Figure 9.2: *Region of attraction and computation of t_{cr}*

These methods are:

1. Lowest energy u.e.p method [99]

   $V_{cr} = V(x^{uc})$, where x^{uc} is the unstable equilibrium point (u.e.p) resulting in the lowest value of V_{cr} among the u.e.p's lying on the stability boundary of (9.5). This requires the computation of many u.e.p's of the postfault system and, hence, is not computationally attractive. Moreover, it gives conservative results. Reference [99] was the first systematic application of Lyapunov's method for transient stability.

2. Potential Energy Boundary Surface (PEBS) method

   $V_{cr} =$ maximum value of the potential energy component of $V(x)$ along the trajectory of (9.4). This is known as the potential energy boundary surface (PEBS) method [100].

3. Controlling u.e.p method

   $V_{cr} = V(x^u)$, at which x^u is the relevant or controlling u.e.p, i.e., the u.e.p closest to the point where the fault-on trajectory of (9.4) exits the region of attraction of (9.5). This is called the controlling u.e.p method first proposed in [97]. The boundary-controlling u.e.p (BCU) method [98] (also called the exit-point method) is an efficient technique to compute this controlling u.e.p.

We will discuss the PEBS method (2) and the controlling u.e.p method (3) in detail. The computation of t_{cr} involves the following steps.

1. Computing x_s, the stable equilibrium point of the postfault system given by (9.5).

2. Formulating $V(x)$ for (9.5). This is not a difficult step. Generally, $V(x)$ is the sum of kinetic and potential energies of the postfault system, i.e., $V(x) = V_{KE} + V_{PE}$.

3. Computation of V_{cr}.

   In the PEBS method (2), V_{cr} is obtained by integrating the faulted trajectory in (9.4) until the potential energy part V_{PE} of $V(x)$ reaches a maximum $V_{PE}^{\max}$. This value is taken as V_{cr} in the PEBS method. In the controlling u.e.p method (3) we integrate (9.4) for a short interval followed by either a minimization problem to get the controlling u.e.p, or integration of a reduced-order postfault system after the PEBS is reached (BCU method) [98] to get the controlling u.e.p x^u. V_{cr} is given by $V(x^u) = V_{PE}(x^u)$, since V_{KE} is zero at an u.e.p.

4. Calculating the time instant t_{cr} when $V(x) = V_{cr}$ on the faulted trajectory of (9.4). The faulted trajectory has to be integrated for all the three methods to obtain t_{cr}. In the PEBS method (2), the faulted trajectory is already available while computing V_{cr}. It is also available in the BCU method. The computation time is least for the PEBS method (2). The relative merits of these various methods and their variations are discussed extensively in the literature [101].

9.3 Lyapunov's Method

In 1892, A. M. Lyapunov, in his famous Ph.D. dissertation [118], proposed that the stability of the equilibrium point of a nonlinear dynamic system of dimension n

$$\dot{x} = f(x), \; f(0) = 0 \tag{9.6}$$

can be ascertained without numerical integration. He said that if there exists a scalar function $V(x)$ for (9.6) that is positive-definite, i.e., $V(x) > 0$ around the equilibrium point "0" and the derivative $\dot{V}(x) < 0$, then the equilibrium is asymptotically stable. $\dot{V}(x)$ is obtained as $\sum_{i=1}^{n} \frac{\partial V}{\partial x_i} \dot{x}_i = \sum_{i=1}^{n} \frac{\partial V}{\partial x_i} f_i(x) = \bigtriangledown V^T \cdot f(x)$ where n is the order of the system in (9.6). Thus, $f(x)$ enters directly in the computation of $\dot{V}(x)$. The condition $\dot{V}(x) < 0$ can be relaxed to $\dot{V}(x) \leq 0$, provided that $\dot{V}(x)$ does not vanish along any other solution with the exception of $x = 0$.

$V(x)$ is actually a generalization of the concept of the energy of a system. Since 1948, when the results of Lyapunov appeared in the English language together with potential applications, there has been extensive literature surrounding this topic. Application of the energy function method to power system stability began with the early work of Magnusson [102] and Aylett [103], followed by a formal application of the more general Lyapunov's method by El-Abiad and Nagappan [99]. Reference [99] provided an algorithmic procedure to compute the critical clearing time. It used the lowest energy u.e.p method to compute V_{cr}. Although many different Lyapunov functions have been tried since then, the first integral of motion, which is the sum of kinetic and potential energies, seemed to have provided the best result. In the power literature, Lyapunov's method has become synonymous with the transient energy function (TEF) method and has been applied successfully [93, 98]. Today, this technique has proved to be a practical tool in dynamic security assessment. To make it a practical tool, it is necessary to compute the region of stability of the equilibrium point of (9.5). In physical systems, it is finite and not the whole state-space. An estimate of the region of stability or attraction is characterized by an inequality of the type $V(x) < V_{cr}$. The computation of V_{cr} remained a formidable barrier for a long time. In the case of a multimachine classical model with loads being treated as constant impedances, there are well-proved algorithms. Extensions to multimachine systems with detailed models have been made [104, 106].

9.4   Modeling Issues

In applying the TEF technique, we must consider the model in two time frames, as follows:

1. Faulted system

$$\dot{x} = f^F(x(t)) , \ 0 < t \leq t_{c\ell} \tag{9.7}$$

2. Postfault system

$$\dot{x} = f(x(t)) , \ t > t_{c\ell} \tag{9.8}$$

In reality, the model is a set of differential-algebraic equations (DAE), i.e.,

$$\dot{x} \;\; = \;\; f^F(x(t), y(t)) \tag{9.9}$$

$$0 \;\; = \;\; g^F(x(t), y(t)) \,, \; 0 < t \le t_{c\ell} \tag{9.10}$$

and

$$\dot{x} \;\; = \;\; f(x(t), y(t)) \tag{9.11}$$

$$0 \;\; = \;\; g(x(t), y(t)), \; t > t_{c\ell} \tag{9.12}$$

The function g represents the nonlinear algebraic equations of the stator and the network, while the differential equations represent the dynamics of the generating unit and its controls. In Chapter 7, the modeling of equations in the form of (9.9) and (9.10) or (9.11) and (9.12) has been covered extensively. Reduced-order models, such as a flux-decay model and a classical model, have also been discussed. In the classical model representation, we can either preserve the network structure (structure-preserving model) or eliminate the load buses (assuming constant impedance load) to obtain the internal-node model. These have also been discussed in Chapter 7. Structure-preserving models involve nonlinear algebraic equations in addition to dynamic equations, and can incorporate nonlinear load models leading to the concept of structure-preserving energy function (SPEF) $V(x, y)$, while models consisting of differential equations lead only to closed-form types of energy functions $V(x)$. The work on SPEF by Bergen and Hill [104] has been extended to more detailed models in [105]–[108].

It is not clear at this stage whether a more detailed generator or load model will lead to more accurate estimates of t_{cr}. What appears to be true, however, from extensive simulation studies by researchers is that, for the so-called first-swing stability (i.e., instability occurring in 1 to 2 sec interval), the classical model with the loads represented as constant impedance will suffice. This results in only differential equations, as opposed to DAE equations. Both the PEBS and BCU methods give satisfactory results for this model. We first discuss this in the multimachine context. The swing equations have been derived in Section 7.9.3 (using $P_{mi} = T_{Mi}$) as

$$\frac{2H_i}{\omega_s} \frac{d^2\delta_i}{dt^2} + D_i \frac{d\delta_i}{dt} = P_{mi} - P_{ei} \,, \; i = 1, \dots, m \tag{9.13}$$

where

$$P_{ei} = E_i^2 G_{ii} + \sum_{\substack{j=1 \\ \neq i}}^{m} (C_{ij} \sin \delta_{ij} + D_{ij} \cos \delta_{ij}) \tag{9.14}$$

Denoting $\frac{2H_i}{\omega_s} \triangleq M_i$ and $P_i \triangleq P_{mi} - E_i^2 G_{ii}$, we get

$$M_i \frac{d^2 \delta_i}{dt^2} + D_i \frac{d\delta_i}{dt} = P_i - \sum_{j=1 \neq i}^{m} (C_{ij} \sin \delta_{ij} + D_{ij} \cos \delta_{ij}) \tag{9.15}$$

which can be written as

$$M_i \frac{d^2 \delta_i}{dt^2} + D_i \frac{d\delta_i}{dt} = P_i - P_{ei}(\delta_i \ldots \delta_m), \quad i = 1, \ldots, m \tag{9.16}$$

Let α_i be the rotor angle with respect to a fixed reference. Then $\delta_i = \alpha_i - \omega_s t$. $\dot{\delta_i} = \frac{d\alpha_i}{dt} - \omega_s \triangleq \omega_i - \omega_s$, where ω_i is the angular velocity of the rotor and ω_s is the synchronous speed in radians per sec. Thus, both δ_i and $\dot{\delta_i}$ are expressed with respect to a synchronously rotating reference frame. Equation (9.16) is converted to a set of first-order differential equations by introducing the state variables δ_i and ω_i:

$$\dot{\delta_i} = \omega_i - \omega_s \tag{9.17}$$

$$\dot{\omega_i} = \frac{1}{M_i}(P_i - P_{ei}(\delta_1, \ldots, \delta_m) - D_i(\omega_i - \omega_s)) \quad i = 1, \ldots, m \tag{9.18}$$

Equations (9.17) and (9.18) are applicable both to the faulted state and the postfault state, with the difference that P_{ei} is different in each case, because the internal node admittance matrix is different for the faulted and postfault system. The model corresponding to (9.17)–(9.18) is known as the internal-node model since the physical buses have been eliminated by network reduction.

9.5 Energy Function Formulation

Prior to 1979, there was considerable research in constructing a Lyapunov function for the system (9.15) using the state-space model given by (9.17) and (9.18) [109]–[112]. However, analytical Lyapunov functions can be constructed only if the transfer conductances are zero, i.e., $D_{ij} \equiv 0$. Since these terms have to be accounted for properly, the first integrals of motion of the system are constructed, and these are called energy functions. We have two options, to use either the relative rotor angle formulation or the center of inertia formulation. We use the latter, since there are some advantages. Since the angles are referred to a center of inertia, the resulting energy function is called the transient energy function (TEF).

In this formulation, the angle of the center of inertia (COI) is used as the reference angle, since it represents the "mean motion" of the system. Although the resulting energy function is identical to $V(\delta, \omega)$ (using relative rotor angles), it has the advantage of being more symmetric and easier to handle in terms of the path-dependent terms. Synchronous stability of all machines is judged by examining the angles referenced only to COI instead of relative rotor angles. Modern literature invariably uses the COI formulation. The energy function in the COI notation, including D_{ij} terms (transfer conductances), was first proposed by Athay et al. [97].

We derive the transient energy function for the conservative system (assuming $D_i \equiv 0$). The center of inertia (COI) for the whole system is defined as

$$\delta_o = \frac{1}{M_T} \sum_{i=1}^{m} M_i \delta_i \text{ and the center of speed as } \omega_o = \frac{1}{M_T} \sum_{i=1}^{m} M_i \omega_i \quad (9.19)$$

where $M_T = \sum_{i=1}^{m} M_i$. We then transform the variables δ_i, ω_i to the COI variables as $\theta_i = \delta_i - \delta_o$, $\tilde{\omega}_i = \omega_i - \omega_o$. It is easy to verify

$$\dot{\theta}_i = \dot{\delta}_i - \dot{\delta}_o$$
$$= \omega_i - \omega_o$$
$$\triangleq \tilde{\omega}_i$$

The swing equations (9.15) with $D_i = 0$ become (omitting the algebra):

$$M_i \frac{d^2 \theta_i}{dt^2} = P_i - \sum_{\substack{j=1 \\ \neq i}}^{n} (C_{ij} \sin \theta_{ij} + D_{ij} \cos \theta_{ij}) - \frac{M_i}{M_T} P_{COI}$$

$$\triangleq f_i(\theta) \quad i = 1, \ldots, m \quad (9.20)$$

where

$$P_i = P_{mi} - E_i^2 G_{ii} \; ; \; P_{COI} = \sum_{i=1}^{m} P_i - 2\sum_{i=1}^{m}\sum_{j=i+1}^{m} D_{ij}\cos\theta_{ij}$$

If one of the machines is an infinite bus, say, m whose inertia constant M_m is very large, then $\frac{M_i}{M_T}P_{COI} \approx 0$ $(i \neq m)$ and also $\delta_o \approx \delta_m$ and $\omega_o \approx \omega_m$. The COI variables become $\theta_i = \delta_i - \delta_m$ and $\tilde{\omega}_i = \omega_i - \omega_m$. In the literature where the BCU method is discussed [98], δ_m is simply taken as zero. Equation (9.20) is modified accordingly, and there will be only $(m-1)$ equations after omitting the equation for machine m.

We consider the general case in which all M_i's are finite. Corresponding to the faulted and the postfault states, we have two sets of differential equations,

$$M_i \frac{d\tilde{\omega}_i}{dt} = f_i^F(\theta) \quad 0 < t \leq t_{cl}$$

$$\frac{d\theta_i}{dt} = \tilde{\omega}_i \; , \quad i = 1, 2, \ldots, m \tag{9.21}$$

and

$$M_i \frac{d\tilde{\omega}_i}{dt} = f_i(\theta) \quad t > t_{cl}$$

$$\frac{d\theta_i}{dt} = \tilde{\omega}_i \; , \quad i = 1, 2, \ldots, m \tag{9.22}$$

Let the postfault system given by (9.22) have the stable equilibrium point at $\theta = \theta^s$, $\tilde{\omega} = 0$. θ^s is obtained by solving the nonlinear algebraic equations

$$f_i(\theta) = 0 \; , \; i = 1, \ldots, m \tag{9.23}$$

Since $\sum_{i=1}^{m} M_i \theta_i = 0$, θ_m can be expressed in terms of the other θ_i's and substituted in (9.23), which is then equivalent to

$$f_i(\theta_1, \ldots, \theta_{m-1}) = 0 \; , \; i = 1, \ldots, m - 1 \tag{9.24}$$

The basic procedure for computing the critical clearing time consists of the following steps:

1. Construct an energy or Lyapunov function $V(\theta, \tilde{\omega})$ for the system (9.22), i.e., the postfault system.

2. Find the critical value of $V(\theta, \tilde{\omega})$ for a *given* fault denoted by V_{cr}.

3. Integrate (9.21), i.e., the faulted equations, until $V(\theta, \tilde{\omega}) = V_{cr}$. This instant of time is called the critical clearing time t_{cr}.

While this procedure is common to all the methods, they differ from one another in steps 2 and 3, i.e., finding V_{cr} and integrating the swing equations. There is general agreement that the first integral of motion of (9.22) constitutes a proper energy function and is derived as follows [94].

From (9.22) we have, for $i = 1, \ldots, m$

$$dt = \frac{M_1 d\tilde{\omega}_1}{f_1(\theta)} = \frac{d\theta_1}{\tilde{\omega}_1} = \frac{M_2 d\tilde{\omega}_2}{f_2(\theta)} = \frac{d\theta_2}{\tilde{\omega}_2} = \cdots = \frac{M_m d\tilde{\omega}_m}{f_m(\theta)} = \frac{d\theta_m}{\tilde{\omega}_m} \quad (9.25)$$

Integrating the pairs of equations for each machine between $(\theta_i^s, 0)$, the postfault s.e.p to $(\theta_i, \tilde{\omega}_i)$ results in

$$V_i(\theta, \tilde{\omega}) = \frac{1}{2} M_i \tilde{\omega}_i^2 - \int_{\theta_i^s}^{\theta_i} f_i(\theta) d\theta_i , \quad i = 1, \ldots, m \quad (9.26)$$

This is known in the literature as the individual machine energy functon [113]. Adding these functions for all the machines, we obtain the first integral of motion for the system as (omitting the algebra):

$$V(\theta, \tilde{\omega}) = \frac{1}{2} \sum_{i=1}^{m} M_i \tilde{\omega}_i^2 - \sum_{i=1}^{m} \int_{\theta_i^s}^{\theta_i} f_i(\theta) d\theta_i \quad (9.27)$$

$$= \frac{1}{2} \sum_{i=1}^{m} M_i \tilde{\omega}_i^2 - \sum_{i=1}^{m} P_i(\theta_i - \theta_i^s) - \sum_{i=1}^{m-1} \sum_{j=i+1}^{m} \Big[C_{ij}(\cos\theta_{ij} - \cos\theta_{ij}^s)$$

$$- \int_{\theta_i^s+\theta_j^s}^{\theta_i+\theta_j} D_{ij} \cos\theta_{ij} d(\theta_i + \theta_j) \Big] \quad (9.28)$$

$$= V_{KE}(\tilde{\omega}) + V_{PE}(\theta) \quad (9.29)$$

since

$$\sum_{i=1}^{m} \frac{M_i}{M_T} \int_{\theta_i^s}^{\theta_i} P_{COI} d\theta_i = 0$$

Note that (9.28) contains path-dependent integral terms. In view of this, we cannot assert that V_i and V are positive-definite. If $D_{ij} \equiv 0$, it can be shown that $V(\theta, \tilde{\omega})$ constitutes a proper Lyapunov function [93, 109, 110].

9.6   Potential Energy Boundary Surface (PEBS)

We first discuss the PEBS method because of its simplicity and its natural relationship to the equal-area criterion.

Ever since it was first proposed by Kakimoto et al. [100] and Athay et al. [97], the method has received wide attention by researchers because it avoids computing the controlling (relevant) u.e.p and requires only a quick fault-on system integration to compute V_{cr}. We can even avoid computing the postfault s.e.p., as discussed in Section 9.6.5. In this section, we will first motivate the method through application to a single-machine infinite-bus system, establish the equivalence between the energy function and the equal-area criterion, and, finally, explain the multimachine PEBS method.

9.6.1   Single-machine infinite-bus system

Consider a single-machine infinite-bus system (Figure 9.3). Two parallel lines each having a reactance of X_1 connect a generator having transient reactance of X'_d through a transformer with a reactance of X_t to an infinite bus whose voltage is $E_2 \angle 0°$. A three-phase fault occurs at the middle of one of the lines at $t = 0$, and is subsequently cleared at $t = t_{cl}$ by opening the circuit breakers at both ends of the faulted line. The prefault, faulted, and postfault configurations and their reduction to a two-machine equivalent are shown in Figures 9.4, 9.5, and 9.6. The electric power P_e during prefault, faulted, and postfault states are $\frac{E_1 E_2}{X^I} \sin \delta$, $\frac{E_1 E_2 \sin \delta}{X^F}$, and $\frac{E_1 E_2 \sin \delta}{X}$, respectively. The computation of X^I for the prefault system and X for the postfault system is straightforward, as shown in Figures 9.4 and 9.6.

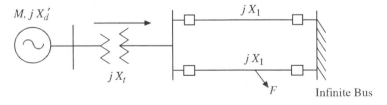

Figure 9.3: *Single-machine infinite-bus system*

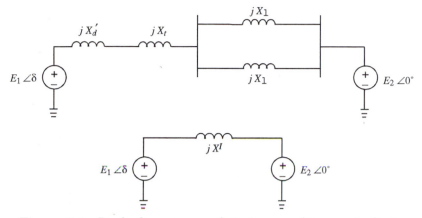

Figure 9.4: *Prefault system and its two-machine equivalent*

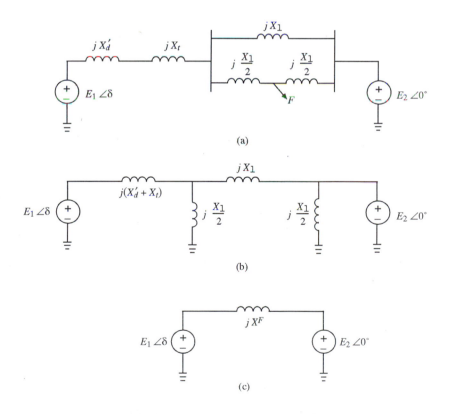

Figure 9.5: *Faulted system and its two-machine equivalent*

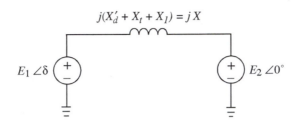

Figure 9.6: *Postfault system and its two-machine equivalent*

Example 9.1

Compute X^F for the faulted system in Figure 9.3. X^F is the reactance between the internal node of the machine and the infinite bus. It can be computed for this network by performing a $Y - \Delta$ transformation in Figure 9.5(b). It is more instructive to illustrate a general method that is applicable to the multimachine case as well. Figure 9.5(a) is redrawn after labeling the various nodes (Figure 9.7). The point at which the fault occurs is labeled node 4. $X_g = X'_d + X_t$.

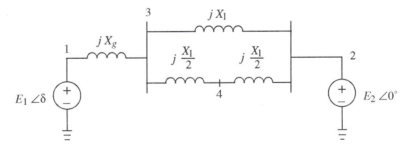

Figure 9.7: *Faulted system*

There are current injections at nodes 1, 2, and 4, and none at node 3. The nodal equation is

$$
\begin{bmatrix} I_1 \\ I_2 \\ 0 \\ I_4 \end{bmatrix} = \begin{bmatrix} Y_{11} & 0 & Y_{13} & 0 \\ 0 & Y_{22} & Y_{23} & Y_{24} \\ Y_{31} & Y_{32} & Y_{33} & Y_{34} \\ 0 & Y_{42} & Y_{42} & Y_{44} \end{bmatrix} \begin{bmatrix} E_1 \\ E_2 \\ V_3 \\ V_4 \end{bmatrix}
$$

Since the fault is at node 4 with the fault impedance equal to zero, $V_4 \equiv 0$. Hence, we can delete row 4 and column 4, resulting in

$$
\begin{bmatrix} I_1 \\ I_2 \\ 0 \end{bmatrix} = \begin{bmatrix} Y_{11} & 0 & Y_{13} \\ 0 & Y_{22} & Y_{23} \\ Y_{31} & Y_{32} & Y_{33} \end{bmatrix} \begin{bmatrix} E_1 \\ E_2 \\ V_3 \end{bmatrix}
$$

Node 3 is eliminated by expressing V_3 from the third equation in terms of E_1 and E_2 and substituting in the first two equations. This results in (omitting the algebra):

$$
\begin{bmatrix} I_1 \\ I_2 \end{bmatrix} = \begin{bmatrix} Y_{11} - Y_{13}Y_{33}^{-1}Y_{31} & -Y_{13}Y_{33}^{-1}Y_{32} \\ -Y_{23}Y_{33}^{-1}Y_{31} & Y_{22} - Y_{23}Y_{33}^{-1}Y_{32} \end{bmatrix} \begin{bmatrix} E_1 \\ E_2 \end{bmatrix}
$$

X^F can be computed from the off-diagonal entry as

$$
\frac{1}{jX^F} = Y_{13}Y_{33}^{-1}Y_{32}
$$

Now

$$
Y_{13}Y_{33}^{-1}Y_{32} = \left(\frac{-1}{jX_g} \right) \left(\frac{1}{jX_g} + \frac{1}{jX_1} + \frac{1}{j\frac{X_1}{2}} \right)^{-1} \left(\frac{-1}{jX_1} \right)
$$

$$
= \frac{1}{j(X_1 + 3X_g)}
$$

Hence

$$
X^F = X_1 + 3X_g
$$

We do not require the other elements of the reduced Y matrix.

□

9.6.2 Energy function for a single-machine infinite-bus system

The energy function is *always* constructed for the postfault system. In the SMIB case, the postfault equations are

$$M \frac{d^2\delta}{dt^2} = P_m - P_e^{\max} \sin\delta \tag{9.30}$$

where $P_e^{\max} = \frac{E_1 E_2}{X}$, δ is the angle relative to the infinite bus, and $\frac{d\delta}{dt} = \omega$ is the relative rotor-angle velocity. The right-hand side of (9.30) can be written as $\frac{-\partial V_{PE}}{\partial \delta}$, where

$$V_{PE}(\delta) = -P_m\delta - P_e^{\max} \cos\delta \tag{9.31}$$

Multiplying (9.30) by $\frac{d\delta}{dt}$, it can be rewritten as

$$\frac{d}{dt}\left[\frac{M}{2}\left(\frac{d\delta}{dt}\right)^2 + V_{PE}(\delta) \right] = 0$$

i.e.,

$$\frac{d}{dt}\left[\frac{1}{2}M\omega^2 + V_{PE}(\delta) \right] = 0$$

i.e.,

$$\frac{d}{dt}[V(\delta,\omega)] = 0 \tag{9.32}$$

Hence, the energy function is

$$V(\delta,\omega) = \frac{1}{2}M\omega^2 + V_{PE}(\delta) \tag{9.33}$$

It follows from (9.32) that the quantity inside the brackets $V(\delta,\omega)$ is a constant. The equilibrium point is given by the solution of $0 = P_m - P_e^{\max}\sin\delta$, i.e., $\delta^s = \sin^{-1}\left(\frac{P_m}{P_e^{\max}}\right)$. This is a stable equilibrium point surrounded by two unstable equilibrium points $\delta^u = \pi - \delta^s$ and $\hat{\delta}_u = -\pi - \delta^s$. If we make a change of coordinates so that $V_{PE} = 0$ at $\delta = \delta^s$, then (9.31) becomes

$$V_{PE}(\delta, \delta^s) = -P_m(\delta - \delta^s) - P_e^{\max}(\cos\delta - \cos\delta^s) \qquad (9.34)$$

With this, the energy function $V(\delta, \omega)$ can be written as

$$V(\delta, \omega) = \frac{1}{2}M\omega^2 - P_m(\delta - \delta^s) - P_e^{\max}(\cos\delta - \cos\delta^s)$$

$$= V_{KE} + V_{PE}(\delta, \delta^s) \qquad (9.35)$$

where $V_{KE} = \frac{1}{2}M\omega^2$ is the transient kinetic energy and $V_{PE}(\delta, \delta^s) = -P_m(\delta - \delta^s) - P_e^{\max}(\cos\delta - \cos\delta^s)$ is the potential energy. From (9.32) it follows that $V(\delta, \omega)$ is equal to a constant E, which is the sum of the kinetic and potential energies, and remains constant once the fault is cleared since the system is conservative. $V(\delta, \omega)$ evaluated at $t = t_{cl}$ from the faulted trajectory represents the total energy E present in the system at $t = t_{cl}$. This energy must be absorbed by the system once the fault is cleared if the system is to be stable. The kinetic energy is always positive, and is the difference between E and $V_{PE}(\delta, \delta^s)$. This is shown graphically in Figure 9.8, which is the potential energy curve.

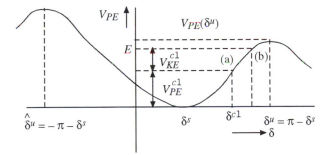

Figure 9.8: *Potential energy "well"*

At $\delta = \delta^s$, the postfault s.e.p, both the V_{KE} and the V_{PE} are zero, since $\omega = 0$ and $\delta = \delta^s$ at this point. Suppose that, at the end of the faulted period $t = t_{cl}$, the rotor angle is $\delta = \delta^{cl}$, and the velocity is ω^{cl}. Then

$$V^{cl}(\delta^{cl}, \omega^{cl}) = \frac{1}{2}M\omega_{cl}^2 - P_m(\delta^{cl} - \delta^s) - P_e^{\max}(\cos\delta^{cl} - \cos\delta^s)$$

$$= V_{KE}^{cl} + V_{PE}^{cl} \qquad (9.36)$$

This is the value of E. There are two other equilibrium points of the system (9.30), namely, $\delta^u = \pi - \delta^s$ and $\hat{\delta}^u = -\pi - \delta^s$. Both of these are unstable and, in fact, are Type 1 (saddle-type) e.p.'s. Type 1 e.p.'s are characterized by the fact that the linearized system at that e.p. has one real eigenvalue in the right-half plane. The potential energy is zero at $\delta = \delta^s$ and has a relative maximum at $\delta = \delta^u$ and $\delta = \hat{\delta}^u$. At the point (a), δ^{cl} and ω^{cl} are known from the faulted trajectory; hence, $V(\delta^{cl}, \omega^{cl}) = E$ is known. This is shown as point (b). If $E < V_{PE}(\delta^u)$, then since the system is conservative, the cleared system at point (a) will accelerate until point (b), and then start decelerating. If $E > V_{PE}(\delta^u)$, then the cleared system will accelerate beyond δ^u and, hence, the system is unstable. $V_{PE}(\delta^u)$ is obtained from (9.34) as $-P_m(\pi - 2\delta^s) + 2P_e^{\max} \cos \delta^s$. If δ decreases due to deceleration for $t > 0$, then the system is unstable if $E > V(\hat{\delta}^u)$. The points δ^u and $\hat{\delta}^u$ constitute the zero-dimensional PEBS for the single-machine system. Some researchers restate the above idea by saying that if the V_{PE} is initialized to zero at δ^{cl}, V_{KE}^{cl} represents the excess kinetic energy injected into the system. Then stability of the system is determined by the ability of the postfault system to absorb this excess kinetic energy (i.e., the system is stable if $V_{PE}(\delta^u) - V_{PE}(\delta^{cl}) > V_{KE}^{cl}$).

Most of the stability concepts can be interpreted as if the moment of inertia M is assumed as a particle that slides without friction within a "hill" with the shape $V_{PE}(\delta)$. Motions within a potential "well" are bounded and, hence, stable. It is interesting to relate the potential "well" concept to the stability of equilibrium points for small disturbances. Using (9.31), (9.30) can be written as

$$M\frac{d^2\delta}{dt^2} = -\frac{\partial V_{PE}(\delta)}{\partial \delta} \tag{9.37}$$

We can expand the right-hand side of (9.37) in a Taylor series about an equilibrium point δ^*, i.e., $\delta = \delta^* + \Delta\delta$, and retain only the linear term. Then

$$M\frac{d^2\Delta\delta}{dt^2} = -\frac{\partial^2 V_{PE}(\delta)}{\partial \delta}\bigg|_{\delta^*} \Delta\delta \tag{9.38}$$

i.e.,

$$M \frac{d^2 \Delta \delta}{dt^2} + \left. \frac{\partial^2 V_{PE}(\delta)}{\partial \delta} \right|_{\delta^*} \Delta \delta = 0 \tag{9.39}$$

If $\left. \frac{\partial^2 V_{PE}}{\partial \delta^2} \right|_{\delta^*} < 0$, the equilibrium is unstable. If $\left. \frac{\partial^2 V_{PE}}{\partial \delta^2} \right|_{\delta^*} > 0$, then it is an oscillatory system, and the oscillations around δ^* are bounded. Since there is always some positive damping, we may call it stable. In the case of (9.30), it can be verified that δ^s is a stable equilibrium point and that both δ^u and $\hat{\delta}^u$ are unstable equilibrium points using this criterion.

The energy function, Lyapunov function, and the PEBS are thus all equivalent in the case of a single-machine infinite-bus system. It is in the case of multimachine systems and nonconservative systems that each method gives only approximations to the true stability boundary! In the case of multimachine systems, the second derivative of V_{PE} is the Hessian matrix.

Example 9.2

Consider an SMIB system whose postfault equation is given by

$$0.2 \frac{d^2 \delta}{dt^2} = 1 - 2 \sin \delta - 0.02 \frac{d\delta}{dt}$$

The equilibrium points are given by $\delta^s = \sin^{-1}\left(\frac{1}{2}\right)$. Hence, $\delta^s = \frac{\pi}{6}$, $\delta^u = \pi - \frac{\pi}{6} = \frac{5\pi}{6}$, $\hat{\delta}^u = -\pi - \frac{\pi}{6} = \frac{-7\pi}{6}$. Linearizing around an equilibrium point "0" results in

$$0.2 \frac{d^2 \Delta \delta}{dt^2} = -2 \cos \delta^\circ \Delta \delta - 0.02 \frac{d\Delta \delta}{dt}$$

This can be put in the state-space form by defining $\Delta \delta$, $\Delta \omega = \Delta \dot{\delta}$ as the state variables.

$$\begin{bmatrix} \Delta \dot{\delta} \\ \Delta \dot{\omega} \end{bmatrix} = \begin{bmatrix} 0 & 1 \\ -10 \cos \delta^\circ & -0.1 \end{bmatrix} \begin{bmatrix} \Delta \delta \\ \Delta \omega \end{bmatrix}$$

For $\delta^\circ = \delta^s = \frac{\pi}{6}$, eigenvalues of this matrix are $\lambda_{1,2} = -0.05 \pm j2.942$. It is a stable equilibrium point called the focus. For $\delta^\circ = \delta^u$ or $\hat{\delta}^u$, the eigenvalues are $\lambda_1 = 2.993$ and $\lambda_2 = $ -2.893. Both are saddle points. Since there is only one eigenvalue in the right-half plane, it is called a Type 1 u.e.p.

□

Example 9.3

Construct the energy function for Example 9.2. Verify the stability of the equilibrium points by using (9.39).

The energy function is constructed for the undamped system, i.e., the coefficient of $\frac{d\delta}{dt}$ is set equal to zero. $M = 0.2$, $P_m = 1$, $P_e^{\max} = 2$, $\delta^s = \frac{\pi}{6}$. The energy function is

$$V(\delta,\omega) = \frac{1}{2}(0.2)\omega^2 - 1(\delta - \pi/6) - 2(\cos\delta - \cos\frac{\pi}{6})$$

$$= 0.1\omega^2 - (\delta - \frac{\pi}{6}) - 2(\cos\delta - (0.866))$$

$$V_{PE}(\delta,\delta^s) = -(\delta - \frac{\pi}{6}) - 2(\cos\delta - 0.866)$$

$$\frac{\partial^2 V_{PE}(\delta,\delta^s)}{\partial\delta} = 2\cos\delta$$

At $\delta = \delta^s = \pi/6$, $2\cos\delta > 0$; hence δ^s is a stable equilibrium point. At $\delta = \delta^u = \frac{5\pi}{6}$ or $\hat{\delta}^u = \frac{-7\pi}{6}$, $2\cos\delta < 0$. Hence, both δ^u and $\hat{\delta}^u$ are unstable equilibrium points.

□

9.6.3 Equal-area criterion and the energy function

The prefault, faulted, and postfault power angle curves P_e for the single-machine infinite-bus system are shown in Figure 9.9. The system is initially at $\delta = \delta^o$. We shall now show that the area A_1 represents the kinetic energy injected into the system during the fault, which is the same as V_{KE}^{cl} in Figure 9.8. A_2 represents the ability of the postfault system to absorb this energy. In terms of Figure 9.8, A_2 represents $V_{PE}(\delta^u) - V_{PE}(\delta^{cl})$. By the equal-area criterion, the system is stable if $A_1 < A_2$.

Let the faulted and postfault equations, respectively, be

$$M\frac{d^2\delta}{dt^2} = P_m - P_e^F \sin\delta \tag{9.40}$$

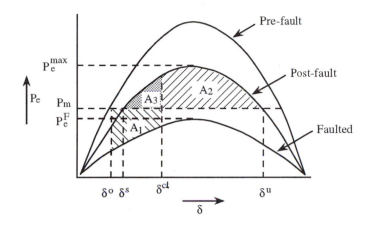

Figure 9.9: *Equal-area criterion for the SMIB case*

and

$$M\frac{d^2\delta}{dt^2} = P_m - P_e^{\max}\sin\delta \qquad (9.41)$$

where

$$P_e^F = \frac{E_1 E_2}{X^F}$$

and

$$P_e^{\max} = \frac{E_1 E_2}{X}$$

The area A_1 is given by

$$
\begin{aligned}
A_1 &= \int_{\delta^o}^{\delta_{c\ell}} \left(P_m - P_e^F \sin\delta \right) d\delta \\
&= \int_{\delta^o}^{\delta_{c\ell}} M\frac{d\omega}{dt} d\delta \\
&= \int_{\delta^o}^{\delta_{c\ell}} M\frac{d\omega}{dt}\omega\, dt \\
&= \int_{o}^{\omega_{c\ell}} M\omega\, d\omega = \frac{1}{2}M(\omega_{c\ell})^2 \qquad (9.42)
\end{aligned}
$$

Hence, A_1 is the kinetic energy injected into the system due to the fault. Area A_2 is given by

$$A_2 = \int_{\delta^{cl}}^{\delta^u} (P_e^{\max} \sin \delta - P_m) d\delta = -P_e^{\max}(\cos \delta^u - \cos \delta^{cl})$$

$$-P_m(\delta^u - \delta^{cl})$$

$$= V_{PE}(\delta^u) - V_{PE}(\delta^{cl})$$

from (9.34). If we add area A_3 to both sides of the criterion $A_1 < A_2$, the result is

$$A_1 + A_3 < A_2 + A_3 \tag{9.43}$$

Now

$$A_3 = \int_{\delta^s}^{\delta_{cl}} (P_e^{\max} \sin \delta - P_m) d\delta$$

$$= -P_m \left(\delta^{cl} - \delta^s \right) - P_e^{\max} \left(\cos \delta^{cl} - \cos \delta^s \right) \tag{9.44}$$

Changing δ^{cl}, ω^{cl} to any δ, ω and adding A_1 to A_3, gives

$$A_1 + A_3 = \frac{1}{2} M\omega^2 - P_m(\delta - \delta^s) - P_e^{\max}(\cos \delta - \cos \delta^s) \tag{9.45}$$

This is the same as $V(\delta, \omega)$ as in (9.35). Now, from Figure 9.9:

$$A_2 + A_3 = \int_{\delta^s}^{\pi - \delta^s} (P_e^{\max} \sin \delta - P_m) d\delta$$

$$= 2P_e^{\max} \cos \delta^s - P_m(\pi - 2\delta^s) \tag{9.46}$$

The right-hand side of (9.46) is also verified to be the sum of the areas A_2 and A_3, for which analytical expressions have been derived. It may be verified from (9.35) that

$$V(\delta, \omega) \big|_{\substack{\delta = \delta^u \\ \omega = 0}} = -P_m(\pi - 2\delta^s) + 2P_e^{\max} \cos \delta^s = A_2 + A_3$$

$$= V_{PE}(\delta^u)$$

$$\triangleq V_{cr} \tag{9.47}$$

Thus, the equal-area criterion $A_1 < A_2$ is equivalent to $A_1 + A_2 < A_2 + A_3$, which in turn is equivalent to

$$V(\delta, \omega) < V_{cr} \tag{9.48}$$

where $V_{cr} = V_{PE}(\delta^u)$. Note that δ, ω are obtained from the faulted equation.

Example 9.4

For Example 9.2, (1) state analytically the equal-area criterion. (2) If the faulted system is given by

$$\frac{0.2 d^2 \delta}{dt^2} = 1 - \sin \delta - 0.02 \frac{d\delta}{dt}$$

compute t_{cr} using the results of part (1). Note that the energy function is for a conservative system, while the faulted system is not conservative.

The energy function is given by

$$V(\delta, \omega) = \frac{1}{2}(0.2)\omega^2 - 1(\delta - \frac{\pi}{6}) - 2(\cos \delta - \cos \frac{\pi}{6})$$

$$= 0.1\omega^2 - (\delta - \frac{\pi}{6}) - 2 \cos \delta + 1.732$$

$$= 0.1\omega^2 - \delta - 2 \cos \delta + 2.256$$

$$V(\delta^u, 0) = -P_m(\pi - 2\delta^s) + 2P_e^{\max} \cos \delta^s$$

$$= - \left(\pi - \frac{2\pi}{6} \right) + 4 \cos \frac{\pi}{6}$$

$$= -2.09 + 3.464 = 1.374$$

Hence, the equal-area criterion is

$$V(\delta, \omega) < 1.374$$

where δ, ω are calculated from the fault-on trajectory. To obtain t_{cr}, the faulted equations are integrated and $V(\delta, \omega)$ are computed at each time point. When $V(\delta, \omega) = 1.374$, the time instant is t_{cr}.

□

9.6.4   Multimachine PEBS

In the previous section, we mentioned that the points δ^u and $\hat{\delta}^u$ were the zero-dimensional PEBS for the SMIB system. In the case of multimachine systems, the PEBS is quite complex in the rotor-angle space. A number of unstable equilibrium points surround the stable equilibrium point of the postfault system. The potential energy boundary surface therefore constitutes a multidimensional surface passing through the u.e.p's. The theory behind the characterization of the PEBS is quite detailed, and is dealt with in the literature [93, 94]. We can extend the concept of computing V_{cr} using the PEBS method for a multimachine system as follows.

In the previous section, we showed that in the SMIB case $V_{cr} = V_{PE}(\delta^u)$, i.e., $V(\delta, \omega)$ is evaluated at the nearest equilibrium point $(\delta^u, 0)$ if the machine loses synchronism by acceleration. δ^u is therefore not only the nearest, but also the relevant (or controlling), u.e.p in this case. In the case of the multimachine system, depending on the location and nature of the fault, the system may lose synchronism by one or more machines going unstable. Hence, each disturbance gives rise to what is called a mode of instability (MOI) [114]. Associated with each MOI is an u.e.p that we call the controlling u.e.p for that particular disturbance. A number of u.e.p's surround the s.e.p of the postfault system. Mathematically, these are the solutions of (9.23). From the prefault s.e.p, if the faulted system is integrated and cleared critically, then the postfault trajectory approaches a particular u.e.p depending on the mode of instability. This u.e.p is called the controlling u.e.p for that disturbance. In the multimachine PEBS, we can visualize a multidimensional potential "well" analogous to Figure 9.8 for the SMIB case. For a three-machine system, one such "well" is shown in Figure 9.10 where the axes are the COI-referenced rotor angles θ_1, θ_2 of two machines. The vertical axis represents $V_{PE}(\theta)$. Equipotential contours are shown, as well as the three nearby u.e.p's U_1, U_2, U_3. The dotted line connecting these u.e.p's is orthogonal to the equipotential curves and is called the PEBS. If at the instant of fault-clearing the system state in the angle space has crossed the PEBS, the system will be unstable. If the fault is cleared early enough, then the postfault trajectory in the angle space will tend to return to equilibrium eventually because of the damping in the system. The critical clearing time t_{cr} is defined to be the time instant such that the postfault trajectory just stays within the "well." It is a conjecture that the critically cleared trajectory passes "very close" to the controlling u.e.p. This is called the "first swing" stable phenomenon. To find the critical value of $V(\delta, \omega)$, the fault-on trajectory is monitored until it crosses the PEBS at a point θ^*. In many cases, θ^u, the controlling u.e.p, is close to θ^*, so that $V_{PE}(\theta^u) \approx V_{PE}(\theta^*) \triangleq V_{cr}$. This is the essence of the PEBS method. A key question here is the detection of the PEBS crossing. This crossing is also approximately the point at which $V_{PE}(\theta)$ is maximum along the faulted trajectory. Hence, V_{cr} can be taken as equal to $V_{PE}^{max}(\theta)$ along the faulted trajectory. It can be shown [97] that the PEBS crossing is also the point at which $f^T(\theta) \cdot (\theta - \theta^s) = 0$. $f(\theta)$ is the accelerating power in the postfault system. That this is the same point at which $V_{PE}(\theta)$ is maximum has been shown to be true for a conservative system [93]. In recent years, this PEBS crossing method has been the basis of improved algorithms such as the "second kick" method [115]. We now explain the basic PEBS algorithm in detail. It will help in understanding the newer algorithms.

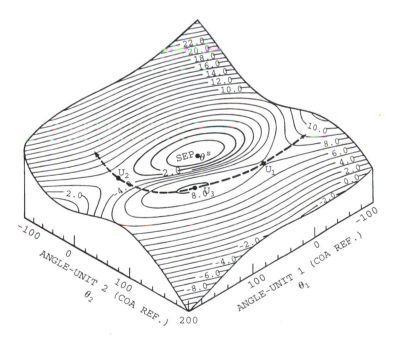

Figure 9.10: *The potential energy boundary surface (reproduced from [97])*

The energy function given by (9.29) is repeated here:

$$V(\theta, \tilde{\omega}) = \frac{1}{2} \sum_{i=1}^{m} M_i \tilde{\omega}_i^2 - \sum_{i=1}^{m} P_i(\theta_i - \theta_i^s) - \sum_{i=1}^{m} \sum_{j=i+1}^{m} [C_{ij}(\cos\theta_{ij} - \cos\theta_{ij}^s)$$

$$- \int_{\theta_i^s + \theta_j^s}^{\theta_i + \theta_j} D_{ij} \cos\theta_{ij} \, d(\theta_i + \theta_j)] \qquad (9.49)$$

The last term on the right-hand side of (9.49) denoted by $V_d(\theta)$ is a path-dependent term. This can be evaluated using trapezoidal integration as

$$V_d(\theta) = \sum_{i=1}^{m-1} \sum_{j=i+1}^{m} I_{ij} \qquad (9.50)$$

where at the kth step

$$I_{ij}(k) = I_{ij}(k-1) + \frac{1}{2} D_{ij}[\cos(\theta_i(k) - \theta_j(k)) + \cos(\theta_i(k-1) - \theta_j(k-1))]$$

$$[\theta_i(k) + \theta_j(k) - \theta_i(k-1) - \theta_j(k-1)] \qquad (9.51)$$

with

$$I_{ij}(0) = 0$$

This evaluation of $I_{ij}(0)$ is correct when the postfault system is the same as the prefault system, but is somewhat inaccurate if there is line-switching. This is explained in the next section. Equation (9.49) is rewritten as

$$V(\theta, \tilde{\omega}) = V_{KE}(\tilde{\omega}) + V_p(\theta) + V_d(\theta) \qquad (9.52)$$

where

$$V_{PE}(\theta) = V_p(\theta) + V_d(\theta) \qquad (9.53)$$

$$V_p(\theta) = -\sum_{i=1}^{m} P_i(\theta_i - \theta_i^s) - \sum_{i=1}^{m} \sum_{j=i+1}^{m} C_{ij}(\cos\theta_{ij} - \cos\theta_{ij}^s)$$

and $V_d(\theta)$ is given by (9.50). It can be shown [97] that the points θ on the PEBS are defined by $\sum_{i=1}^{m} f_i(\theta)(\theta_i - \theta_i^s) = 0$. This is the characterization of the PEBS. In vector form, this can be written as $f^T(\theta) \cdot (\theta - \theta^s) = 0$. Denoting $\theta - \theta^s = \hat{\theta}$, we can show by analogy to the zero transfer conductance case that inside the PEBS $f^T(\theta) \cdot \hat{\theta} < 0$, and outside the PEBS it is > 0 [93]. In the absence of transfer conductances, $f(\theta) = -\frac{\partial V_{PE}(\theta)}{\partial \theta}$. When θ is away from θ^s, within the potential multidimensional "well," $\frac{\partial V_{PE}(\theta)}{\partial \theta}$ (which is the gradient of the potential energy function) and $\hat{\theta}$ (i.e., $(\theta - \theta^s)$) are both > 0. Hence, $f^T(\theta) \cdot \hat{\theta} < 0$ inside the "well." Outside the "well," $\theta - \theta^s$ is > 0 and $\frac{\partial V_{PE}(\theta)}{\partial \theta} < 0$, resulting in $f^T(\theta) \cdot \hat{\theta} > 0$. On the PEBS, the product $f^T(\theta) \cdot \hat{\theta}$ is equal to zero.

The steps to compute t_{cr} using the PEBS method are as follows:

1. Compute the postfault s.e.p θ^s by solving (9.23).

2. Compute the fault-on trajectory given by (9.21).

3. Monitor $f^T(\theta) \cdot \hat{\theta}$ and $V_{PE}(\theta)$ at each time step. The parameters in $f(\theta)$ and $V_{PE}(\theta)$ pertain to the postfault configuration.

4. Inside the potential "well" $f^T(\theta) \cdot \hat{\theta} < 0$. Continue steps 2 and 3 until $f^T(\theta) \cdot \hat{\theta} = 0$. This is the PEBS crossing $(t^*, \theta^*, \tilde{\omega}^*)$. At this point, find $V_{PE}(\theta^*)$. This is a good estimate of V_{cr} for that fault.

5. Find when $V(\tilde{\theta}, \tilde{\omega}) = V_{cr}$ from the fault-on trajectory. This gives a good estimate of t_{cr}.

One can replace steps 3 and 4 by monitoring when $V_{PE}^{max}(\theta)$ is reached, and taking it as V_{cr}. There will be some error in either of the algorithms.

9.6.5 Initialization of $V_{PE}(\theta)$ and its use in PEBS method

In this section, we outline a further simplification of the PEBS method that works well in many cases, particularly when θ^s is "close" to θ^o. While integrating the faulted trajectory given by (9.21), the initial conditions on the states are $\theta_i(0) = \theta_i^o$ and $\tilde{\omega}_i(0) = 0$. In the energy function (9.29), the reference angle and velocity variables are θ_i^s and $\tilde{\omega}_i(0) = 0$. Thus, at $t = 0$, we evaluate $V_{PE}(\theta)$ in (9.29) as

$$V_{PE}(\theta^o) = -\sum_{i=1}^{m} \int_{\theta_i^s}^{\theta_i^o} f_i(\theta)d\theta_i$$

$$= -P_i(\theta_i^o - \theta_i^s) - \sum_{i=1}^{m-1}\sum_{j=i+1}^{m}$$

$$\left[C_{ij}(\cos\theta_{ij}^o - \cos\theta_{ij}^s) - \int_{\theta_i^s+\theta_j^s}^{\theta_i^o+\theta_j^o} D_{ij}\cos\theta_{ij}d(\theta_i + \theta_j) \right]$$

$$= K \text{ (a constant)} \tag{9.54}$$

The path-dependent integral term in (9.54) is evaluated using the trapezoidal rule:

$$I_{ij}(0) = \frac{1}{2}D_{ij}\left[\cos(\theta_i^o - \theta_j^o) + \cos(\theta_i^s - \theta_j^s)\right]\left[(\theta_i^o + \theta_j^o) - (\theta_i^s + \theta_j^s)\right] \tag{9.55}$$

If the postfault network is the same as the prefault network, then $K = 0$. Otherwise, this value of K should be included in the energy function.

If one uses the potential energy boundary surface (PEBS) method, then when the postfault network is not equal to the prefault network, this term can be subtracted from the energy function, i.e.,

$$V(\theta, \tilde{\omega}) = V_{KE}(\tilde{\omega}) + V_{PE}(\theta) - V_{PE}(\theta^o) \tag{9.56}$$

Hence, the potential energy can be defined, with θ^o as the datum, as

$$\hat{V}_{PE}(\theta) \triangleq V_{PE}(\theta) - V_{PE}(\theta^o) = -\left[\sum_{i=1}^{m}\int_{\theta_i^s}^{\theta_i} f_i(\theta)d\theta_i - \sum_{i=1}^{m}\int_{\theta_i^s}^{\theta_i^o} f_i(\theta)d\theta_i\right]$$

$$= -\sum_{i=1}^{m}\int_{\theta_i^o}^{\theta_i} f_i(\theta)d\theta_i \tag{9.57}$$

If the path-dependent integral term in (9.57) is evaluated, using trapezoidal integration as in (9.51), $I_{ij}(0) = 0$. At the PEBS crossing θ^*, $\hat{V}_{PE}(\theta^*)$ gives a good approximation to V_{cr}. The PEBS crossing has been shown as approximately the point at which the potential energy V_{PE} reaches a maximum value. Hence, one can directly monitor $\hat{V}_{PE}$ and thus avoid having to monitor the dot product $f^T(\theta) \cdot (\theta - \theta^s)$ as in step 4 of the previous section. This leads to the important advantage of not having to compute θ^s. In fast screening of contingencies, this could result in a significant saving of computation. On large-scale systems, this has not been investigated in the literature so far.

Example 9.5

Compute the $\overline{Y}_{int}$ for Example 7.1, using the classical model for the fault at bus 7 followed clearing lines of 7–5. Using the PEBS method, compute t_{cr}. Use $f^T(\theta) \cdot \hat{\theta}$ as the criterion for PEBS crossing.

$\underline{\overline{Y}_{int}}$ with fault at bus 7 (faulted system)

In the Y_{aug}^{new} matrix of Example 7.6, since $V_5 \equiv 0$, we delete the row and column corresponding to bus 5. Then eliminate all buses except the internal nodes 10, 11, and 12. The result is

$$\overline{Y}_{int}^F = \begin{bmatrix} 0.6588 - j3.8175 & 0.0000 - j0.0000 & 0.0714 + j0.6296 \\ 0.0000 - j0.0000 & 0.0000 - j5.4855 & 0.0000 - j0.0000 \\ 0.0714 + j0.6296 & 0.0000 - j0.0000 & 0.1750 - j2.7966 \end{bmatrix}$$

$\underline{\overline{Y}_{int}}$ with lines 7–5 cleared (postfault system)

$\overline{Y}_{bus} = \overline{Y}_N$ is first computed with lines 7–5 removed, and the rest of the steps are as in Example 7.6. Buses 1 to 9 are eliminated, resulting in

$$\overline{Y}_{int} = \begin{bmatrix} 1.1411 - j2.2980 & 0.1323 + j0.7035 & 0.1854 + j1.0611 \\ 0.1323 + j0.7035 & 0.3810 - j2.0202 & 0.1965 + j1.2031 \\ 0.1854 + j1.0611 & 0.1965 + j1.2031 & 0.2723 - j2.3544 \end{bmatrix}$$

The initial rotor angles are $\delta_1(0) = 0.0396$ rad, $\delta_2(0) = 0.344$ rad, and $\delta_3(0) = 0.23$ rad. The COA is calculated as $\delta_o = \frac{1}{M_T} \sum_{i=1}^{3} M_i \delta_i(0) = 0.116$ rad, where $M_T = M_1 + M_2 + M_3$. Hence, we have $\theta_1(0) = \delta_1(0) - \delta_o = -0.0764$ rad, $\theta_2(0) = \delta_2(0) - \delta_o = 0.229$ rad, and $\theta_3(0) = \delta_3(0) - \delta_o = 0.114$ rad, $\tilde{\omega}_1(0) = \tilde{\omega}_2(0) = \tilde{\omega}_3(0) = 0$. The postfault s.e.p is calculated as $\theta_1^s = -0.1649$, $\theta_2^s = 0.4987$, $\theta_3^s = 0.2344$. The steps in computing t_{cr} are given below.

1. From the entries in Y_{int} for faulted and postfault systems, the appropriate C_{ij} and D_{ij}'s are calculated to put the equations in the form of (9.21) and (9.22).

2. $V(\theta, \omega)$ is given by $V_{KE} + V_{PE}(\theta)$, where $V_{KE} = \frac{1}{2}M_i\tilde{\omega}_i^2$ and $V_{PE}(\theta)$ is given by (9.49). The path-integral term is evaluated as in (9.51), with $I_{ij}(0) = 0$, and the term (9.55) is added to $V_{PE}(\theta)$.

3. The faulted system corresponding to (9.21) is integrated and at each time step $V(\theta, \tilde{\omega})$ as well as $V_{PE}(\theta)$ are computed. Also the dot product $f^T(\theta) \cdot \hat{\theta}$ is monitored. The plots of $V(\theta, \tilde{\omega})$ and $V_{PE}(\theta)$ are shown in Figure 9.11. Figure 9.12 shows the plot of $f^T(\theta) \cdot \hat{\theta}$.

4. $V_{PE}^{max} = 1.3269$ is reached at approximate by 0.36 sec. Note from Figure 9.12 that the zero crossing of $f^T(\theta) \cdot \hat{\theta}$ occurs at approximately the same time.

5. From the graph for $V(\theta, \tilde{\omega})$, $t_{cr} = 0.199$ sec when $V(\theta, \tilde{\omega}) = 1.3269$.

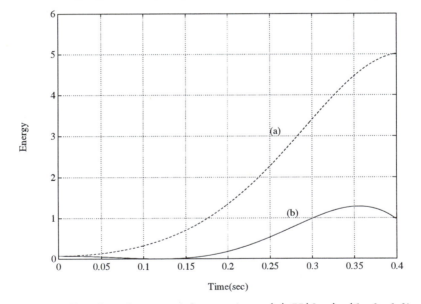

Figure 9.11: *Total and potential energies: (a) $V(\delta, \omega)$ (dashed line); (b) $V_{PE}(\theta)$ (solid line)*

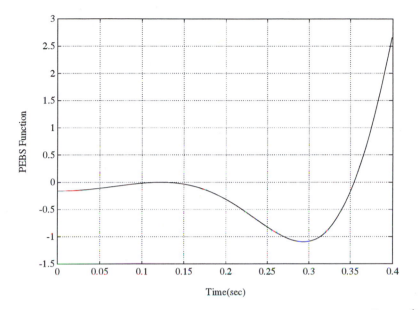

Figure 9.12: *The monitoring of the PEBS crossing by* $f^T(\theta) \cdot \dot{\theta}$

□

9.7 The Boundary Controlling u.e.p (BCU) Method

This method [98] provided another breakthrough in applying energy function methods to stability analysis after the work of Athay et al. [97], which originally proposed the controlling u.e.p method. The equations of the postfault system (9.22) can be put in the state-space form as

$$
\begin{aligned}
\dot{\theta} &= \tilde{\omega}_i \\
M_i \dot{\tilde{\omega}}_i &= f_i(\theta) \\
&= -\frac{\partial V_{PE}(\theta)}{\partial \theta_i} \quad i = 1, \ldots, m
\end{aligned}
\tag{9.58}
$$

Now

$$
\dot{V}(\theta, \tilde{\omega}) = \frac{\partial V}{\partial \theta}\dot{\theta} + \frac{\partial V}{\partial \tilde{\omega}}\dot{\tilde{\omega}}
$$

$$= \frac{\partial V_{PE}(\theta)}{\partial \theta} \dot{\theta} + \frac{\partial V_{KE}(\tilde{\omega})}{\partial \tilde{\omega}} \dot{\tilde{\omega}}$$

$$= -\sum_{i=1}^{m} f_i(\theta)\dot{\theta}_i + \sum_{i=1}^{m} M_i\tilde{\omega}_i\dot{\tilde{\omega}}_i$$

$$= \sum_{i=1}^{m} \tilde{\omega}_i(-f_i(\theta) + M_i\dot{\tilde{\omega}}_i)$$

$$= 0 \tag{9.59}$$

Hence, $V(\theta, \tilde{\omega})$, is a valid energy function.

The equilibrium points of (9.58) lie on the subspace $\theta, \tilde{\omega}$ such that $\theta \epsilon R^m$, $\tilde{\omega} = 0$. In the previous section, we have qualitatively characterized the PEBS as the hypersurfaces connecting the u.e.p's. We make it somewhat more precise now. Consider the gradient system

$$\dot{\theta} = \frac{-\partial V_{PE}(\theta)}{\partial \theta} \tag{9.60}$$

Note that the gradient system has dimension m, which is half the order of the system (9.58). It has been shown by Chiang et al. [98] that the region of attraction of (9.58) is the union of the stable manifolds of u.e.p's lying on the stability boundary. If this region of attraction is projected onto the angle space, it can be characterized by

$$\partial A(\theta^s) = \cup_i W^s(\theta_i^{ub}) \tag{9.61}$$

where θ_i^{ub} is an u.e.p on the stability boundary in the angle space. The stable manifold $W^s(\theta_i^{ub})$ of θ_i^{ub} is defined as the set of trajectories that converge to θ_i^{ub} as $t \to +\infty$. Since the gradient of $V_{PE}(\theta)$ is a vector orthogonal to the level surfaces $V_{PE}(\theta) = $ constant in the direction of increased values of $V_{PE}(\theta)$, the PEBS in the direction of decreasing values of $V_{PE}(\theta)$ can be described by the differential equations $\dot{\theta} = -\frac{\partial V_{PE}(\theta)}{\partial \theta} = f(\theta)$. Hence, when the fault-on trajectory reaches the PEBS at $\theta = \theta^*$ corresponding to $t = t^*$, we can integrate the set of equations for $t > t^*$ as

$$\dot{\theta} = f(\theta), \ \theta(t^*) = \theta^* \tag{9.62}$$

where $f(\theta)$ pertains to the postfault system. This will take $\theta(t)$ along the PEBS to the saddle points (u.e.p's U_1 or U_2 in Figure 9.10 depending on

θ^*). The integration of (9.62) requires very small time steps since it is "stiff." Hence, we stop the integration until $\| f(\theta) \|$ is minimum. At this point, let $\theta = \theta_{\text{app}}^u$. If we need the exact θ^u, we can solve for $f(\theta) = 0$ in (9.23) using θ_{app}^u as an initial guess. The BCU method is now explained in an algorithmic manner.

Algorithm

1. For a given contingency that involves either line switching or load/generation change, compute the postdisturbance s.e.p. θ^s as follows.

   The s.e.p and u.e.p's are solutions of the real power equations

   $$f_i(\theta) = P_i - P_{ei}(\theta) - \frac{M_i}{M_T} P_{COI}(\theta) = 0 \quad i = 1, \ldots, m \qquad (9.63)$$

   Since $\theta_m = \frac{-1}{M_m} \sum_{i=1}^{m-1} M_i \theta_i$, it is sufficient to solve for

   $$f_i(\theta) = 0 \quad i = 1, \ldots, m - 1 \qquad (9.64)$$

   with θ_m being substituted in (9.64) in terms of $\theta_1, \ldots, \theta_{m-1}$. Generally, the s.e.p. θ^s is close to θ^o the prefault e.p. Hence, using θ^o as the starting point, (9.64) can be solved using the Newton-Raphson method.

2. Next, compute the controlling u.e.p. θ^u as follows:

   (a) Integrate the faulted system (9.21) and compute $V(\theta, \tilde{\omega}) = V_{KE}(\tilde{\omega}) + V_{PE}(\theta)$ given in (9.52) at each time step. As in the PEBS algorithm of the previous section, determine when the PEBS is crossed at $\theta = \theta^*$ corresponding to $t = t^*$. This is best done by finding when $f^T(\theta) \cdot (\theta - \theta^s) = 0$.

   (b) After the PEBS is crossed, the faulted swing equations are no longer integrated. Instead, the gradient system (9.62) of the postfault system is used. This is a reduced-order system in that only the θ dynamics are considered as explained earlier, i.e., for $t > t^*$

   $$\dot{\theta} = f(\theta) , \ \theta(t^*) = \theta^* \qquad (9.65)$$

Equation (9.65) is integrated while looking for a minimum of

$$\sum_{i=1}^{m} | f_i(\theta) | \tag{9.66}$$

At the first minimum of the norm given by (9.66), $\theta = \theta^u_{app}$ and $V_{PE}(\theta^u_{app}) = V_{cr}$ is a good approximation to the critical energy of the system. The value of θ^u_{app} is almost the relevant or the controlling u.e.p.

(c) The exact u.e.p can be obtained by solving $f(\theta) = 0$ and using θ^u_{app} as a starting point to arrive at θ^u. Note that since $f(\theta)$ is nonlinear, some type of minimization routine must be used to arrive at θ^u. Generally, θ^u_{app} is so close to θ^u that it makes very little difference in the value of V_{cr} whether θ^u or θ^u_{app} is used.

3. V_{cr} is approximated as $V_{cr} = V(\theta^u, 0) = V_{PE}(\theta^u)$.

Because of the path-dependent integral term in V_{PE}, this computation also involves approximation. Unlike computing $V_{PE}(\theta)$ from the faulted trajectory where θ was known, here we do not know the trajectory from the full system. Hence, an approximation has to be used. The most convenient one is the straight-line path of integration. $V_{PE}(\theta^u)$ is evaluated as [97]:

$$V_{PE}(\theta^u) = -\sum_{i=1}^{m} P_i(\theta^u_i - \theta^s_i) - \sum_{i=1}^{m-1} \sum_{j=i+1}^{m} \left[C_{ij} \cos \left(\theta^u_{ij} - \cos \theta^s_{ij} \right) \right.$$

$$\left. -\frac{(\theta^u_i - \theta^s_i) + (\theta^u_j - \theta^s_j)}{(\theta^u_i - \theta^s_i) - (\theta^u_j - \theta^s_j)} D_{ij} \left(\sin \theta^u_{ij} - \sin \theta^s_{ij} \right) \right] \tag{9.67}$$

We derive the third term of (9.67) as follows. Assume a ray from θ^s_i to θ^u_i and then any point on the ray is $\theta_i = \theta^s_i + p(\theta^u_i - \theta^s_i)(0 \le p \le 1)$. Thus, $d(\theta_i + \theta_j) = dp(\theta^u_i - \theta^s_i + \theta^u_j - \theta^s_j)$. The path-dependent term $V_d(\theta)$ in (9.49) is now evaluated at θ^u as $V_d(\theta) = \sum_{i=1}^{m-1} \sum_{j=i+1}^{m} I_{ij}$.

$$I_{ij} = [(\theta^u_i - \theta^s_i) + (\theta^u_j - \theta^s_j)] \int_0^1 D_{ij} \cos[(\theta^s_i - \theta^s_j)]$$

$$+p((\theta_i^u - \theta_i^s) - (\theta_j^u - \theta_j^s))]dp$$

$$= \frac{(\theta_i^u - \theta_i^s) + (\theta_j^u - \theta_j^s)}{(\theta_i^u - \theta_i^s) - (\theta_j^u - \theta_j^s)} D_{ij} \left[\sin \left(\theta_i^s - \theta_j^s \right) \right.$$

$$\left. + p \left((\theta_i^u - \theta_i^s) - (\theta_j^u - \theta_j^s) \right) \right] \Big|_{p=0}^{p=1}$$

$$= \frac{(\theta_i^u - \theta_i^s) + (\theta_j^u - \theta_j^s)}{(\theta_i^u - \theta_i^s) - (\theta_j^u - \theta_j^s)} D_{ij} \left[\sin \theta_{ij}^u - \sin \theta_{ij}^s \right] \qquad (9.68)$$

4. To compute t_{cr}, we go back to the already-computed value of $V(\theta, \tilde{\omega})$ from the fault-on trajectory in the case of a fault, or the postdisturbance trajectory in the case of load/generation change, and find the time when $V(\theta, \tilde{\omega}) = V_{cr}$. In the case of a fault this time instant gives t_{cr} and the system is stable if the fault is set to clear at a time $t < t_{cr}$. In the case of load/generation change, the system is stable if $V(\theta, \tilde{\omega}) < V_{cr}$ for all t.

Example 9.6

Compute t_{cr} using the BCU method for Example 9.5. Use θ_{app}^u instead of the exact controlling u.e.p θ^u to compute V_{cr}.

The PEBS crossing is computed as in Example 9.5. t^* is obtained as 0.3445 sec and $\theta_1^* = $ -0.7319 rad, $\theta_2^* = 2.413$ rad and $\theta_3^* = 0.618$ rad.

The gradient system for $t \geq t^*$ is given by

$$\dot{\theta} = f(\theta), \ \theta(t^*) = \theta^*$$

where $f(\theta)$ is given by (9.30) with the parameters C_{ij} and D_{ij} pertaining to the postfault system. The equations are now integrated until $\| f(\theta) \| = \sum_{i=1}^{3} | f_i(\theta) |$ is minimum (Fig. 9.13). The minimum obtained is $\| f \| = 0.985$. At this point, $\theta_{app1}^u = $ -0.9894 rad, $\theta_{app2}^u = 2.124$ rad, and $\theta_{app3}^u = 1.164$ rad. With this, $V_{cr} = V_{PE}(\theta_{app}^u)$ is calculated using (9.67) as 1.3198. Hence, from the fault-on trajectory in Figure 9.11, $t_{cr} = $ is between 0.196 and 0.199 sec. In this case, t_{cr} by both the PEBS and BCU method is about the same. The effect of loading, as well as the closed-form approximation to the path-dependent integral on t_{cr}, is discussed in [120].

□

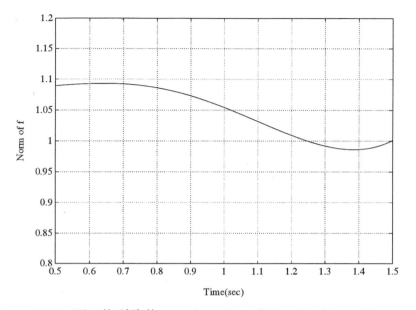

Figure 9.13: *Plot* $\| f(\theta) \|$ *as a function of time in the gradient system*

9.8   Structure-Preserving Energy Functions

The structure-preserving model (7.201)–(7.204) is reproduced below.

$$\dot{\theta}_i = \omega_i - \omega_s \tag{9.69}$$

$$M_i \dot{\omega}_i = T_{Mi} - \sum_{j=1}^{n+m} V_i V_j B_{ij} \sin(\theta_i - \theta_j)$$

$$i = n + 1, \ldots, n + m \tag{9.70}$$

$$P_{Li}(V_i) = \sum_{j=1}^{n+m} V_i V_j B_{ij} \sin(\theta_i - \theta_j) \quad i = 1, \ldots, n \tag{9.71}$$

$$Q_L(V_i) = - \sum_{j=1}^{n+m} V_i V_j B_{ij} \cos(\theta_i - \theta_j) \quad i = 1, \ldots, n \tag{9.72}$$

Note that the rotor angles δ_i's are also denoted as θ_i's. Here, we are assuming constant real power loads so that $P_{Li}(V_i) = P_{Li}$, and reactive power as nonlinear voltage-dependent loads. If the angles are referred to the COI δ_o $= \frac{1}{M_T} \sum_{i=1}^{m} M_i \theta_i$, then the angles referenced to COI become $\tilde{\theta}_i = \theta_i - \delta_o$ ($i = 1, \ldots, n+m$). It can be shown [94] that the energy function is given by

$$V(\tilde{\omega}, \tilde{\theta}, V) = V_{KE}(\tilde{\omega}) + V_{P1}(\tilde{\theta}, V) + V_{P2}(\tilde{\theta}) \qquad (9.73)$$

where

$$V_{KE}(\tilde{\omega}) = \frac{1}{2} \sum_{i=1}^{m} M_i \tilde{\omega}_i^2$$

$$V_{p1}(\tilde{\theta}, V) = - \sum_{i=n+1}^{n+m} T_{Mi}(\tilde{\theta}_i - \tilde{\theta}_i^s) + \sum_{i=1}^{n} \int_{V_i^s}^{V_i} \frac{Q_{Li}(V_i)}{V_i} dV_i \qquad (9.74)$$

$$\frac{1}{2} \sum_{1}^{n} B_{ii}(V_i^2 - (V_i^s)^2) \qquad (9.75)$$

$$- \sum_{i=1}^{n+m-1} \sum_{j=i+1}^{n+m} B_{ij}(V_i V_j \cos \tilde{\theta}_{ij} - V_i^s V_j^s \cos \tilde{\theta}_{ij}^s) \qquad (9.76)$$

$$V_{p2}(\tilde{\theta}) = - \sum_{1}^{n} P_{Li}(\tilde{\theta}_i - \tilde{\theta}_i^s) \qquad (9.77)$$

This energy function has been used for both transient-stability and voltage-stability analyses. In the second case, only the PE term is used, along with the concepts of high- and low-power flow solutions [117].

9.9 Conclusion

In this chapter, we have discussed in detail the transient energy function method for angle stability. The basis of the method has been shown to be the famous Lyapunov's direct method [118]. The equivalence of the energy function to the equal-area criterion has been shown for the single-machine case. For the multimachine case, the PEBS and the BCU have been explained in detail. The TEF method can be used to act as a filter to screen out contingencies in a dynamic security assessment framework [119]. The BCU method is known to be sensitive with respect to the PEBS crossing θ^*.

The method can be made more robust by tracking the stable manifold to the controlling u.e.p using what is called "shadowing" method [121]. Finally, the structure-preserving energy function has been derived. The TEF area is an active area of research.

9.10   Problems

9.1 A single machine connected to an infinite bus has the following faulted and postfault equations.

Faulted

$$0.0133\frac{d^2\delta}{dt^2} = 0.91 \quad 0 < t \le t_{cl}$$

Postfault

$$0.0133\frac{d^2\delta}{dt^2} = 0.91 - 3.24\sin\delta \quad t > t_{cl}$$

The prefault system is the same as the postfault system.

(a) Find $V(\delta,\omega)$ and V_{cr} using the u.e.p formulation.

(b) Explain stability test of (a) using the equal-area criterion. Sketch the areas A_1, A_2, A_3.

(c) Find t_{cr} using V_{cr}.

(d) Find t_{cr} using PEBS method.

9.2 For the 3-machine system of Example 7.1, a fault occurs at bus 7 and is cleared at t_{cl} with no line switching.

(a) Based on prefault load flow and using the classical model, compute the voltages behind transient reactances and the rotor angles at $t = 0^-$.

(b) Find $\overline{Y}_{\text{int}}$ for the faulted and the postfault cases.

(c) Write the energy function $V(\theta, \tilde{\omega})$, assuming $G_{ij} = 0 (i \ne j)$.

(d) Write the faulted equations in COI notation, together with the initial conditions.

(e) Compute t_{cr} using the PEBS method.

(f) Repeat (c), (d), and (e), assuming $G_{ij} \ne 0$.

9.3 For Problem 8.1 using the classical model, compute $\overline{Y}_{\text{int}}$ for prefault and $\overline{Y}_{\text{int}}$ for a fault at bus 112.

9.4 Compute for Problem 8.1 the voltages behind transient reactances and the rotor angles at $t = 0^-$.

    (a) Using results of Problem 9.3, write the energy function $V(\theta, \tilde{\omega})$. Assume that the fault at bus 112 is self-clearing.

    (b) Using the PEBS method, compute t_{cr}.

9.5 Use the BCU method to compute t_{cr} for Problems 9.2 and 9.4. Do this first for $G_{ij} = 0$ $(i \neq j)$ and then for $G_{ij} \neq 0$. Compare the results with the PEBS method.

Appendix A

Integral Manifolds for Model Reduction

A.1   Manifolds and Integral Manifolds

The term "manifold" in this chapter refers to a functional relationship between variables. For example, a manifold for z as a function of x is simply another term for the expression

$$z = h(x) \tag{A.1}$$

When x is a scalar, the manifold is a line when plotted in the z, x space. When x is two-dimensional, the manifold is a surface that might appear as in Figure A.1

To define an integral manifold, we have to introduce a multidimensional dynamic model of the form

$$\frac{dx}{dt} = f(x, z) \qquad\qquad x(o) = x^o \tag{A.2}$$

$$\frac{dz}{dt} = g(x, z) \qquad\qquad z(o) = z^o \tag{A.3}$$

An integral manifold for z as a function of x is a manifold

$$z = h(x) \tag{A.4}$$

which satisfies the differential equation for z. Thus, $h(x)$ is an integral manifold of (A.2)–(A.3) if it satisfies

$$\frac{\partial h}{\partial x} f(x, h) = g(x, h) \tag{A.5}$$

323

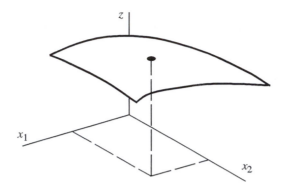

Figure A.1: *Geometrical interpretation of a manifold*

If the initial conditions on x and z lie on the manifold $(z^o = h(x^o))$, then the integral manifold is an exact solution of the differential equation (A.3), and the following reduced-order model is exact:

$$\frac{dx}{dt} = f(x, h(x)) \qquad\qquad x(o) = x^o \qquad\qquad (A.6)$$

A.2   Integral Manifolds for Linear Systems

The concept of integral manifolds is illustrated in this section through a series of examples. Consider the oversimplified system

$$\frac{dx}{dt} = -x + z \qquad x(o) = x^o \qquad\qquad (A.7)$$

$$\frac{dz}{dt} = -10z + 10 \qquad z(o) = z^o \qquad\qquad (A.8)$$

We say that z is predominantly *fast* because of the 10 multiplying the right side. We say that x is predominantly *slow* because it has only 1 multiplying the right side. For comparison later, let's solve for the exact response. Because it is a linear time-invariant system, the solution will be of the form

$$x(t) = c_1 e^{\lambda_1 t} + c_2 e^{\lambda_2 t} + c_3 \qquad\qquad (A.9)$$
$$z(t) = c_4 e^{\lambda_1 t} + c_5 e^{\lambda_2 t} + c_6 \qquad\qquad (A.10)$$

where λ_1 and λ_2 are the two eigenvalues. These are found from the system-state matrix as follows:

$$\left[\begin{array}{c} \frac{dx}{dt} \\ \frac{dz}{dt} \end{array} \right] = \underbrace{\left[\begin{array}{cc} -1 & 1 \\ 0 & -10 \end{array} \right]}_{A} \left[\begin{array}{c} x \\ z \end{array} \right] + \left[\begin{array}{c} 0 \\ 10 \end{array} \right] \tag{A.11}$$

The eigenvalues are the solutions of

$$\text{determinant } [\lambda I - A] = \left| \begin{array}{cc} \lambda + 1 & -1 \\ 0 & \lambda + 10 \end{array} \right| = 0 \tag{A.12}$$

or $(\lambda + 1)(\lambda + 10) - 0 = 0$. The roots are $\lambda_1 = -1$, $\lambda_2 = -10$. The constants c_3 and c_6 are the steady-state solution

$$\left. \begin{array}{c} 0 = -x_{ss} + z_{ss} \\ 0 = -10z_{ss} + 10 \end{array} \right\} z_{ss} = 1 \quad x_{ss} = 1 \tag{A.13}$$

so

$$x = c_1 e^{-t} + c_2 e^{-10t} + 1 \tag{A.14}$$
$$z = c_4 e^{-t} + c_5 e^{-10t} + 1 \tag{A.15}$$

The other constants are found from the initial conditions:

$$x(o) = c_1 + c_2 + 1 = x^o \tag{A.16}$$
$$\frac{dx}{dt} \Big|_o = -c_1 - 10c_2 = -x^o + z^o \tag{A.17}$$
$$z(o) = c_4 + c_5 + 1 = z^o \tag{A.18}$$
$$\frac{dz}{dt} \Big|_o = -c_4 - 10c_5 = -10z^o + 10 \tag{A.19}$$

Solving

$$c_1 = x^o + \frac{z^o - 10}{9} \quad c_2 = -\frac{z^o - 1}{9} \tag{A.20}$$
$$c_3 = z^o - 1 \quad c_4 = 0 \tag{A.21}$$

the exact solution is

$$x = \left(x^o - \frac{10}{9} + \underbrace{\frac{z^o}{9}}_{\substack{\text{small if } z^o \\ \text{is not large}}} \right) e^{-t} - \underbrace{\frac{z^o - 1}{9}}_{\substack{\text{small if } z^o \\ \text{is not large}}} e^{-10t} + 1 \tag{A.22}$$

If z^o is not large, the major part of x is slow. The fast variation of z contributes only small amounts to x.

To see how we can develop a reduced-order model, look for an integral manifold of the form

$$z = hx + c \tag{A.23}$$

where h and c are constants. Substituting into the z differential equation,

$$h\frac{dx}{dt} = -10(hx + c) + 10 \tag{A.24}$$

or

$$h(-x + hx + c) = -10hx - 10c + 10 \tag{A.25}$$

One solution is $h = 0$, $c = 1$. This means that $z = 1$ is an exact integral manifold for the z variable. That is, if $z = 1$ at any time, then z remains equal to one for all time. Or, more properly stated: "If the initial conditions start on the manifold, then the system remains on the manifold." If this integral manifold is substituted into the x differential equation, the reduced-order model (valid only for $z^o = 1$) is

$$\frac{dx}{dt} = -x + 1 \qquad\qquad x(o) = x^o \tag{A.26}$$

which clearly exhibits the exact slow eigenvalue and the following solution

$$x = (x^o - 1)e^{-t} + 1 \tag{A.27}$$

Compare this to the exact solution of (A.22). If z^o is equal to 1.0, then (A.26) gives the exact response of x for any x^o. If z^o is not equal to 1.0, then (A.26) will not give the exact response of x. For this case, we define the off-manifold variable η as

$$\eta \triangleq z - 1 \tag{A.28}$$

which has the dynamics

$$\frac{d\eta}{dt} = -10\eta \qquad\qquad \eta(o) = z^o - 1 \tag{A.29}$$

Note that $\eta = 0$ is an exact integral manifold because if $\eta = 0$ at any time, $\eta = 0$ for all time. The exact solution, when $\eta(0) \neq 0$, is

$$\eta(t) = (z^o - 1)e^{-10t} \tag{A.30}$$

The integral manifold plus off-manifold solution for z as a function of x and t is

$$z = 1 + (z^o - 1)e^{-10t} \qquad\qquad\qquad (A.31)$$

This gives the exact reduced-order model (for any z^o)

$$\frac{dx}{dt} = -x + 1 + (z^o - 1)e^{-10t} \qquad x(o) = x^o \qquad (A.32)$$

While this may have been obvious from the beginning, the steps that led to this result are important for cases where it is not obvious.

The fast time-varying input into this slow subsystem is somewhat undesirable. An interesting approximation can be made to eliminate this term as follows. Since the time-varying term decays very rapidly (e^{-10t}), this term enters the slow differential equation almost as an impulse. It is $z^o - 1$ at time zero and essentially zero for $t > 0$. Using the following impulse identity,

$$\mathrm{Imp}(t) \triangleq \lim_{a \to o} \frac{1}{a}e^{-\frac{t}{a}} \qquad\qquad (A.33)$$

Equation (A.32) can be approximated by

$$\frac{dx}{dt} \approx -x + 1 + \frac{(z^o - 1)}{10}\mathrm{Imp}(t) \qquad x(o) = x^o \qquad (A.34)$$

This impulse can be eliminated by recognizing that its integral can be included in the initial condition on x

$$x(t) = x^o + \int_o^t (-x + 1)d\hat{t} + \int_o^t \frac{(z^o - 1)}{10}\mathrm{Imp}(\hat{t})d\hat{t} \qquad (A.35)$$

or

$$x(t) = x^o + \frac{z^o - 1}{10} + \int_o^t (-x + 1)d\hat{t} \qquad (A.36)$$

This gives the approximate reduced-order model using the exact integral manifold for z and accounts for the z initial condition off-manifold dynamics through a revised initial condition on x:

$$\frac{dx}{dt} \approx -x + 1 \qquad x(o) = x^o + \frac{(z^o - 1)}{10} \qquad (A.37)$$

To see the impact of this approximation, consider the example in which $x^o = 1$ and $z^o = 0$. The exact response of x is the solution of

$$(a) \quad \frac{dx}{dt} = -x + 1 - e^{-10t} \qquad x(o) = 1.0 \qquad (A.38)$$

The approximate response of x using only the exact integral manifold for z is the solution of

$$(b) \quad \frac{dx}{dt} \approx -x + 1 \quad x(o) = 1.0 \tag{A.39}$$

The improved approximate response of x accounting for the off-manifold dynamics is the solution of

$$(c) \quad \frac{dx}{dt} \approx -x + 1 \quad x(o) = 0.9 \tag{A.40}$$

A comparison of these solutions is given in Figure A.2.

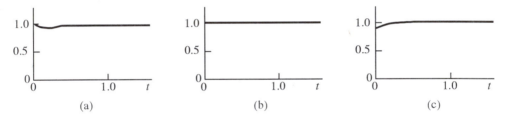

Figure A.2: *Comparison of exact and approximate solutions*

It is important to observe that the basic phenomenon was captured by the slow manifold of (b). The correction for the fast initial condition was a second-order effect approximated fairly well.

Before leaving this example, we return to (A.25) and observe that another integral manifold solution is

$$h = -9 , \ c = 10 \tag{A.41}$$

which gives another exact integral manifold,

$$z = -9x + 10 \tag{A.42}$$

If this solution is used in the original differential equation for x, the reduced-order model is

$$\frac{dx}{dt} = -10x + 10 \quad x(o) = x^{o} \tag{A.43}$$

The solution of this is

$$x = (x^{o} - 1)e^{-10t} + 1 \tag{A.44}$$

Compare this to the exact solution of (A.22). It is exact if $z^o = -9x^o + 10$. This means it is exact for $x^o = 1$, $z^o = 1$. But, if $x^o = 0.5$, $z^o = 1$ (not significantly different), there will be a large error in this reduced-order model, which is good only for a very limited range of initial conditions.

We now extend this concept to the more general case with coupling in both equations as:

$$\frac{dx}{dt} = -x + z \tag{A.45}$$

$$\frac{dz}{dt} = -10x - 10z \tag{A.46}$$

As before, we say that z is predominantly fast and x is predominantly slow. This is due to the 10 multiplying the right side of the z differential equation. The eigenvalues of this system are found from the state matrix

$$A = \begin{bmatrix} -1 & 1 \\ -10 & -10 \end{bmatrix} \tag{A.47}$$

$$|\lambda I - A| = \begin{vmatrix} \lambda + 1 & -1 \\ 10 & \lambda + 10 \end{vmatrix} = (\lambda + 1)(\lambda + 10) + 10 = 0 \tag{A.48}$$

The roots are

$$\lambda_1 = -2.3 \quad \lambda_2 = -8.7 \tag{A.49}$$

so the exact solution is

$$x = c_1 e^{-2.3t} + c_2 e^{-8.7t} + c_3 \tag{A.50}$$
$$z = c_4 e^{-2.3t} + c_5 e^{-8.7t} + c_6 \tag{A.51}$$

The steady-state values of x and z are

$$x_{ss} = c_3 = 0 \quad z_{ss} = c_6 = 0 \tag{A.52}$$

Focusing on x, let's solve for c_1 and c_2.

$$x(o) = c_1 + c_2 = x^o \tag{A.53}$$

$$\frac{dx}{dt}\Big|_o = -2.3c_1 - 8.7c_2 = -x^o + z^o \tag{A.54}$$

Solving

$$c_1 = \frac{7.7x^o + z^o}{6.4}, \quad c_2 = -\frac{1.3x^o + z^o}{6.4} \tag{A.55}$$

so:

$$x = \left(\underbrace{\frac{7.7}{6.4} x^o + \frac{z^o}{6.4}}_{\substack{\text{small if } z^o \\ \text{is not large}}} \right) e^{-2.3t} - \left(\frac{1.3}{6.4} x^o + \underbrace{\frac{x^o}{6.4}}_{\substack{\text{small if } z^o \\ \text{is not large}}} \right) e^{-8.7t} \quad (A.56)$$

As before, if z^o is not large, the x response is dominated by the slow component. If we are interested only in capturing the mode with eigenvalue -2.3, we propose that this mode is associated with the state x, and seek to eliminate z from (A.45)–(A.46) through an integral manifold of the general linear form

$$z = hx + c \quad (A.57)$$

Substituting into (A.46) (and using (A.45)) gives

$$h(-x + hx + c) = -10x - 10hx - 10c \quad (A.58)$$

For arbitrary x, this has the solutions

$$c = 0 \qquad\qquad h = -1.3 \quad (A.59)$$

$$c = 0 \qquad\qquad h = -7.7 \quad (A.60)$$

Thus, there are two integral manifolds for z as before:

$$IM\ 1: \quad z = -1.3x \quad (A.61)$$

$$IM\ 2: \quad z = -7.7x \quad (A.62)$$

Substitution of IM 1 into (A.45) gives

$$\frac{dx}{dt} = -2.3x \quad (A.63)$$

and substitution of IM 2 into (A.45) gives the faster subsystem

$$\frac{dx}{dt} = -8.7x \quad (A.64)$$

This second result is somewhat disturbing, since we had originally proposed that the variable x was associated with the -2.3 mode. Several important points are illustrated here. First, integral manifolds can be used to decouple systems and find eigenvalues. Second, integral manifolds are not unique (one for each eigenvalue, it appears!). Third, the choice of the integral manifold determines the phenomena retained in the reduced-order model. To understand more about this technique, consider the same system with the introduction of the small parameter ϵ (1/10 in the above example), and the inclusion of initial conditions

$$\frac{dx}{dt} = -x + z \qquad\qquad x(o) = x^o \qquad\qquad \text{(A.65)}$$

$$\epsilon\frac{dz}{dt} = -x - z \qquad\qquad z(o) = z^o \qquad\qquad \text{(A.66)}$$

Again we seek a linear manifold

$$z = h(\epsilon)x + c(\epsilon) \qquad\qquad \text{(A.67)}$$

where we presume that h and c would normally depend on the parameter ϵ. Substitution of (A.67) into (A.65)–(A.66) gives

$$\epsilon h(\epsilon)(-x + h(\epsilon)x + c(\epsilon)) = -x - h(\epsilon)x - c(\epsilon) \qquad\qquad \text{(A.68)}$$

Again, for arbitrary x, the solutions are

$$c(\epsilon) = 0, \quad h(\epsilon) = -\frac{1-\epsilon}{2\epsilon} + \frac{1}{2\epsilon}\sqrt{(1-\epsilon)^2 - 4\epsilon}, \quad h(o) = -1 \quad \text{(A.69)}$$

$$c(\epsilon) = 0, \quad h(\epsilon) = -\frac{1-\epsilon}{2\epsilon} - \frac{1}{2\epsilon}\sqrt{(1-\epsilon)^2 - 4\epsilon}, \quad h(o) = -\infty \quad \text{(A.70)}$$

For positive ϵ, these solutions exist only for

$$0 \le \epsilon \le 0.1715 \qquad\qquad \text{(A.71)}$$

This makes sense, since the original system has complex eigenvalues for ϵ less than 1 but greater than 0.1715. It would be impossible to capture a complex mode from a single-state equation.

   This simple example illustrates that integral manifolds may not exist and may not be unique. If we are interested in the "slow mode," which we propose is associated with the variable x, we need a systematic way to compute the correct integral manifold. If z is infinitely fast ($\epsilon = 0$), the integral manifold of interest is $z = -x$ ($h = -1$). When ϵ is near zero, we propose that the integral manifold of interest should be near -1. To systematically compute h for ϵ near zero, we expand $h(\epsilon)$ in a power series in ϵ as

$$h(\epsilon) = h_o + \epsilon h_1 + \epsilon^2 h_2 + \ldots \tag{A.72}$$

and return to the example by substituting into (A.66) and (A.67) (using $c(\epsilon) = 0$)

$$\epsilon(h_o + \epsilon h_1 + \epsilon^2 h_2 + \ldots)(-x + (h_o + \epsilon h_1 + \epsilon^2 h_2 + \ldots)x)$$

$$= -x - (h_o + \epsilon h_1 + \epsilon^2 h_2 + \ldots)x \tag{A.73}$$

For arbitrary x, we solve for $h_o, h_1, h_2, \ldots$ by equating coefficients of powers of ϵ:

$$\epsilon^o : \quad 0 = -1 - h_o \text{ or } h_o = -1 \tag{A.74}$$

$$\epsilon^1 : \quad h_o(-1 + h_o) = -h_1 \text{ or } h_1 = h_o - h_o^2 = -2 \tag{A.75}$$

etc.

Stopping with these terms,

$$h(\epsilon) \approx -1 - 2\epsilon \tag{A.76}$$

This approximates the exact integral manifold of (A.65)–(A.66) as

$$z \approx -(1 + 2\epsilon)x \tag{A.77}$$

Using this in (A.65) gives the approximate reduced-order model (valid when the initial conditions satisfy $z^o = h(\epsilon)x^o$)

$$\frac{dx}{dt} \approx -(2 + 2\epsilon)x \qquad\qquad x(o) = x^o \tag{A.78}$$

This model could be improved to any degree of accuracy by including additional terms of $h(\epsilon)$. Since these terms were computed from a power series near the integral manifold of interest (slow manifold), the correct mode has been captured. Note that it is necessary to consider only the off-manifold dynamics due to the initial conditions if the reduced order model is going to be used in a simulation. If only eigenvalues are of interest, the off-manifold dynamics due to initial conditions are not relevant. For $\epsilon = 0.1$, as in the previous example, the slow eigenvalue (-2.3) is approximated in (A.78) by

-2.2. As a warning, it is important to note that in this example the integral manifold always exists when ϵ is actually zero. When ϵ is small but not zero, the infinite series of (A.72) can always be computed, but its convergence and, hence, its validity depend on the size of the actual ϵ. In this example, we would expect that this series converges only when ϵ is less than 0.1715. Thus, while terms of the integral manifold series can be computed for any ϵ, it should be used only when ϵ is less than 0.1715.

While computation of the integral manifold is the primary task in model reduction, it may not make sense to find a good approximation and then ignore the possibility that z does not start on the manifold ($z^o \neq h(\epsilon)x^o$). As in the earlier example, it is possible to approximate this impact on x by computing the off-manifold correction. To do this, we define the off-manifold variable as (with $c(\epsilon) = 0$)

$$\eta \overset{\triangle}{=} z - h(\epsilon)x \qquad (A.79)$$

which has dynamics (recalling (A.68) with $c(\epsilon) = 0$)

$$\epsilon\frac{d\eta}{dt} = -(1 + \epsilon h(\epsilon))\eta \qquad \eta(o) = z^o - h(\epsilon)x^o \qquad (A.80)$$

Clearly, $\eta = 0$ is also an integral manifold because if $\eta = 0$ at any time, then $\eta = 0$ for all time. The exact off-manifold dynamics are the solution of (A.80):

$$\eta(t) = (z^o - h(\epsilon)x^o)e^{-\left(\frac{1+\epsilon h(\epsilon)}{\epsilon}\right)t} \qquad (A.81)$$

If $h(\epsilon)$ is known exactly, then the exact slow subsystem is

$$\frac{dx}{dt} = -(1 - h(\epsilon))x + (z^o - h(\epsilon)x^o)e^{\left(\frac{1+\epsilon h(\epsilon)}{\epsilon}\right)t}$$

$$x(o) = x^o \qquad (A.82)$$

As before, this fast input can be approximated by an impulse and eliminated by modification of the initial condition to obtain

$$\frac{dx}{dt} \approx -(1 - h(\epsilon))x$$

$$x(o) = x^o + \left(\frac{\epsilon}{1 + \epsilon h(\epsilon)}\right)(z^o - h(\epsilon)x^o) \qquad (A.83)$$

Clearly, if $z^o = h(\epsilon)x^o$ (on the slow manifold), this result is exact when $h(\epsilon)$ is exact. In our example, $h(\epsilon)$ was never found exactly. Using the approximation $h(\epsilon) \approx -1 - 2\epsilon$ with $\epsilon = 1/10$ gives the following slow subsystem approximate model (approximate for two reasons: $h(\epsilon)$ is not exact and the off-manifold dynamics are not exact):

$$\frac{dx}{dt} \approx -2.2x \qquad x(o) = \frac{100}{88}x^o + \frac{10}{88}z^o \tag{A.84}$$

The basic result of all this is summarized as follows. Consider a linear system in standard two-time-scale form

$$\frac{dx}{dt} = Ax + Bz \tag{A.85}$$

$$\epsilon\frac{dz}{dt} = Cx + Dz \tag{A.86}$$

where A, B, C, C are of order 1 (not big) and D is nonsingular (D^{-1} exists).

A first-order approximation of the slow dynamics of x is obtained by simply setting $\epsilon = 0$.

$$\frac{dx}{dt} = Ax + Bz \tag{A.87}$$

$$0 = Cx + Dz \tag{A.88}$$

which gives the first approximation of the slow manifold (h_o)

$$z = -D^{-1}Cx \tag{A.89}$$

and the first approximation of the slow dynamics of x

$$\frac{dx}{dt} = (A - BD^{-1}C)x \tag{A.90}$$

Using this approximate model will introduce two errors. One is due to the fact that the integral manifold is not exact (h_o is only the largest part of $h(\epsilon)$). The second error is due to the fact that the initial condition on z may not satisfy the integral manifold.

It can be shown that if the initial condition on z is not big, then the total error of these two approximations is small. The term "not big" means that z^o should be on the order of magnitude of 1 or less. The term "small" means on the order of magnitude of ϵ. References [66] and [67] discuss this in detail.

In most cases, the first-order approximation of the slow manifold coincides exactly with the steady-state relationship between z and x in the z differential equation, because the result of setting ϵ to zero is the same as the result of setting $\frac{dz}{dt}$ to zero. There is, however, a profound theoretical difference. Setting $\frac{dz}{dt} = 0$ implies $z = $ constant. This would be contradicted by the change of z when x changes ($z = h_o x$). Setting $\epsilon = 0$ does not imply $z = $ constant, and thus is not contradicted by the change of z when x changes.

A.3 Integral Manifolds for Nonlinear Systems

While there are many reduction techniques that can be applied to linear systems, the primary advantage of the integral manifold approach is its straightforward extension to nonlinear systems. We begin by considering the general form

$$\frac{dx}{dt} = f(x, z) \qquad x(o) = x^o \tag{A.91}$$

$$\frac{dz}{dt} = g(x, z) \qquad z(o) = z^o \tag{A.92}$$

To analyze the dynamics of x, it is necessary to also compute the dynamics of z. A reduced-order model involving only x requires the elimination of z from (A.91). An integral manifold for z as a function of x has the form

$$z = h(x) \tag{A.93}$$

and must satisfy (A.92)

$$\frac{\partial h}{\partial x} f(x, h) = g(x, h) \tag{A.94}$$

While, in general, it is very difficult to find such an integral manifold, there are several very important cases in which h can be either found exactly or approximated to any degree of accuracy. We begin with a generic example, which closely resembles the single synchronous machine connected to an infinite bus with stator transients

$$\frac{dx_1}{dt} = x_2 - 1 \tag{A.95}$$

$$\frac{dx_2}{dt} = f(x_1, x_2, z_1, z_2) \tag{A.96}$$

$$\frac{dz_1}{dt} = -\sigma z_1 + x_2 z_2 + V \sin x_1 \tag{A.97}$$

$$\frac{dz_2}{dt} = -\sigma z_2 - x_2 z_1 + V \cos x_1 \tag{A.98}$$

We suppose that only x_1 and x_2 are of interest, and look for an integral manifold of the form

$$z_1 = h_1(x_1, x_2) \tag{A.99}$$

$$z_2 = h_2(x_1, x_2) \tag{A.100}$$

This two-dimensional integral manifold must satisfy

$$\frac{\partial h_1}{\partial x_1}(x_2 - 1) + \frac{\partial h_1}{\partial x_2} f(x_1, x_2, h_1, h_2) =$$
$$-\sigma h_1 + x_2 h_2 + V \sin x_1 \tag{A.101}$$

$$\frac{\partial h_2}{\partial x_1}(x_2 - 1) + \frac{\partial h_2}{\partial x_2} f(x_1, x_2, h_1, h_2) =$$
$$-\sigma h_2 - x_2 h_1 + V \cos x_1 \tag{A.102}$$

These partial differential equations can be solved by first assuming that h_1 and h_2 are independent of x_2 and then equating coefficients of x_2 to give

$$\frac{\partial h_1}{\partial x_1} = h_2, \quad \frac{\partial h_1}{\partial x_1} = \sigma h_1 - V \sin x_1 \tag{A.103}$$

$$\frac{\partial h_2}{\partial x_1} = -h_1, \quad \frac{\partial h_2}{\partial x_1} = \sigma h_2 - V \cos x_1 \tag{A.104}$$

Eliminating the partials gives the solution

$$h_1 = V \cos \alpha \cos(\alpha - x_1) \tag{A.105}$$

$$h_2 = V \cos \alpha \sin(\alpha - x_1) \tag{A.106}$$

where $\tan \alpha = \sigma$. Thus, if the initial conditions on z_1, z_2, and x_1 satisfy (A.105)–(A.106), then substitution of (A.105)–(A.106) into (A.96) gives an exact reduced-order model. When the initial conditions on z_1, z_2, and x_1 do not satisfy (A.105)–(A.106), it is necessary to introduce the "off-manifold" variables

$$\eta_1 \stackrel{\triangle}{=} z_1 - V \cos \alpha \cos(\alpha - x_1) \tag{A.107}$$

$$\eta_2 \stackrel{\triangle}{=} z_2 - V \cos \alpha \sin(\alpha - x_1) \tag{A.108}$$

These off-manifold variables have the following dynamics:

$$\frac{d\eta_1}{dt} = -\sigma\eta_1 + x_2\eta_2 \tag{A.109}$$

$$\frac{d\eta_2}{dt} = -\sigma\eta_2 - x_2\eta_1 \tag{A.110}$$

This system has the explicit solution

$$\eta_1 = c_1 e^{-\sigma t}\cos(t + x_1 - c_2) \tag{A.111}$$

$$\eta_2 = -c_1 e^{-\sigma t}\sin(t + x_1 - c_2) \tag{A.112}$$

where c_1 and c_2 are found from initial conditions by solving

$$z_1^o - V\,\cos\alpha\,\cos(\alpha - x_1^o) = c_1\,\cos(x_1^o - c_2) \tag{A.113}$$

$$z_2^o - V\,\cos\alpha\,\sin(\alpha - x_1^o) = -c_1\,\sin(x_1^o - c_2) \tag{A.114}$$

This result leads to an exact reduced-order model in x_1 and x_2 by using the following in (A.96):

$$z_1 = V\,\cos\alpha\,\cos(\alpha - x_1) + c_1 e^{-\sigma t}\cos(t + x_1 - c_2) \tag{A.115}$$

$$z_2 = V\,\cos\alpha\,\sin(\alpha - x_1) - c_1 e^{-\sigma t}\sin(t + x_1 - c_2) \tag{A.116}$$

Such exact integral manifolds and exact off-manifold solutions are rare in dynamic systems. The synchronous machine stands out as a unique device with this property [64].

A very broad class of systems in which integral manifolds can often be found or approximated to any degree of accuracy is the class of two-time-scale systems of the form

$$\frac{dx}{dt} = f(x, z) \quad x(o) = x^o \tag{A.117}$$

$$\epsilon\frac{dz}{dt} = g(x, z) \quad z(o) = z^o \tag{A.118}$$

These systems are called two-time scales because when ϵ is small, the z variables are predominantly fast and the x variables are predominantly slow. This is clear because the derivative of z with respect to time is proportional to $1/\epsilon$, which is large for small ϵ. As in the linear case of the last section, we propose an integral manifold for z as a function of x and ϵ:

$$z = h(x, \epsilon) \tag{A.119}$$

We assume that ϵ is sufficiently small so that the manifold can be expressed as a power series in ϵ:

$$h(x, \epsilon) = h_o(x) + \epsilon h_1(x) + \epsilon^2 h_2(x) + \ldots \qquad (A.120)$$

Substitution into (A.119) and then (A.118) gives

$$\epsilon \left(\frac{\partial h_o}{\partial x} + \epsilon \frac{\partial h_1}{\partial x} + \ldots \right) f(x, h) = g(x, h) \qquad (A.121)$$

Expanding f and g about $\epsilon = 0$,

$$f(x, h) = f(x, h_o) + \epsilon \frac{\partial f}{\partial z} |_{z=h_o} h_1 \ldots \qquad (A.122)$$

$$g(x, h) = g(x, h_o) + \epsilon \frac{\partial g}{\partial z} |_{z=h_o} h_1 + \ldots \qquad (A.123)$$

the partial differential equation to be solved is

$$\epsilon \left(\frac{\partial h_o}{\partial x} + \epsilon \frac{\partial h_1}{\partial x} + \ldots \right) \left(f(x, h_o) + \epsilon \frac{\partial f}{\partial z} |_{z=h_o} h_1 + \ldots \right) =$$

$$g(x, h_o) + \epsilon \frac{\partial g}{\partial z} |_{z=h_o} h_1 + \ldots \qquad (A.124)$$

Equating coefficients of powers of ϵ produces a set of algebraic equations to be solved for $h_o, h_1, h_2 \ldots$:

$$\epsilon^o : \qquad 0 = g(x, h_o) \qquad (A.125)$$

$$\epsilon^1 : \qquad \frac{\partial h_o}{\partial x} f(x, h_o) = \frac{\partial g}{\partial z} |_{z=h_o} h_1 \qquad (A.126)$$

$$\text{etc.}$$

Clearly, the most important equation is (A.125), which requires the solution of the nonlinear equation for h_o. Once this is found, the solution for h_1 simply requires nonsingular $\partial g/\partial z$. Normally, if (A.125) can be solved, the nonsingularity of $\partial g/\partial z$ follows.

As in the linear case, the use of an integral manifold in a reduced-order model can give exact results only if it is found exactly and if the initial conditions start on it. If the initial conditions do not start on the manifold (do not satisfy (A.119)), an error will be introduced. To eliminate this error, it is necessary to compute the off-manifold dynamics. This is done by introducing the off-manifold variables

$$\eta \overset{\triangle}{=} z - h(x, \epsilon) \tag{A.127}$$

with the following dynamics

$$\epsilon \frac{d\eta}{dt} = g(x, \eta + h) - \epsilon \frac{\partial h}{\partial x} f(x, \eta + h) \tag{A.128}$$

and the initial condition

$$\eta(o) = z^o - h(x^o, \epsilon) \tag{A.129}$$

These off-manifold dynamics normally are difficult to compute because they require x. As a first approximation, (A.128) could be solved using x as a constant equal to its initial condition. This is a reasonably good approximation because the off-manifold dynamics should decay (if they are stable) before x changes significantly. A geometric illustration of the integral manifold and the off-manifold dynamics is shown in Figure A.3.

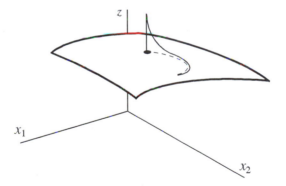

Figure A.3: *Off-manifold dynamics*

If z starts off the surface $z = h(x, \epsilon)$, it should decay rapidly to the surface (if it is stable), as shown in the solid line. The dotted line shows the trajectory of z if off-manifold dynamics are neglected and z is forced to begin on the surface.

It can be shown that, if z is stable, using h_o as an approximation for h and neglecting off-manifold dynamics only introduces "order ϵ" error into the slow variable x response. If further accuracy is desired and h is approximated by $h_o + \epsilon h_1$, there will still be "order ϵ" error if the off-manifold dynamics are neglected. To reduce the error to "order ϵ^2," it is necessary to include h_1 and approximate η to order ϵ^2. This can be done by approximating the off-manifold dynamics as

$$\epsilon \frac{d\eta}{dt} \approx g(x^o, \eta + h_o + \epsilon h_1) - \epsilon \frac{\partial h_o}{\partial x} f(x^o, \eta + h_o) \qquad (A.130)$$

with

$$\eta(o) = z^o - h_o(x^o) - \epsilon h_1(x^o) \qquad (A.131)$$

and h_o, h_1, $\partial h_o/\partial x$ evaluated at $x = x^o$. Additional illustrations of integral manifolds and off-manifold dynamics are given in [63]–[69].

Bibliography

[1] IEEE Working Group on Dynamic System Performance of the System Planning Subcommittee, IEEE Power Engineering Society Power System Engineering Committee, "Symposium on Adequacy and Philosophy of Modeling: Dynamic System Performance," *IEEE Publication 75 CHO 970-4-PWR IEEE*, New York, 1975.

[2] J. H. Chow, Ed., *Time-Scale Modeling of Dynamic Networks with Applications to Power Systems*, Springer-Verlag Publishers, New York, 1983.

[3] North American Electric Reliability Council, "Reliability Assessment 1995-2004," *NERC*, Sept. 1995.

[4] U. S. Energy Research and Development Administration, "Systems Engineering for Power: Status and Prospects," *Proceedings Engineering Foundation Conference, CONF-750867*, Henniker, NH, Aug. 17–22, 1975.

[5] Electric Power Research Institute, "Power System Planning and Operations: Future Problems and Research Needs," *EPRI Special Report EL-377-SR*, Feb. 1977.

[6] E. G. Gate, K. Hemmaplardh, J. W. Manke, and D. P. Gelopulos, "Time Frame Notion and Time Response of the Models in Transient, Mid-Term and Long-Term Stability Programs," *IEEE Trans. Power Appar. Syst., PAS-103*, 1, Jan. 1984, 143–150.

[7] E. Kuffel and M. Abdullah, *High-Voltage Engineering*, Pergamon Press, Oxford, England, 1970.

[8] Electric Power Research Institute, "User Survey and Implementation Plan for EMTP Enhancements," *EPRI Report EL-3668*, Palo Alto, CA, Sept. 1984.

[9] Electric Power Research Institute, "Electromagnetic Transients Program (EMTP) Primer," *EPRI Report EL-4202*, Palo Alto, CA, Sept. 1985.

[10] J. J. Grainger and W. D. Stevenson, *Power System Analysis*, McGraw-Hill, New York, 1994.

[11] C. A. Gross, *Power System Analysis*, John Wiley & Sons, Inc., New York, 1986.

[12] A. R. Bergen, *Power System Analysis*, Prentice-Hall, Inc., Englewood Cliffs, NJ, 1986.

[13] S. Ramo, J. R. Whinnery, and T. Van Duzer, *Fields and Waves in Communication Electronics*, John Wiley & Sons, Inc., New York, 1965.

[14] H. W. Dommel, "Digital Computer Solution of Electromagnetic Transients in Single and Multiphase Networks," *IEEE Trans. PAS*, 88, 4, April 1969, 388–399.

[15] G. W. Stagg and A. H. El-Abiad, *Computer Methods in Power System Analysis*, McGraw-Hill Book Co., New York, 1968.

[16] M. A. Pai, *Computer Techniques in Power System Analysis*, Tata McGraw-Hill, New Delhi, India, 1979.

[17] L. O. Chua and P. M. Lin, *Computer-Aided Analysis of Electronic Circuits*, Prentice-Hall, Inc., Englewood Cliffs, NJ, 1975.

[18] G. T. Heydt, *Computer Analysis Methods for Power Systems*, Macmillan Publishing Co., New York, 1986.

[19] IEEE PES Engineering Education Committee, "IEEE Tutorial Course: Digital Simulation of Electrical Transient Phenomena," *IEEE Tutorial 81 EHO 173-5-PWR*, IEEE, New York, 1980.

[20] P. C. Krause, *Analysis of Electric Machinery*, McGraw-Hill, New York, 1986.

[21] A. E. Fitzgerald, C. Kingsley, and S. D. Umans, *Electric Machinery*, McGraw-Hill Book Co., New York, 1983.

[22] E. W. Kimbark, *Power System Stability: Synchronous Machines*, Dover Publications, Inc., New York, 1956.

[23] B. Adkins, *The General Theory of Electrical Machines*, Chapman and Hall Ltd., London, 1962.

[24] F. P. DeMello and L. N. Hannett, "Determination of Synchronous Machine Stability Study Constants," *EPRI Report EL-1424, 3*, Electric Power Research Institute, Palo Alto, CA, June 1980.

[25] D. P. Gelopulos, "Midterm Simulation of Electric Power Systems," *EPRI Report EL-596*, Electric Power Research Institute, Palo Alto, CA, June 1979.

[26] D. W. Olive, "New Techniques for the Calculation of Dynamic Stability," *IEEE Trans. Power Appar. Sys., PAS-85*, 7, July 1966.

[27] C. Concordia, *Synchronous Machine*, John Wiley & Sons, New York, 1951.

[28] A. W. Rankin, "Per-unit Impedance of Synchronous Machines," *AIEE Trans., 64*, Aug. 1945.

[29] M. R. Harris, P. J. Lawrenson, and J. M. Stephenson, *Per-unit Systems with Special Reference to Electrical Machines*, Cambridge University Press, Cambridge, England, 1970.

[30] I. M. Canay, "Causes of Discrepancies on Calculation of Rotor Quantities and Exact Equivalent Diagrams of the Synchronous Machine," *IEEE Trans. Power Appar. Syst., PAS-88*, 7, July 1969.

[31] H. H. Woodson and J. R. Melcher, *Electromechanical Dynamics Part I: Discrete Systems*, John Wiley & Sons, Inc., New York, 1968.

[32] T. S. Peterson, *Calculus with Analytic Geometry*, Harper & Row Publishers, Inc., New York, 1960.

[33] R. G. Bartle, *The Elements of Real Analysis*, John Wiley & Sons, Inc., New York, 1976.

[34] E. Kreyszig, *Advanced Engineering Mathematics*, John Wiley & Sons, Inc., New York, 1972.

[35] C. C. Young, "The Synchronous Machine," *IEEE Tutorial Course on Modern Concepts of Power System Dynamics, 70 M 62-PWR*, Institute of Electrical and Electronics Engineers, New York, 1970.

[36] P. M. Anderson and A. A. Fouad, *Power System Control and Stability*, Iowa State University Press, Ames, IA, 1977.

[37] F. P. DeMello and L. N. Hannett, "Representation of Saturation in Synchronous Machines," *IEEE Trans. Power Syst., PWRS-1*, 4, Nov. 1986, 8–18.

[38] P. L. Dandeno et al, "Current Usage and Suggested Practices in Power System Stability Simulations for Synchronous Machines," *IEEE Trans. Energy Conversion, EC-1*, 1, Mar. 1986, 77–93.

[39] G. Shackshaft, "Model of Generator Saturation for Use in Power System Studies," *Proc. IEE*, 126, 8, 759–763.

[40] G. Xie and R. S. Ramshaw, "Nonlinear Model of Synchronous Machines with Saliency," *Paper 85 SM 347-0*, IEEE/PES 1985 Summer Meeting, Vancouver, BC, July 14–19, 1985.

[41] K. Prabhashankar and W. Janischewsyj, "Digital Simulation of Multi-machine Power Systems for Stability Studies," *IEEE Trans. on Power Appar. Syst., PAS-87*, 1, Jan. 1968, 73–79.

[42] J. E. Brown, K. P. Kovacs, and P. Vas, "A Method of Including the Effects of Main Flux Path Saturation in the Generalized Equations of A.C. Machines," *IEEE Trans. Power Appar. Syst., PAS-102*, 1, Jan. 1983, 96–102.

[43] I. M. Canay, "Determination of Model Parameters of Synchronous Machines," *IEE Proc., 130*, Part B, 2, Mar. 1983, 86-94.

[44] W. B. Jackson and R. L. Winchester, "Direct and Quadrature Axis Equivalent Circuits for Solid-Rotor Turbine Generators," *IEEE Trans. Power Appar. Syst., PAS-88*, 7, July 1969, 1121–1136.

[45] P. L. Dandeno, P. Kundur, and R. P. Schulz, "Recent Trends and Progress in Synchronous Machine Modeling in the Electric Utility Industry," *IEEE Proc., 62*, 7, July 1974, 941–950.

[46] S. H. Minnich et al, "Saturation Functions for Synchronous Generators from Finite Elements," *Paper 87 WM 207-4*, IEEE/PES 1987 Winter Meeting, New Orleans, Feb. 1–6, 1987.

[47] S. H. Minnich, "Small Signal, Large Signals and Saturation in Generator Modeling," *IEEE Trans. Energy Conversion, EC-1*, 1, Mar. 1986, 94–102.

[48] T. J. Hammons and D. J. Winning, "Comparisons of Synchronous Machine Models in the Study of the Transient Behaviour of Electrical Power Systems," *Proc. IEEE, 118*, 10, Oct. 1971, 1442–1458.

[49] IEEE, "Symposium on Synchronous Machine Modeling for Power System Studies," *83TH0101-6-PWR*, Tutorial at IEEE/PES 1983 Winter Meeting, New York, 1983.

[50] J. D. Ojo and T. A. Lipo, "An Improved Model for Saturated Salient Pole Synchronous Motors," *Paper 88 SM 614-0*, IEEE/PES 1988 Summer Meeting, Portland, OR, July 24–29, 1988.

[51] P. W. Sauer, "Constraints on Saturation Modeling in AC Machines," *IEEE Trans. Energy Conversion, 7*, 1, Mar. 1992, 161–167.

[52] M. E. Coultes and W. Watson, "Synchronous Machine Models by Standstill Frequency Response Tests," *IEEE Trans. Power Appar. Syst., PAS-100*, 4, Apr. 1981, 1480–1489.

[53] IEEE Std 115A, "IEEE Trial Use Standard Procedures for Obtaining Synchronous Machine Parameters by Standstill Frequency Response Testing," *Supplement to ANSI/IEEE Std. 115-1983*, IEEE, New York, 1984.

[54] A. E. Fitzgerald and C. Kingsley, *Electric Machinery*, McGraw-Hill Book Co., New York, 1961.

[55] IEEE Committee report, "Computer Representation of Excitation Systems," *IEEE Trans. Power Appar. Syst., PAS-87*, 6, June 1968, 1460–1464.

[56] IEEE Committee report, "Excitation System Models for Power System Stability Studies," *IEEE Trans. Power Appar. Syst., PAS-100*, 2, Feb. 1981, 494–509.

[57] D. G. Ramey and J. W. Skooglund, "Detailed Hydrogovernor Representation for System Stability Studies," *IEEE Trans. Power Appar. and Syst., PAS-89*, 1, Jan. 1970, 106–112.

[58] C. C. Young, "Equipment and System Modeling for Large Scale Stability Studies," *IEEE Trans. Power Appar. and Syst., PAS-91*, 1, Jan./Feb. 1972, 99–109.

[59] IEEE Committee report, "Dynamic Models for Steam and Hydro Turbines in Power System Studies," *IEEE Trans. Power Appar. and Syst., PAS-92*, 6, Nov/Dec 1973, 1904–1915.

[60] Systems Control Inc., "Development of Dynamic Equivalents for Transient Stability Studies," *Final Report EPRI EL-456*, Electric Power Research Institute, Palo Alto, CA, Apr. 1977.

[61] O. I. Elgerd, *Electric Energy System Theory, An Introduction*, McGraw-Hill Book Co., New York, 1982.

[62] C. C. Young and R. M. Webler, "A New Stability Program for Predicting the Dynamic Performance of Electric Power Systems," *Proceedings of the American Power Conference, 29*, Chicago, 1967, 1126–1138.

[63] P. W. Sauer, D. J. LaGesse, S. Ahmed-Zaid and M. A. Pai, "Reduced Order Modeling of Interconnected Multimachine Power Systems Using Time-Scale Decomposition," *IEEE Trans. Power Syst., PWRS-2*, 2, May 1987, 310–320.

[64] P. W. Sauer, S. Ahmed-Zaid, and P. V. Kokotovic, "An Integral Manifold Approach to Reduced Order Dynamic Modeling of Synchronous Machines," *IEEE Trans. Power Syst., 3*, 1, Feb. 1988, 17–23.

[65] P. V. Kokotovic and P. W. Sauer, "Integral Manifold as a Tool for Reduced Order Modeling of Nonlinear Systems: A Synchronous Machine Case Study," *IEEE Trans. Circuits Syst., 36*, 3, Mar. 1989, 403–410.

[66] J. H. Chow, Ed., "Time-Scale Modeling of Dynamic Networks with Applications to Power Systems," vol. 46 of lecture notes in *Control and Information Sciences*, Springer-Verlag, New York, 1982.

[67] P. V. Kokotovic, H. K. Khalil, and J. O'Reilly, *Singular Perturbation Methods in Control: Analysis and Design*, Academic Press Inc., London, 1986.

[68] P. W. Sauer, "Reduced Order Dynamic Modeling of Machines and Power Systems," *Proceedings of the American Power Conference, 49*, Chicago, 1987, 789–794.

[69] P. W. Sauer, S. Ahmed-Zaid, and M. A. Pai, "Systematic Inclusion of Stator Transients in Reduced Order Synchronous Machine Models," *IEEE Trans. Power Appar. Syst., PAS-103*, 6, June 1984, 1348–1354.

[70] "Extended Transient-Midterm Stability Package (ETMSP)," *Final Report EPRI EL-4610*, Electric Power Research Institute, Palo Alto, Jan. 1987.

[71] P. W. Sauer and M. A. Pai, "Simulation of Multimachine Power System Dynamics," *Control and Dynamic System* (C. T. Leondes, Ed.), *43*, Part 3, Academic Press, San Diego, 1991.

[72] B. Stott, "Power System Dynamic Response Calculations," *Proc. IEEE, 67*, Feb. 1979, 219–241.

[73] "Power System Dynamic Analysis-Phase I," *EPRI Report EL-484*, Electric Power Research Institute, July 1977.

[74] P. M. Anderson and A. A. Fouad, "Power System Control and Stability," Iowa State University Press, Ames, IA, 1977.

[75] "Interactive Power Flow (IPFLOW)," *Final Report TR-103643*, 1, May 1994.

[76] P. Kundur, *Power System Stability and Control*, McGraw-Hill, New York, 1994.

[77] E. V. Larsen and D. A. Swann, "Applying Power System Stabilizers, Part I: General Concepts, Part II: Performance Objectives and Tuning Concepts, Part III: Practical Considerations," *IEEE Power Appar. Syst., PAS-100*, 12, Dec. 1981, 3017–3046.

[78] F. P. DeMello and C. Concordia, "Concepts of Synchronous Machine Stability as Affected by Excitation Control," *IEEE Trans., PAS-88*, Apr. 1969, 316–329.

[79] P. W. Sauer and M. A. Pai, "Power System Steady State Stability and the Load Flow Jacobian," *IEEE Trans. Power Appar. Syst., PAS-5*, Nov. 1990, 1374–1383.

[80] R. K. Ranjan, M. A. Pai, and P. W. Sauer, "Analytical Formulation of Small Signal Stability Analysis of Power Systems with Nonlinear Load Models," in *Sadhana, Proc. in Eng. Sci.*, Indian Acad. Sci., Bangalore, India, *18*, 5, Sept. 1993, 869–889, Errata, *20*, 6, Dec. 1995, 971.

[81] G. C. Verghese, I. J. Perez-Arriaga and F. C. Scheweppe, "Selective Modal Analysis with Applications to Electric Power Systems, Part I and II," *IEEE Trans. Power Appar. Syst.*, PAS-101, Sept. 1982, 3117–3134.

[82] C. Rajagopalan, B. Lesieutre, P. W. Sauer, and M. A. Pai, "Dynamic Aspects of Voltage/Power Characteristics," *IEEE Trans. Power Syst.*, 7, Aug. 1992, 990–1000.

[83] B. C. Lesieutre, P. W. Sauer, and M. A. Pai, "Sufficient Conditions on Static Load Models for Network Solvability," in *Proc. 24th Annu. North Amer. Power Symp.*, Reno, Oct. 1992, 262–271.

[84] P. W. Sauer and B. C. Lesieutre, "Power System Load Modeling," *System and Control Theory for Power Systems* (J. H. Chow et al., Eds.), Springer-Verlag, *64*, 1995.

[85] V. Venkatasubramanian, H. Schattler, and J. Zaborszky, "Voltage Dynamics: Study of a Generator with Voltage Control, Transmission and Matched MW Load," *IEEE Trans. Autom. Contr.*, *37*, Nov. 1992, 1717–1733.

[86] C. Rajagopalan, "Dynamic of Power Systems at Critical Load Levels," Ph.D. thesis, University of Illinois, Urbana-Champaign, 1989.

[87] E. H. Abed and P. Varaiya, "Nonlinear Oscillations in Power Systems," *Int. J. Elect. Power Energy Syst.*, *6*, 1984, 37–43.

[88] W. G. Heffron and R. A. Phillips, "Effects of Modern Amplidyne Voltage Regulator in Underexcited Operation of Large Turbine Generators," *AIEE Trans.*, PAS-71, Aug. 1952, 692–697.

[89] A. R. Bergen, "Analytical Methods for the Problem of Dynamic Stability," *Proc. International Symposium on Circuits and Systems*, 1977.

[90] Y. N. Yu, *Electric Power System Dynamics*, Academic Press, New York, 1983.

[91] P. Kundur, M. Klein, G. J. Rogers, and M. S. Zywno, "Application of Power System Stabilizer for Enhancement of Overall System Stability," *IEEE Trans.*, PWRS-4, May 1989, 614–626.

[92] F-P. DeMello, P. J. Nolan, T. F. Laskowski and J. M. Undrill, "Co-ordinated Application of Stabilizers in Multimachine Power Systems," *IEEE Trans. Power Appar. Syst., PAS-99*, May-June 1980, 892–901.

[93] M. A. Pai, *Power System Stability*, North Holland Publishing Co., New York, 1981.

[94] M. A. Pai, *Energy Function Analysis for Power System Stability*, Kluwer Academic Publishers, Boston, 1989.

[95] A. A. Fouad and V. Vittal, *Power System Transient Stability Analysis Using the Transient Energy Function Method*, Prentice Hall, Englewood Cliffs, 1991.

[96] M. Pavella and P. G. Murthy, *Transient Stability of Power Systems: Theory and Practice*, John Wiley & Sons, Inc., New York, 1994.

[97] T. Athay, R. Podmore, and S. Virmani, "A Practical Method for Direct Analysis of Transient Stability," *IEEE Trans. Power Appar. Syst., PAS-98*, 2, Mar./Apr. 1979, 573–584.

[98] H. D. Chiang, F. F. Wu, and P. P. Varaiya, "Foundations of Direct Methods for Power System Transient Stability Analysis," *IEEE Trans. Circuits Syst., CAS-34*, Feb. 1987, 160–173.

[99] A. H. El-Abiad and K. Nagappan, "Transient Stability Region of Multi-Machine Power Systems," *IEEE Trans. Power Appar. Syst., PAS-85*, 2, Feb. 1966, 169–178.

[100] N. Kakimoto, Y. Ohsawa, and M. Hayashi, "Transient Stability Analysis of Electric Power Systems Via Lure Type Lyapunov Functions, Parts I and II," *Trans. IEE of Japan, 98*, 5/6, May/June 1978.

[101] F. A. Rahimi, M. G. Lauby, J. N. Wrubel, and K. W. Lee, "Evaluation of the Transient Energy Function Method for On-line Dynamic Security Analysis," *IEEE Transactions on Power Systems, 8*, 2, May 1993, 497–506.

[102] P. C. Magnusson, "Transient Energy Method of Calculating Stability," *AIEE Trans., 66*, 1947, 747–755.

[103] P. D. Aylett, "The Energy-Integral Criterion of Transient Stability Limits of Power Systems," *Proc. of the Institution of Electrical Engineers (London), 105C*, 8, Sept. 1958, 257–536.

[104] A. R. Bergen and D. J. Hill, "Structure Preserving Model for Power System Stability Analysis," *IEEE Trans. Power Appar. Syst., PAS-100*, 1, Jan. 1981.

[105] K. R. Padiyar and H. S. Y. Sastry, "Topological Energy Function Analysis of Stability of Power Systems," *Int. Electr. Power and Energy Syst.*, *9*, 1, Jan. 1987, 9–16.

[106] K. R. Padiyar and K. K. Ghosh, "Direct Stability Evaluation of Power Systems with Detailed Generator Models Using Structure Preserving Energy Functions," *Int. J. Electr. Power and Energy Syst., 11*, 1, Jan. 1989, 47–56.

[107] P. W. Sauer, A. K. Behera, M. A. Pai, J. R. Winkelman, and J. H. Chow, "Trajectory Approximations for Direct Energy Methods That Use Sustained Faults with Detailed Power System Models," *IEEE Transactions on Power Systems, 4*, 2, May 1988, 499–506.

[108] A. R. Bergen, D. J. Hill, and C. L. DeMarco, "A Lyapunov Function for Multi-Machine Power Systems with Generator Flux Decay and Voltage Dependent Loads," *Int. J. of Electric Power and Energy Syst., 8*, 1, Jan. 1986, 2–10.

[109] M. A. Pai, "Power System Stability Studies by Lyapunov-Popov Approach," *Proc. 5th IFAC World Congress*, Paris, 1972.

[110] J. Willems, "Direct Methods for Transient Stability Studies in Power System Analysis," *IEEE Trans. Autom. Control, AC-16*, 4, July-Aug. 1971, 1469–81.

[111] M. Ribbens-Pavella and F. J. Evans, "Direct Methods for Studying of the Dynamics of Large Scale Electric Power Systems - A Survey," *Automatica, 21*, 1, 1985, 1–21.

[112] A. A. Fouad and S. E. Stanton, "Transient Stability of Multi-Machine Power Systems, Part I and II," *IEEE Trans. Power Appar. Syst., PAS-100*, 1951, 3408–3424.

[113] A. N. Michel, A. A. Fouad, and V. Vittal, "Power System Transient Stability Using Individual Machine Energy Functions," *IEEE Trans. Circuits Syst., CAS-30*, 5, May 1983, 266–276.

[114] M. A. El-Kady, C. K. Tang, V. F. Carvalho, A. A. Fouad, and V. Vittal, "Dynamic Security Assessment Utilizing the Transient Energy Function Method," *IEEE Trans. Power Syst., PWRS1*, 3, Aug. 1986, 284–291.

[115] Y. Mansour, E. Vaahedi, A. Y. Chang, B. R. Corns, B. W. Garrett, K. Demaree, T. Athay, and K. Cheung, "B. C. Hydro's On-Line Transient Stability Assessment (TSA): Model Development, Analysis and Post-Processing," *IEEE Trans. Power Syst., 10*, 1, Feb. 1995, 241–253.

[116] H. D. Chiang, "Analytical results on direct methods for power system transient stability assessment," *Control and Dynamic Systems* (C. T. Leondes, Ed.), *43*, Academic Press, San Diego, 1991, 275–334.

[117] C. L. DeMarco and T. J. Overbye, "An Energy Based Measure for Assessing Vulnerability to Voltage Collapse," *IEEE Trans. Power Syst., 5*, 2, May 1990, 582–591.

[118] A. M. Lyapunov, *The General Problem of the Stability of Motion* (in Russian), The Mathematical Society of Kharkov, Russia, 1892. English translation, Taylor and Francis Ltd., London, 1992.

[119] A. A. Fouad and V. Vittal, "Power System Transient Stability Assessment Using the Transient Energy Function Method," *Control and Dynamic Systems* (C. T. Leondes, Ed.), *43*, Academic Press, San Diego, 1991, 115–184.

[120] M. A. Pai, M. Laufenberg, and P. W. Sauer, "Some Clarification in the Transient Energy Function Method," *Int. Journal of Elec. Power and Energy Systems, 18*, 1, 1996, 65–72.

[121] R. T. Treinen, V. Vittal, and W. Kliemann, "An Improved Technique to Determine the Controlling Unstable Equilibrium Point in a Power System," *IEEE Transactions on Circuits and Systems-I: Fundamental Theory and Application, 43*, 4, Apr. 1996, 313–323.

Index